Patrick Cohendet · Marc J. Ledoux
Ehud Zuscovitch

New Advanced Materials

Economic Dynamics and European Strategy

A Report from the FAST Programme of the Commission of the European Communities

English Version Edited by: Marc J. Ledoux

In Cooperation with:
Isabelle Boudoux, Philippe Capdevielle, Michèle Forte Frédéric Heran, Jean-Alain Heraud, Danièle Kieny, Patrick Llerena, Rémy Risser, Renate Schwarz and Marc Willinger

With the Assistance of:
Yves Ayrault, Laurent Bach, Godefroy Dang Nguyen and Jean Troton

Translated by: M. Bettison, I. Smith and R. Tanner

Springer-Verlag
Berlin Heidelberg New York
London Paris Tokyo

Professor Dr. Patrick Cohendet

Université Louis Pasteur, Bureau d'Economie Théorique et Appliquée (BETA), 4 rue Blaise Pascal, F-67070 Strasbourg Cedex, France

Dr. Marc J. Ledoux

CNRS - Université Louis Pasteur, Bureau d'Economie Théorique et Appliquée (BETA), 4 rue Blaise Pascal, F-67070 Strasbourg Cedex, France

Dr. Ehud Zuscovitch

Université Louis Pasteur, Bureau d'Economie Théorique et Appliquée (BETA), 4 rue Blaise Pascal, F-67070 Strasbourg Cedex, France

With 44 Figures

This book has been published in a French edition under the title:

Les Materiaux Nouveaux

Dynamique économique et stratégie européenne

Economica

ISBN 2-7178-1295-4

ISBN 3-540-19414-2 Springer-Verlag Berlin Heidelberg New York Tokyo
ISBN 0-387-19414-2 Springer-Verlag New York Heidelberg Berlin Tokyo

Printed in Germany

Typeset by Heinz-Dieter Ecker on TEX, under assistance of Sfb 303 at the University of Bonn.
Printing: Weihert-Druck GmbH & Co. KG, Darmstadt
Bookbinding: J. Schäffer GmbH u. Co. KG., Grünstadt
2142/7130-543210

Introducing the Research Team which Carried out the Study

Study directors and editors of the final report

Patrick COHENDET,
Professor of economic sciences, co-director of BETA (Bureau d'Economie Théorique et Appliquée) ;

Marc-Jacques LEDOUX,
CNRS (Centre National de Recherche Scientifique - national scientific research centre) research officer, Chemistry Department, Université Louis Pasteur ;

Ehud ZUSCOVITCH,
Lecturer in economic sciences, Université Louis Pasteur.

Assisted with the study and the drafting of the report

Isabelle BOUDOUX,
Documentalist at the Université Louis Pasteur (general documentation for the study) ;

Philippe CAPDEVIELLE,
Researcher with Cia-CEREQ/BETA (co-drafting of Chapters I.2, II.3, III.3 and drafting of the technical appendix on steels in French edition) ;

Michèle FORTE,
Visiting lecturer, Strasbourg III University (co-drafting of Chapter V.4) ;

Frédéric HERAN,
Researcher with Cia-CEREQ/BETA (co-drafting of Chapter V.4) ;

Jean-Alain HERAUD,
Lecturer in economic sciences, Université Louis Pasteur (co-drafting of Chapter IV.2) ;

Danièle KIENY,
Studies manager at the Strasbourg Chamber of Commerce and Industry (general documentation for the study ; preparation of the technical appendix on ceramics in French edition) ;

Rémy RISSER,

Researcher with BETA (co-drafting of Chapters I.2, II.3, III.1, III.3 and technical appendix on plastics in French edition);

Patrick LLERENA,

Researcher with BETA, Docteur d'Etat (drafting of Chapter II.5) ;

Renate SCHWARZ,

Researcher with BETA, (co-drafting of the technical appendix on plastics in French edition and contribution to section V.2.3) ;

Marc WILLINGER,

Visiting lecturer, Strasbourg III University (drafting of Chapters II.2 and II.4, co-drafting of Chapter III.1 and drafting of the technical appendix on composites in French edition).

__Assisted with the drafting of the final report__

Yves AYRAULT,

Director of the Strasbourg School of Architecture (Chapter III.3);

Laurent BACH

Researcher with BETA (section V.3.1) ;

Godefroy DANG NGUYEN,

Lecturer, CNET (Centre National d'Etudes des Télécommunications - national telecommunications research centre) (section IV.1.2) ;

Jean TROTON,

Researcher with BETA (section V.3.1).

__Layout and typing of the final report__

Christine DEMANGE and Sandra ZEVIAN,

BETA Secretariat.

__Supervision of the study in the Commission__

__For the FAST Project__

Riccardo PETRELLA,

FAST Project Director.

Werner WOBBE.

For the EURAM Project

J.G. WURM,

Head of EURAM Programme on Materials Technology.

The BETA team wishes to thank all those who contributed through their ideas and support to the considerations presented in this report, and in particular :

Jean-Claude BERNIER,

Professor, Strasbourg, Ecole des Hautes Etudes Européennes des Industries Chimiques, for many useful discussions on the future of engineering ceramics ;

Catherine GARDETTE,

(Cabinet Arthur Andersen, Strasbourg), who initiated the group's consideration of packaging materials ;

Sonia GEORGENTHUM,

(Alsace Regional Authority), who initiated the group's consideration of semiconductors ;

Pierre-François GOBIN,

Professor, materials sciences, INSA (Centre National de la Recherche Scientifique - national scientific research centre), Lyons, many of whose ideas are embodied in the text ;

D. EVANS and D.M. JACOBSON,

(GEC Hirst Research Centre, Wembley, United Kingdom) for their contribution on metal materials ;

Dominique MUSTER,

(research laboratory of the Stomatology and Maxillo-Facial Surgery Department, University Louis Pasteur, for his contribution on biomaterials ;

Pierre PANISSOD,

Research Director CNRS, LMSES, Université Louis Pasteur ;

Paul SIFFERT,

Professor, PHASE laboratory director, for his unfailing support, particularly during the MRS Colloquy ;

Pierre WIPPLER,

Professor, Charles Sadron Laboratory, CNRS-Université Louis Pasteur.

Various events (colloquies, congresses, seminars, talks, etc ...) also brought the members of the research team into contact with people whose ideas have been incorporated into this report. BETA owes special thanks to Jean-Pierre CAUSSE, Deputy Managing Director, responsible for research (Saint-Gobain Company), Catherine BLUNDEN (Chargée de mission, CGP-Commissariat Général du Plan), Pierre FRECHET, Director of Technology (Rhône-Poulenc) and Jean HANUS, Professor, director of the CNRS Projet Interdisciplinaire Mécanique des Matériaux, for profitable exchanges during the preparation of "Forecast 2005" (CGP-CNRS) ; all members of the Lyons ECT team whom they met at a seminar in Arc-et-Senans and who, headed by Henri JACOT, are working on a further materials project.

Lastly, the Materials Centre in Strasbourg provided the group with a most fruitful and stimulating atmosphere in which to carry out this task.

Foreword

This report on materials is not a sequel to the five or six high-quality reports published in certain Community countries over the last few years, nor does it attempt to summarize them. Nor is it a technical summary of the state of the art in new materials. It is rather to be seen as a survey of economic dynamics and strategy, carried out for the purpose of prompting political and industrial leaders throughout the European Community to reflect in some depth on the subject of materials.

The report is arranged in five parts : the first is concerned with a definition of materials, the second with structural materials and their influence on the reconception of industrial processes, and this part is complemented by the third, which looks at the three main sectors using structural materials (transport, packaging, building and public works). The fourth part analyzes functional materials and the way in which they affect basic components in the technological system (information, energy, biotechnology). The fifth and last part makes a strategic analysis for Europe (in the spheres of industrial strategies, the role of the authorities and developments in employment and skills).

The paper is based on an investigation of the many reports on materials drawn up in the last few years, and on some hundred or so direct interviews carried out in the different European countries by the BETA team (Bureau d'Economie Théorique et Appliquée - office of theoretical and applied economics) with industrialists, research workers and officials. The intermediary findings were outlined at a conference held in December 1985 at the Council of Europe, Strasbourg, when an indication was given of the main thrust of the report.

Introduction

Official priorities in most western countries, after shifting from information technology in the early 70s to new energy sources in the mid-70s and then to biomaterials technologies in the early 80s, are now focused on the development of materials. This strategy bears witness to a major alteration in the technology system, as Bertrand GILLE sees it, in that its four mainstays - information, energy, biotechnology and materials - are being revolutionized in turn.

The change in the materials sector is not, however, of recent origin. Each technological breakthrough has depended on the development of a specific material ; silicon for information technology, optical fibres for opto-electronics, composite materials for space exploration, ceramics for high-temperature conditions and so on. Yet materials so far seem to have been the "dark side of technical breakthroughs" ; since they account for only part of overall added value, they have often been regarded as a secondary aspect of a broader technological revolution.

There are perhaps two main reasons for the change which has now taken place, and the first is that the fundamental importance of materials has no doubt been realized. Although they represent only a tiny fraction of a product's added value, their intrinsic properties are critical factors in the success of the product. The second and perhaps chief reason is that technological progress in the materials sector has spawned an extremely wide variety of materials, an "overchoice". Consequently, a given product no longer relies on a given material ; instead, several materials compete to assume a given function at a specified cost and for a certain space of time. In the automotive industry, for instance, there is keen competition among fine cast iron, aluminium, plastics and composites. In a way, this great diversity heralds the possibility of an inversion in the logics of production. Hitherto, the initial stage (material) was a determinant for the final stage (products incorporating the material) whereas today all the right conditions seem to exist for selecting or fashioning the ideal material according to the specifications for a given product.

The overchoice of materials has profound repercussions on the production system, as the current revolution in materials affects the production method as well as society's lifestyles and opens new avenues for creativity. A major part of this publication will be devoted to discussing these phenomena, but at this point we can already draw attention to two repercussions at either end of the production system :

At the upper end of production systems, the hyperchoice of materials seems to have allayed fears of a supply scarcity. The concept of a "critical material" or "strategic material" vital to defence or to the economy still stands, but is less crucial than it was a few years ago. Of course, threats continue to overshadow the supply of chromium, manganese, platinum, but the prospect of quicker discovery of substitutes does much to relieve the pressure where these materials are concerned. Furthermore, it is much more a question of "critical products" than of the "critical materials" on which they rely, for a breakdown in the supply of one such material would have a very severe, albeit passing, effect on the products involved. At all events, what now seems strategic is "grey matter". Carbon itself is not strategic, but carbon fibre is. This example clearly illustrates that a given material can be strategic without being a physical rarity.

At the other end, further down the production system, the consumers confronted with an overchoice of materials are less and less capable of discerning which ones constitute the products around them, a situation which means that the end phase of the production system tends more and more to convey its requirements in terms of functions to be fulfilled, specifications to be met and capabilities to be realized, instead of type of material to be used. It also means that supply, in order to adapt to the highly problematic consumer choice aspect, will have to attribute more importance (and probably a substantial proportion of available jobs) to ascertaining customer requirements and making the manufactured products known.

NEW MATERIALS / TRADITIONAL MATERIALS

The development of "new materials", those whose anticipated growth rate up to the end of the century exceeds the average growth rate of the industrialized countries (about 3 % per year), is doubtless the most spectacular expression of the materials revolution. In this category can be placed technical plastics, composite materials, engineering ceramics, electronic materials and a few groups of other materials including amorphous metals and rare earths. Among these materials, a closer analysis distinguishes several categories according to stage of development (given that almost 20 years elapse between the research phase and the production phase, it can be assumed that the materials which will go into production in the year 2000 are already known) :

- materials now in a phase of recent industrialization (titanium, fibre reinforced plastics in widespread use) ;
- materials now in a pre-industrial phase (high-performance composites, advanced ceramics) ;
- materials in the development phase (aluminium-lithium) ;
- materials in the research phase (multilayer and amorphous materials).

The application of these materials is one of the keys to the development of new sectors (eg electronics and telecommunications), new activities (sport, recreation, household appliances) and the remodelling of production processes (automotive and aeronautics industries), and in these terms their evolution may be considered strategic.

However, to confine the strategic considerations to the so-called "new" materials alone would no doubt be a serious error. Indeed, a host of reasons urge caution in examining the economic impact of these materials :

- "New" materials today represent a mere 5 % of the global turnover in materials. Even though their comparative growth rate guarantes that the proportion will increase significantly by the close of the century, new materials will indubitably have only limited importance compared to other materials.
- Under the growing stimulus of advances in new materials, "<u>traditional</u>" materials (steel, wood, glass, commodity plastics, concretes) are making a more and more "organized" riposte. Consequently, although per capital and volume consumption of the older materials is on the downward trend in the most highly industrialized countries, at the same time their <u>quality</u> has noticeably improved. For instance, aluminium's "service to the economy" over the last decade, ie the development of its uses per constant kilogram, shows an "adjusted" real cumulative growth rate in the region of 50 %. Improved refining processes, control of impurities, the use of computer-aided techniques and weight reduction give aluminium many features of a new material, and observation of sales volume factors alone can prove misleading.

The "new materials"/"traditional materials" antithesis is therefore largely artificial (cf. Part I : A trial definition of new materials : towards a definition to assist the understanding of the current changes). Anticipated growth rate cannot be taken as the sole criterion for drawing the distinction. We can therefore dispense with this "old versus new" dichotomy and suggest a definition of materials which will be of some use in devising a strategy, by putting forward the following considerations :

- It is becoming increasingly difficult at the industrial stage to dissociate materials, processes and products. To be more precise, what seems really "new" is the association of a material with a working-up process to turn out a given product or industrial article. The example of single-crystal silicon in micro-electronics clearly reveals that a material does not establish itself in industry on the strength of its intrinsic value alone, but does so because a series of well-controlled processes permits its production and shaping on a competitive basis.
- A scrutiny of all new options in materials reveals that the progress made stems from the general advances in microscopic mastery of matter. The distinctive feature of recent times is in fact the gradual transition from a civilization with macroscopic control over materials (progressive physical reshaping of matter) to microscopic mastery (allowing "tailored" approaches in particular). This new departure is one of the keys to the transition from an old-style production system, founded on standardization and efficiency of scale, to a new system founded on diversity of products and variety-based economics.
- The technological trajectories and industrial strategies applying to the various materials contrast spharply according as the materials considered are functional (used for their physical properties and serving the functions of electricity or light transmission, magnetization, catalysis of major chemical reactions, etc ...) or structural (determining the forms of industrial articles, eg car bodies, packaging bags and building structures). The former category includes materials which determine the limits to the possibilities of a technology system at a given point in time and are generally fundamental to the creation of new technological sectors. These materials belong to a highly competitive international market. The second category consists of materials which are generally incorporated into existing industrial objects, thus allowing their production processes to be remodelled units. This class of materials is less dependent on the main international competitive markets but on the

other hand is particularly prone to the restrictions imposed by demand and most of all to official environmental restrictions.

The foregoing considerations show that today the development of materials cannot be summed up as a straighforward substitution of "new" materials for traditional ones. The radical alteration in materials as a whole, resulting from advances in microscopic mastery of matter, is the source of a considerable and highly diversified contribution to economic growth. For instance, through their weight-saving effect on the structures of industrial objects, the new materials have made their contribution to energy saving ; through the miniaturization of electronic parts, semiconductor materials have made the decentralization of information technologies feasible ; structural materials, through the redeployment of shaping processes, are responsible for increased company productivity ; similarly, the novel technical solutions provided by new materials (eg optical fibre endoscopy in medical analysis) have revolutionized methods in certain traditional sectors ; lastly, new materials pave the way for the development of new products or new applications (eg sailboards in the leisure activities sector). Generally speaking, the huge variety of materials gives rise to new opportunities for creation and design in all fields of activity.

STRUCTURAL MATERIALS : REDESIGN OF PROCESSES

The structural materials determining the form of industrial objects play a considerable part in remodelling production processes (cf. Part II : Structural materials). In having to reconcile the increasing range of consumer requirements with the growing variety of technical options made available by materials, firms will be more and more frequently confronted with the major phenomenon of increasing complexity in the management of production processes. How indeed can viable industrial responses be found to reconcile growth in the variety of demand with growth in the constantly regenerated possibilities of materials supply ? How can production be organized to cater for the dynamics of change ? The method commonly suggested for mastering this increase in complexity is the use of information technologies (CAD, CAM/CAD, robotics, etc ...). Yet whatever the importance of information technologies, the application of new processes for the manufacture and application of structural materials is at least as important as micro-electronics in offering workable industrial approaches to the mastery of complexity and consequently to the operation of the new variety-based system.

Processes which can be expected to undergo major development in the next decade (powder metallurgy, precision casting, superplastic forming, diffusion welding, magnetic moulding, ion implantation and laser treatments, bonding technologies, non-destructive testing methods, etc ...) are not only subtended by the traditional constants of technological progress in industrial application (economy of material, energy, labour, etc ...) but, above and beyond these familiar phenomena, are also alike in offering a solution to the problem of increasing complexity in production. The solution takes the form of integration principles. Integration basically means a simplification of the production process and an appreciable reduction in assembly costs. In practice, integration can take the following forms :

- Integration of manufacturing steps through reduction of the number of phases in a process (precision casting, for instance, obviates such operations as milling, tooling and finishing).
- Integration of the number of parts (the adoption of plastic in place of steel sheet has made it possible to incorporate in three parts only, in the rear hatch of the Citroën BX, all the functions of the former 27 parts in the Visa's rear hatch).
- Integration of production time (the use of a thermoplastic instead of a phenolic polyester for an electric light bulb socket cuts assembly time from 2.5 minutes to 16 seconds).

Thus the upshot of the integration phenomenon is a simplification of the overall process. Conversely, the material itself is visibly tending to gain in complexity by "integrating" several functions at once (the new packaging films can combine the heat sealability of polyethylene, the rigidity of cardboard and the gasproofness of aluminium). This integration of functions reflects the will to respond to the complexity of the technical environments to which industrially produced articles and structures will be sujected.

The chief applications of structural materials (cf. Part III : Structural materials : main sectors of application) essentially concern three sectors : transport, packaging and building/public works. While these sectors have a number of distinguishable common features (keen competition among the various structural materials ; redeployment of production processes ; major role of governments in setting technical and consumer norms and making environmental regulations), they do not all play the same part in the dissemination of new materials. Today it is plain that the automotive sector (mass-produced materials) and the aircraft industry ("high-performance" materials) are acting as catalysts in the implementation of new industrial approaches.

In transport, the emphasis over the last decade has been on energy savings which can be achieved in the operation of vehicles (cars, trains, aircraft, etc ...), which is why materials allowing weight reduction in the existing systems have found favour. For instance, a gain of 100 kg on a medium-capacity car cuts full consumption by about 0.5 litres/100 km, hence the use of parts made of polyester composites reinforced with aluminium fibre for the Renault 9's (Alliance) wraparound bumper, or the use of plastics instead of steel or aluminium. Their use is further warranted by the lower energy content of plastics compared to aluminium - less than half - and the smaller energy requirements of polymer processing. However, taking into account the recovery phase for polymers, the opposite applies since aluminium is easier to recover.

While weight reduction is now one of the main criteria for materials, it will not necessarily be so in years to come. In certain road vehicles (cars) it is already approaching the limits of tolerance imposed by safety requirements. Moreover, the preoccupation with a feature like weight reduction may gradually raise contradictions with other forms of energy expenditure. For instance, the use of composite materials in cars may ultimately raise the problem of their recovery (very difficult at present) and environmental problems. For all these reasons, other features (aerodynamics, engine performance, etc ...) will probably acquire greater importance in the next few years.

Packaging materials must at once absorb the technical advances of products and processes, the demands of expansion in distribution and transport, changing consumer tastes, environmental protection requirements and the need to supply information material on product type and quality. Fulfilling this combination of requirements produces fierce competition among materials which tends to disrupt the former segmentation of packaging whereby a specific material (glass for bottles, cardboard for cartons and so on) used to dominate each separate market. Although the outcome of this competition still remain largely undecided, the packaging sector more than others indicates how far the "plastics solution" succeeds in meeting a very large number of concurrent demands and accordingly competes consistently with every present or planned solution.

The "building and public works" sector's apparent traditionalism masks radical changes in the use of materials. Economic recession and above all population decline in the industrialized countries have steered trends towards the satisfaction of two major necessities : energy saving (achieved mainly through improved insulation) and rejuvenation of old materials. The process has given the sector a wider variety of materials (plastics in particular), significantly increased the importance of finishing compared to structural operations (the introduction of information technologies into the "home" or "domestic" market is no doubt a promising trend) and caused the authrorities to adopt a more flexible approach to regulations in order to foster the development of innovations.

FUNCTIONAL MATERIALS : THE EMERGENCE OF NEW BRANCHES OF TECHNOLOGY

Functional materials are vehicles for the physical properties of technology (cf. Part IV : Functional materials) and so mark the boundaries within which technology can evolve at a given time. Above and beyond each specific application (electronics, magnetism, catalysis, etc ...), functional materials are closely involved in establishing the basis of the technology system as conceived by Bertrand GILLE ; an understanding of its radical alteration depends on an understanding of the interactions between functional materials and the other mainstays of the system - information, energy and life, the watchword being more than ever "less energy, less material and more information". In this framework, certain functional materials constitute the point of origin of technological trajectories which, as in the case of micro-electronics, can help to restructure an entire sector of economic activity, and which govern the production of other materials "compatible" with these basic materials in making the system a coherent whole.

The importance of new materials to information technologies is particularly well illustrated by single-crystal silicon in micro-electronics. Two main steps can be distinguished in the development of micro-electronics :

- A first step reaching from the early 60s to the mid-80s and coinciding with the constitution of silicon technology (growing of the crystal and production of integrated circuits) and of allied industries (electronics and information technology). The main vector of this development was the effort towards more and more advanced integration in integrated circuity, which was attended by a spectacular fall in production costs. This integration process is not yet complete, and silicon is destined to remain a dominant material for more than a decade ahead, but it is now realized that the ceaseless expansion of requirements in the electronics sector will eventually come up against the theoretical limits to the use of silicon.

- The second stage began not long ago in the mid-80s and is expected to feature a growing convergence between information and telecommunication technologies. The henceforth inevitable linkage of these two fields is due largely to the development of micro-electronics ; with its growing decentralizing effect on data processing through microprocessors and microcomputers, it became patently urgent to interconnect data processing units. Data transmission has become indissociable from data processing, and the need to speed up its transmission may eventually create a bias in certain uses towards data processing and the application of semiconductors other than silicon, in particular gallium arsenide.

The relationship between materials and energy presents increasing degrees of complexity depending on the stage of industrial sophistication :

- The most elementary relation makes materials and energy indissociable, as in the use of primary energy fuels such as uranium. The use of wood and coal as an energy source is another typical elementary relation. The correspondence is also expressed by the expediency of destructive/combustive recycling of the materials incorporated into advanced structures, eg used packaging and demolition materials and in general all flammable structural materials whose primary use is finished.
- A more complex relation characterizes the energy needed to manufacture a given material (eg the number of kilowatts to prepare an aluminium sheet) as well as the quantity and quality of the materials required to produce some form of energy (power station turbine blade as a typical structural material, or refining catalyst as a typical functional material).
- A still more complex degree of interrelation between materials and energy cand be perceived within an industrial system or object. This is the type of relation taken into account in initial research efforts to lighten materials in order to save energy in the operation of the various modes of transport. In this case, mainly structural materials

were involved. However, this type of relation can also be considered in terms of functional materials, for instance with new permanent magnets improving electric motor efficiency, or heat-resistant self-lubricating ceramics which could make the diesel engine extremely competitive.

- An ultimate degree of complexity is reached when the relations between materials, energy and environment are simultaneously taken into account. In this overall perspective, catalytic converters for cars and filters or membranes in scrubbing processes are examples clearly illustrating the propensity of materials to make specific forms of energy usable by virtue of their function in pollution control processes.

Biotechnology materials occur in the following main sectors :

- prostheses and artificial organs ;
- throwaway products ;
- sensors and analytical products ;
- membranes used in biology and biotechnology.

Prostheses exploit a very wide range of materials. For joint implants, non-corrosive metals and alloys are supplemented or rivalled by high-performance plastics, composites - especially of the carbon-carbon type - or certain ceramics which the organism tolerates better.

As well as biological heart valves, there are mechanical valves made from titanium, silicone or carbon. Artificial arteries are based on polyester or fluorocarbon resins. Silicones are often used in permanent implants or "social" prostheses (artificial hands, noses, etc ...). Metals in dentures are being replaced by composites and ceramics.

Artificial organs, eg for haemodialysis, make use of a whole range of sophisticated plastics.

The likely areas of change are the selection of materials and the improvement of their biocompatibility. New materials remain to be devised for new prostheses that do not yet exist, eg artificial nerves and tendons, implantable organs etc ...

Throwaway products used in health and hygiene form a very large and disparate group ranging from absorbents (cotton wool, pads, napkins) to syringes, probes, ligatures, etc ... It is a field which concerns the general public but which may require sophisticated materials, eg resorbable synthetics in surgery.

Among the materials used in biological sensors and analytical products, attention should focus not only on grafted materials, usually polymers, but also on minerals with highly specific affinities. This is a rapidly developing sector. "Homing" materials of this type are going to become widespread. There are already plans for microcapsules of actives pharmaceuticals homing on and lodging in their biological target to release their constituents gradually under optimum performance conditions.

Another promising material for biology is the membrane already being used in the artificial kidney machine or the food industry. The base material is diversifiying away from the original cellulose to include various type of polymers or mineral substrates.

The precision of the cut-off zone of these membranes should come under increasingly accurate control. Higher levels of permeability would make it possible to broaden the field of application of these filtration techniques.

A STRATEGY FOR EUROPE

As an aid to strategic planning in the materials field, the analyses presented in this publication suggest the construction of two scenarios for the future.

The first scenario corresponds to the "growing importance of the variety phenomenon". It assumes that the industrial system as a whole is capable of functioning under permanent conditions of variety, ie able to manage an unceasingly diversified supply of goods and services efficiently by means of a range of technical options constantly renewed by an increasing array of possibilities. Where materials are concerned, the paramount concept in a permanent scheme of variety is necessarily that of multimaterials resulting from the association of several types of material.

The second scenario, corresponding to the "irreversible ascent of plastics", is to a large extent a simple projection of the current situation where plastics offer competitive options in almost every field. However, this simple extrapolation is subtended by a major hypothesis about the evolution of the industrial system : a variety-based order is essentially unstable and too complex to manage. The variety observed from the mid-70s was merely a framework of transition between two industrial eras, marking the switch from a civilization focused on "steel and standardization" to one of "plastics and tailored production".

The two scenarios are to serve as a guide in interpreting the following considerations on industrial strategies, the role of the authorities and the problems of jobs and skills.

Changes in materials are liable to challenge the industrial strategies and structures which formerly predominated in a standardized production system.

- The main challenge is probably to the rigid sectoral division between branches specializing in a given material : until recently the roles were evenly distributed so that steel, for instance, was produced by the structural steel industry. This sectoral division, the outcome of strong traditions in training and industrial strategy, may have definite drawbacks where large-scale substitution of materials occurs. Their replacement may then threaten entire sectors (steel industry) unable to branch out into substitutes beyond their own sphere. The development of alternative materials such as composites further complicates this picture. These materials were divised and then put into production in their sectors of origin (space and aeronautics). Thus new firms have gained prominence as materials manufacturers. Furthermore, the strategy of American or Japanese firms reveals another path of development : such firms, particularly those originating in the steelmaking sector, have recently acquired interests in units producing composites, ceramics or optical fibres. This strategy, the outcome of which will be the formation of a multimaterial production sector, is feasible because formerly unrelated materials such as plastics and steel have become readily interchangeable owing to recent advances. But whichever trend finally dominates in relations between firms, attention should be drawn to the driving influence which the automotive sector currently appears to be exerting on the processing of materials. The intermingling of methods and technologies in this sector clearly provides a lead for many other economic activities. Conversely, the building sector, possibly because the relevant industrial activities are spatially dispersed, now fits the profile of a sector following the lead of innovations achieved elsewhere.
- The second tendency in the industrial evolution of materials is the omnipresence of the chemical industry. The big chemicals firms are making a worldwide offensive in all fields related to the production, processing and application of materials (electronics included). They usually operate by successive takeovers enabling them to absorb one new

area of expertise after another. The confrontation between the chemical industry and other industries producing materials will probably be one of the major issues in this field by the end of the century. The relative advantage held by chemicals is dues to the growing need for chemical know-how as microscopic knowledge of matter advances. This situation is a golden opportunity for Europe, which has managed to preserve intact the competitiveness of its chemical industry, still the best in the world.

- Lastly, the evolution which is occurring in materials and related processes may have profound repercussions on inter-company and particularly customer-supplier relations. The choice between "do it yourself" and "have it done" might be affected in a manner more favourable to the concept of "industrial partnership" than to conventional "customer-supplier" relations. A significant example of this developmental process is again the rear hatch of the Citroën BX, an achievement which was only possible thanks to the development of a new process (ZMC) resulting from close cooperation among three sectors, namely a press manufacturer, a tool maker and a fibre producer. This cannot be described as a "customer-supplier" relationship, but rather as an ad hoc technical association of three "partners". The impact of the wide variety of materials and related processes is evident in this case as well. In such circumstances the term "industrial sector" becomes less and less applicable ; actually there are ad hoc combines of firms according to need. This is no doubt why there are so many agreements between firms, a state of affairs which is bound to assume growing importance and whose investigation in qualitative and quantitative terms could well prove a useful yardstick or instrument for understanding the major industrial changes of today.

The three leading countries or groups of countries - USA, Japan and Europe - are competing relentlessly in the new materials field, where their official strategies nevertheless appear very different :

- The USA, thanks to multiple sources of finance from the various agencies or ministries, seems alone in being able to make progress on all fronts at once. However, US dominance is plainly most conspicuous in functional materials where basic research is most crucially important. The existence of a "critical" mass in public and private research, the traditional mobility of researchers and ideas and regular support to major projects are factors accounting for the might of American research.

- Japan operates in a very different way ; since unlike the USA it does not have the critical mass to tackle every area of materials, the Japanese authorities under the aegis of MITI (Ministry of International Trade and Industry) have decided to pursue a policy geared to a number of precise objectives (advanced ceramics, carbon fibres, semiconductors, etc ...). This policy features close cooperation between scientists and industrialists and determination to find immediate mass market outlets (vehicles, electronics, sports, recreation, etc ...). Japanese strategy in these areas generally involves infiltrating bread-and-butter markets with bottom-of-the-range goods and then moving up to increasingly sophisticated goods at the top of the range (it is also founded on principles of efficient production management with the main emphasis on quality).

Japan's forward thrust is most significant in engineering ceramics, particularly advanced ceramics for electronics and engines. This strategy reveals a determination to develop not only materials assured of a high growth rate for the remainder of the century but also materials to fil the slots that are most resistant to competition from other materials, particularly plastics.

- As for Europe, the arguments developed so far converge in a general recommendation not to be ignored by the authorities : at all costs we must equip ourselves so that Europe can catch up in the functional materials sector and retain the lead gained in the design, manufacture and applications of structural materials. US and Japanese dominance does in fact tend to make Europe systematically dependent where the main basic technologies are concerned. Europe's competitive position in structural materials (due largely to the might of its chemical industry) is endangered more and more each day by advances in other countries.

Europe has nevertheless made a tangible if belated response. Each major European country recently devised a materials programme. At Community level, the Commission has adopted a real Community programme dealing specifically with materials. The EURAM programme thus comprises three main research areas : metal materials, engineering ceramics and composite materials. The magnitude of the technological "challenge" from the United States and Japan means that this action must be intensified, although consideration should be given to Europe's specialities (the strength of the chemical industry and of certain others such as glass and aluminium) and to the need for wide distribution of innovations in materials among small and medium enterprises, the major employers in the materials sector.

As well as in directly financing materials research, the authorities clearly have a decisive role in preserving Europe's competitiveness through major public and military programmes (especially where space programmes are concerned, Europe will have to make a prompt decision on the option of developing an ambitious space materials programme), the establishment of norms and standards and environmental and safety regulations.

It is very hard to estimate the employment figures in the materials sector and to see how they are evolving, since materials (like information technologies) constitute a <u>cross-sectoral</u> technology, so that there is no materials sector in the true sence. A rough estimate of jobs in Europe's materials production and processing industries puts the workforce at about 7 million.

Furthermore, any attempt (even were it statistically feasible) to distinguish "new materials" from "old materials" (according to the definition rejected in Part I) would probably lead nowhere or merely to a very restrictive and distorted impression of the industrial reality. The conclusion would be that "new materials" (composites, ceramics, etc ...) represent only a very small proportion of industrial employment (certainly not more than 5 %) and that the jobs it is hoped to create in these sectors barely make up for the high job losses in "old materials" (steel in particular). Such an outlook, conjuring up a straightforward mechanim whereby old materials are superseded by new, disregards not only the full importance of the mechanisms which govern the rethinking of production tasks and the reshaping of skills but also the central role of know-how. Even where statistics are really available, the "balance" between jobs created with "new" materials and jobs which perish with "old" ones is valueless unless the yardstick is a <u>stable</u> one, ie the <u>entire industrial structure must remain unchanged</u>. If it evolves concurrently with materials, <u>it is no longer meaningful to reason in terms of jobs. What does become essential, on the other hand, is how working methods change</u>. That is, attention should centre on the evolution of skills, and this depends largely on which scenario is likely to materialize :

Assuming a stratightforward system of permanent variety (in which materials play a decisive part, as has been demonstrated) with <u>continuous</u> remodelling of production and increasing mastery of complexity, there would be considerable problems in terms of <u>skills</u>. Permanent variety irrevocably breaks with one of the chief aspects of the old order founded on standardization of goods and processes and specialization of tasks because it <u>destroys the traditional mechanisms</u>

whereby know-how is acquired. Under a "fully developed" Taylorian system there is in fact a sharp division of labour between design and execution which tends to negate the idea of learning by the operatives, who are merely required to perform repetitive, specialized tasks geared to standardized production. Yet the permanence of the production system and the accumulation of experience inherent in the specialization of tasks provide scope for successive improvements and paradoxically favour the emergence of expertise specific to workers.

This development can no longer take place under a variety-based system because, depending on the state of the relationship between supply and demand, a rapid transition can be made from one technical approach to another and these traditional learning mechanisms no longer work. This brings us back to the origins of Taylorism, where all knowledge and expertise are located at the production design and organization level.

This situation would appear to be in utter contradiction with the fact that under a system of permanent variety, since competitiveness depends primarily on mastering complexity, the workforce must itself be capable of mastering it by means of suitable tools (expert systems, flexible workshops, etc ...).

On a more general level, difficulties stem from the fact that no degree of complexity attained in a scheme of variety can long be regarded as constant. This prompts the workforce at all levels to update their knowledge constantly through a suitable training system. The ease of adaptation depends on age ; your people with a good standard of basic training appear fortunate compared to other age groups in that their training is recent and comparatively suitable ; as they do not yet have a work routine, they can be trained for versatility and diversified knowledge of materials. Their elders, however, will be less "malleable", irrespective of their level of responsability, and will also need to expand their knowledge if they are to keep their jobs in the long term.

The difficulties inherent in the operation of a system of permanent variety are therefore of at least two kinds :

- such a system presupposes considerable ambitiousness in terms of training ; such being the case, most skills would in fact need to be substantially re-assessed and suitable means of training found ;
- this system naturally raises the problem of social relationships in the firm and more generally the problem of the system's acceptability to society at large.

If variety is imposed without consultation or training, there is a considerable likelihood that a general functional mechanism based on two classes of "agents" would develop in society. On the one hand there would be "designers and organizers" creating added value, enjoying learning opportunities, earning high wages and holding real power in many spheres, and on the other hand "operatives", poorly paid, deprived in practice of any real responsabilities in their work and unable to take part in devising future approaches. This danger may of course become nationwide (aggravation of the hiatus between "high-tech" and other areas), or even international (aggravation of the disparity between "highly developped" countries and the rest). As this is inconceivable, it seems in conclusion that a workable variety-based system calls for a high standard of training and co-responsibility.

The "plastics" scenario has "social cost" implications which differ from those of the variety scenario ; it fits a traditional production logic and in particular maintains the traditional learning mechanisms through its much more gradual remodelling of the industrial system. The "plastics solution" naturally includes technological solutions of extreme complexity requiring skills of the highest standard, but on the whole it follows the conventional pattern of technological change. As we have seen, it involves imitating conventional approaches before gradually replacing them. Furthermore, it is based on a corpus of knowledge which is relatively compatible with existing skills.

Scenarios are merely an ideal, and necessarily provocative, scheme of reality. In exagerating the tendencies, their sole merit is to provide a frame of reference for further consideration. Where employment and qualification problems are concerned, they at least help to locate the point at which the role of skills is central to the general evolution of technology and to the specific evolution of materials. It is also plain that the problem of skills and working conditions does not depend on technological development alone. It lies at the core of the objectives and strategies of a country or group of countries. The fundamental problem, which exceeds the scope of this study, is doubtless to ascertain whether the state of international competition allows different qualification systems in different countries to coexist.

In scenario terms, Japan has issued a challenge by proving its ability to manage a quasi-permanent system of variety, while Europe would appear more inclined to opt for a system committed to the "irreversible ascent of plastics" for many different reasons, eg preponderance of the chemical industry, rejection of undue mobility and slow rate of adaptation. Assuming that such a dichotomy could really exist, would the European "project" be viable against a background of fiercer competition and worldwide markets ? Much more pragmatically, it can be seen by examining the two scenarios that irrespective of the tendencies identified a number of major underlying trends seem to be gaining momentum in Europe in relation to employment and qualification problems in the materials sectors. They concern the importance of forward management of jobs, the importance of the evolutive skills concept and above all the importance of the concept of <u>training combining linked periods of study and work</u>.

PART I

A Trial Definition of Materials : Towards a Definition of Materials to Assist the Understanding of Current Changes

The idea that the materials referred to as "new" (composites, engineering ceramics, etc ...) which are being developed in advanced sectors like aeronautics and electronics will gradually replace "old" materials (steel, glass, wood, etc ...) in certain specific applications still has wide currency, which no doubt explains why it has taken so long to reach the stage of a comprehensive investigation of the role and future of materials. To regard the current changes as a simple matter of new materials superseding the old is in fact misleading and may furthermore have serious repercussions on understanding of current developments, and above all may lead to the wrong strategic moves. Indeed, the materials revolution has now become radical and all-embracing it affects all materials and through them influences design and production styles, while its effect on society's lifestyles is as far-reaching as that of the computer revolution. That is why any strategic analysis in this field must begin by attempting an "operational" definition of materials.

"New" Materials versus "Old" Materials : a Groundless Controversy

A simple and handy method of classifying materials is as follows. With reference to the theory of the life cycle of a product, it suffices in practice to employ the economic criterion of growth rate. In these terms, a "new" material (one whose growth trajectory is in a phase of acceleration) is any material whose anticipated growth rate over the next decade exceeds the average growth rate of western economies, ie about 3 %. Conversely, a material is regarded as "old" if its anticipated growth rate is below the average (cf. Figure I.1.a).

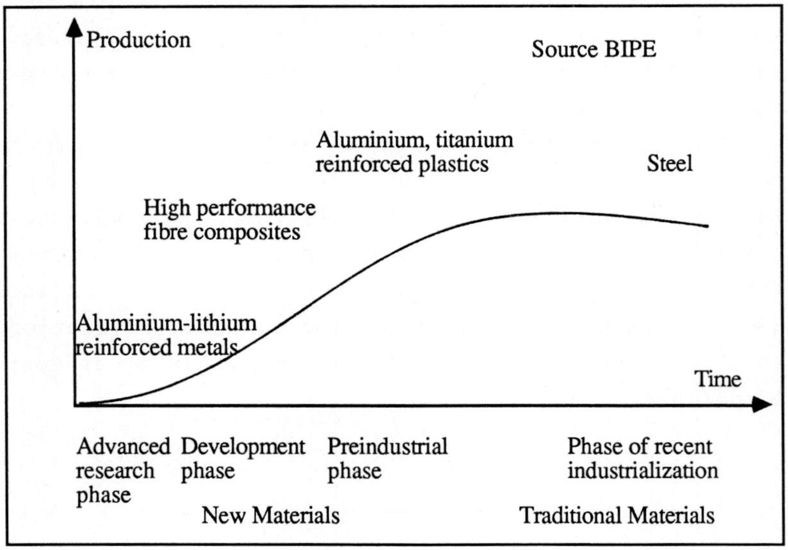

Figure I.1.a : Developmental curve of materials (according to the product lifespan curve).

	WORLD MARKET 1984 (10^6 $)	AVERAGE ANNUAL GROWTH RATE (1984-90)(%)
Single crystals		
- Silicon (wafers)	1400	15 - 20
- Ga-As and InP (semi-insulating wafers)	240	25 - 30
- Ga-As and InP (opto-electronics wafers)	165	25 - 30
- Ga.P (opto-electronics wafers)	30	5
- Piezoelectric quartz (raw)	80	< 6
- Lithium niobate (piezo.)	7	> 15
- Garnets for MBM	30	50
Plastics		
- Technical thermo-plastics	240	10 - 20
- Technical thermosets	300	10 - 20
- Technical adhesives (conductors)	9	12
- Consumer, trade and special laminates	1600	11
Metals		
- Rare earths (separated)	100	15
- Copper (supply and alloys)	1200 - 1500	< 5
- Nickel (pure and alloys)	100 - 200	2 - 3
- Molybdenum	40	4
- Tantalum (powder)	50	0 - 2
- Precious metals incl. electricity)	1600 - 2000	3
- Serigraphic compounds	250	14
Miscellaneous		
- Liquid crystals	10	20
- Metallized glasses (opto-electronics and photomatic)	230	25
- Ceramics (condensers)	1000	15
- Ferrite ceramics	600	< 5

Source BIPE

Table I.1.A : SOME STRATEGIC MATERIALS FOR ELECTRONICS

	WORLD MARKET		ANNUAL GROWTH
	10^3.t(1983)	M$(1983)	(%1963-90)
<u>Technical plastics</u>	<u>4 962</u>	<u>11 160</u>	
- technical thermoplastics	<u>1 882</u>	<u>4 970</u>	<u>9</u>
type : PA	480	1 300	8
POM	195	540	7
TPE - Bottle	420	680	12
TPE - PBT	85	230	8
PC	250	830	9
Top-of-the-range	12	130	> 15
TP elastomers	300	880	7
PPO/PS	140	380	6
- technical thermosets	<u>3 080</u>	<u>6 190</u>	<u>4</u>
type : epoxides	340	850	6
polyurethanes	2 380	4 040	3
silicons	360	1 260	9
polyamides	< 1	40	
<u>Advanced ceramics</u>			
(incl. catal. converter)	<u>3 200</u>	<u>300</u>	<u>20</u>
<u>Glasses</u>		700	
laminated glass	35 Mm^2	600	4
low-emission glasses	3 Mm^2	20	24
composite glasses (protective glazing)		70	5
<u>High-strength composites</u>	<u>27</u>	> <u>500</u>	
Plastic composites			
- glass fibres	21	170	15
- carbon fibres	3	210	26
- aramide fibres	2.5	125	20
- B and SIC fibres	0.015	15	

<u>Table I.1.B</u> : NEW NON-ELECTRONIC MATERIALS

I.1.1. **"NEW MATERIALS"**, by this standard, are organic (especially high-performance plastics and elastomers) and inorganic materials (chiefly engineering ceramics), metals and alloys (with improved composition, structure and purity), composites and most electronics materials (cf. Tables I.1.A and I.1.B). While today these materials as a whole account for only about 5 % of the total materials turnover, their high rate of growth guarantees that by the end of the century they will make up a much larger proportion. Furthermore, given that nearly 20 years often elapse between the research phase and the industrialization phase of materials, it is fair to say that the materials which will be used in the year 2000 are already known.

By observing their stage of development, "new" materials can be placed in the following categories :

- materials in a phase of recent industrialization (titanium, reinforced plastics in widespread use) ;
- materials now in a pre-industrial phase (high-performance composites, advanced ceramics, etc ...) ;
- materials in the development phase (aluminium-lithium) ;
- materials in the research phase (amorphous and microcrystalline metals (cf. Tables I.1.A and I.1.B).

Among new materials in general, emphasis is usually placed on the following main types : composites, engineering ceramics and electronics materials (1) .

- **Composite materials** afford technology a whole range of new physical possibilities ranging from their ability to offer tailor-made properties to scope for the redesign of existing products. What makes these materials special in the combination of properties of an initial substance, the fibre, which gives the material its structure, with those of a second substance, the matrix, which gives it "body". The first composite was produced some forty years ago by associating glass and resin ; such a material is as tough as glass but less fragile.

Composite materials today are classified according to whether they have thermosetting or thermoplastic matrices, glass, carbon or directional polymer reinforcement, metal matrices reinforced by ceramic or carbon fibres, or an inorganic matrix (carbon, ceramics, cements) reinforced by carbon or ceramic fibres or by polymers. These materials, developed in the aerospace sector, are finding more and more

---

(1) Strictly speaking, technical plastics should be added to this group. At several points, reference will nevertheless be made to these materials.

applications in the chemical and transport industries, off-shore drilling and marine construction, recreation and sports, telecommunications, machine tools, building and housing uses, biotechnologies and so on).

They are dealt with more specifically in Chapter II.2.

- **Engineering ceramics** are materials formed at high temperatures from compounds containing silicon, carbon, oxygen, nitrogen, aluminium, zirconium, etc ... They are light, hard, wear-resistant and rustproof, and poor conductors of heat and electricity. They are invaluable for a number of high-tech applications since they can function at higher temperatures than superalloys. Yet they are brittle, non-ductible and highly prone to small imperfections. Design offices must incorporate any ceramic parts into a specially designed structure whose contraints and assembly match the part, and not simply replace a metal part by a ceramic one, especially as ceramics can be thoughened by reinforcement and their reliability enhanced by closer control of their successive manufacturing stages.

The next 20 years will witness gradual penetration by heat-resistant ceramics in engines and turbines, vehicles and engineering and also in low-temperature applications where the characteristics exploited will be hardness and wear resistance, possibly in combination with biocompatibility (for prostheses, for instance).

Ceramics are discussed more specifically in Chapters IV.2 and V.3.

Electronics materials are dominated by one substance, single-crystal silicon, the discovery of whose semiconducting properties gave rise to the development of the computer industry. Its position of near monopoly stems not only from its exceptional qualitities but also largely from a group of fully mastered technologies (gaseous diffusion, epitaxial techniques, masking, photoengraving, etc ...) which have grown up around it. Owing to progress in these techniques it will soon be possible to envisage micro-chips carrying one million transistors. The future position of single-crystal silicon may be threastened by "III-V" materials (particularly gallium arsenide) which already hold large shares of the display, fibre telecommunications and high-speed switching markets. Micro-electronics technology is moreover highly diffusive, being evident in many solid or thin-layer materials, including magnetic and non metallic materials. Thin-layer circuitry on amorphous materials holds considerable prospects for large-surface display, and infra-red imaging circuits based on "II-VI" materials are expected to reach maturity. The commercial distribution of bubble memories has begun, while thin-layer magnetic recording may equal silicon micro-electronics in volume. "Smart" micro-electronic sensors and actuators are arousing keen interest and opening horizons for very extensive development in conjunction with electrotechnical engineering, automotive and consumer applications. This dissemination of basic technologies developed in micro-electronics may recur in other areas of materials. The demand for materials in electronics itself remains large and we shall witness the emergence of advanced chemicals for the electronics branches concerned by semiconductors, ceramics, laser optics materials, magnetic materials and polymers (optics and passivation). Electronics materials are studied in Chapter III.1. In the longer term, other new materials such as amorphous metals (metals whose atoms, as in glass, are not arrayed in crystals) will probably assume growing importance. Amorphous metal alloys produced by fast hardening may play a leading part in electricity distribution, particularly transformers, owing to their exceptional magnetic properties.

I.1.2. **"OLD" MATERIALS** (or traditional materials) are those whose growth rate is under 3 % (or even negative in some cases). They include traditional steels, metals and alloys, glass, wood, concret and even the major "commodity" (PVC, polyurethane, polystyrene).

Statistics thus indicate that the volume consumption of these materials correlated with GNP or population size is generally tending to reach saturation point or even to fall in most industrialized countries.

Figure I.1.b showing Japanese consumption of the principal metals exemplifies this phenomenon. The turning point can actually be placed just after the oil crisis of the mid-70s.

Of all the traditional materials, <u>steel</u> is probably the one which, owing to the sudden crisis in the steel industry following the oil shock, has been most representative of a threatened or declining material. All the industrialized countries have had to make large-scale reductions in capacity (and also in jobs) to restructure the steel industry. In Europe, for instance, the Commission of the European Communities late in 1980 even declared the steel industry to be in an obvious state of crisis (since then, the Commission's measures have reduced capacities by about 15 to 20 % according to product, and cut the workforce by about 30 %). However, steel is not the only material affected by the crisis ; a "young" metal like aluminium, whose annual growth rate prior to the first oil shock was approximately 10 %, is also having to trim its surplus production capacity. World consumption increased by a mere 2.5 % in 1986 (12.9 million tonnes) and the still

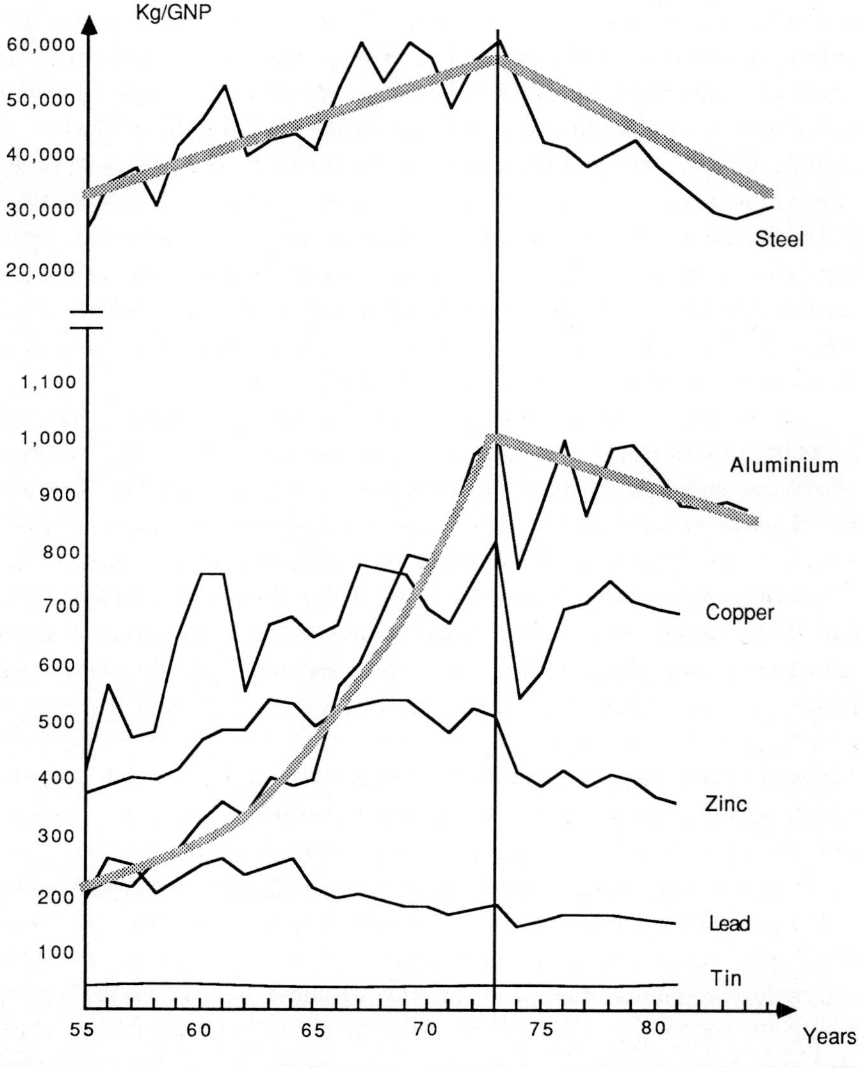

Figure I.1.b: Progress of metals consumption per unit GNP in Japan : reversal of trend after 1973.

uncertain prospects have induced some countries (Japan in particular) to close down a number of production units.

The inarguable decoupling of growth in the industrialized countries from volume consumption of traditional materials, as indicated by the figures for the mid-70s onwards, should nevertheless be interpreted very cautiously. Close examination of the industrial situation in fact shows that occasionally the quality of traditional materials has considerably improved in the last decade, and this has significantly qualified the debate over the decline of "old" materials. In the case of iron and steel, a proportion of the decline in tonnage is due to advances in performance : much less steel is needed for the same structures. Progress in the steel industry lies in the materials and methods. High strength steels are advancing more rapidly than are plastics in the automotive sector, and pre-coated steels resist corrosion. Continuous casting has reduced production costs. Refining processes and control of impurities have made alloys considerably more reliable. Thickness can be reduced through prior calculation of parts and structures and by computer-assisted techniques. The evolution of steels over the last 20 years has been such that "most of the production is now accounted for by steels which might be called new compared to those manufactured at the beginning of the 60s" (2) .

Likewise, aluminium's "service to the economy" from 1975 to 1985, ie the development of its uses per constant kilo, is calculated to have achieved an adjusted cumulative growth rate in the region of 50 %. Furthermore, after having given aeronautics the benefit of their high performance, aluminium alloys are starting to penetrate the automotive industry where they have already wrested a few important positions from cast iron. The new foundry alloys, which have far superior elongation properties to pressure moulded alloys, make it possible to use aluminium in structural parts which are important for vehicle safety.

For these developments to become general, new fabrication processes further removed from aeronautics techniques must emerge. Not all these developments are final ; new generations of cast iron are appearing and can now be used in much thinner sections.

---

(2) A. CONSTANT and G. HENRY "L'évolution des aciers pendant les 20 dernières années", ("Steel developments over the last 20 years", in "Traitement Thermique", n°177, October 1983. For further particulars about steels, see Appendix I.D.

Aluminium-lithium alloys, which will lead to a weight reduction of 10 % in future aircraft structures, may similarly check the invasion by carbon fibre composites.

The foregoing examples show that traditional materials can stay young and hold their own against "new" materials. On the other hand, there are several remarks which serve to qualify the importance of "new" materials.

- Firstly, "new" materials are often as venerable as "old" materials. For instance, ceramics started to be manufactured towards the end of the neolithic age. Closer to our own times, plastics began to appear between 1860 an 1900 (3) . Composites are perhaps the only "new" materials. But if we look back, how are we to regard the pre-1850 "unannealed drawn wires consisting of strands of thin wires and a vegetable fibre core" (4) , or "composite materials prepared from cements, water and aggregates", ie concrete (5) ?

- Moreover, the percolation of "new" materials from the high-tech to the more traditional sectors is far from inevitable. Certain composites or ceramics developed over the past 20 years were produced under defence programmes, but how much longer will it be before the multiple applications envisaged reach the industrial stage ? This points to a basic underestimation of the time factor : "advanced to mass technology" percolation is not automatic. Indeed, plant has to be designed and built to produce and process or shape these materials at cost-performance ratios which make competition with older materials feasible. It is therefore necessary either to create a new sector or to "completely transform the production tools of an entire economic and industrial sector" (6) .

---

(3) Pp. 651-658, Vol.4 of Histoire Générale des Techniques (general history of technology) edited by M. DAUMAS, PUF, 1978. See also pp.898-899 of Histoire des Techniques (history of technology) edited by B. GILLE, Gallimard, 1978.

(4) P.101, Vol.III, Histoire Générale des Techniques, op.cit.

(5) P.67, Report of the materials taskforce (Chairman : J.P. CAUSSE ; rapporteur : P.F. GOBIN, French Ministry of Research and Industry, 1982).

(6) J.H. BOURDOIS "La guerre n'est pas celle que vous croyez", ("The war is not what you think"), in Les Dynasteurs, supplements on materials to "Les Echos", magazine n°14484 of 9/10/1985.

- Finally, even if they are spreading satisfactorily, new materials must be convincing. It is one thing to produce and shape a new material with specific properties, but its reliability in the application planned or achieved during the lifetime of the technical artefact incorporating it is something else(7) . From this angle, old materials long embodied in the technical object perhaps have the advantages of experience, a completed technical and commercial learning process, an established body of information and tried and tested know-how. Of course, in so far as they also are being, or liable to be, "redesigned", "old" materials must also be convincing. However, their attainments in terms of quantity if not quality give them an edge over new materials in their typical applications (which new materials are trying to penetrate).

To sum up, the opposition between new and old materials is of purely statistical value since a closer look at industrial reality indicates that there are no old materials, simply outmoded techniques, processes and plant. Thus the novelty does not arise so much from the material itself as from the "material, process, product" complex. In this respect, the English term "new and improved materials" reflects current changes much better than does the French term "nouveaux matériaux" (8) . The English term clearly emphasizes that any material can be regarded as new and that the basic problem is how a new technical or industrial culture can fit into an old one and what sort of continuities, discontinuities and blockages will occur in the transition from one state of technology to another.

As the "new-old" contrast is inoperative and the problem of novelty is a general one, in that each material has the potential to become "new", another way to describe current changes must be discovered. This is where it becomes worthwhile to demonstrate that a basic transformation has been having radical effects on materials design and manufacturing methods for some 20 years ; in other words we have moved from macroscopic to microscopic industrial mastery of materials. This distinction is all the more important since it corresponds to the transition in economic terms from production based on scale-conditioned efficiency to variety-oriented production.

---

(7) Not to mention maintenance and recycling problems.

(8) The term "new material" is to be taken hereafter in the English sense of "new and improved materials".

CHAPTER I.2

From Macroscopic to Microscopic Industrial Mastery of Materials: the Characteristics of the New Materials

I.2.1. CHARACTERISTICS OF MATERIALS IN THE GROWTH ECONOMY (1950-1970)

The growth economy which lasted from the 50s to the 70s had as its salient feature a macroscopic industrial mastery. As well as presenting this feature, materials had economic characteristics ; a material was basically a mass-produced commodity. These technical and economic characteristics formed a coherent whole tending to restrict the variety of new materials or to relegate them to a marginal position.

I.2.1.1. Technical characteristics : macroscopic mastery of materials

Before the 70s, materials used to undergo homogeneity tests or checks which were made during production, with samples being taken and then analyzed. Macroscopic industrial mastery (9) of materials merely required statistical testing since the sample is considered representative of the whole production. The test, usually destructive, was conducted outside the production and working-up process. Any improvement in the properties of the materials was due mainly to research designed to improve their purity grades and so make them more homogeneous (10) . Materials which had to be industrially controlled at

---

(9) The macroscopic physical properties of a materials include hardness, elasticity, malleability, ductility, transparency, density, viscosity, etc ...

(10) Impurities are thus no longer conceived as elements to be eliminated, but as some fo those elements which may be absolutely .../...

the microscopic level only relate to very limited <u>specific applications</u> (major nuclear or space programmes, for instance).

In these particular cases, cost requirements have little or no influence whereas reliability and safety requirements are paramount in the development, operation and maintenance of complex technical systems such a nuclear power stations, satellites or rockets. In other words, technical constraints are the essential ones for these materials, economic contraints retreating into the background. The materials concerned are produced industrially in limited tonnages ; they are tailor made in factories where skills and know-hows make up a large proportion of the value of the product in which they are incorporated.

I.2.1.2. Economic characteristics

The technical characteristics described above are matched by economic characteristics. As a rule, materials which are <u>macroscopically controlled in industry</u> are <u>commodities</u> produced in factories geared to <u>increasing scale-conditioned efficiency</u> and <u>standardization</u>, so that emphasis is placed on the quantitative aspect. Under these conditions, the <u>supplier of materials</u> usually appears in a <u>strong position</u> in relation to the <u>customer for materials</u>, either because the latter does not have the financial, technical or human resources to state exactly what he wants or because the former is always able to meet a slightly more specific demand with existing equipment. Most materials are produced in the form of commodities to begin with, ie in the form of <u>wide-distribution, low-cost goods</u> used concurrently in a large <u>number</u> of <u>technical objects</u>. Accordingly, three members of the thermoplastics family dominated the period in question,

---

.../...

vital. In other words, they are controlled and then mastered, ie impurities deliberately incorporated into the material are given a specific form and distribution. If the status of impurities is "upgraded", then the status of the noble elements is "downgraded". All elements of the material thus acquire the same status and play an equal part in defining the microstructure and consequently the macroscopic properties of the material.

covering two-thirds of requirements (11) . They were polyvinyl chloride (PVC), polystyrene (PS) and its derivatives, and polyethylene (PE). Similarly, ordinary steels specifically intended to create hollow or static structures (containers, furniture, car bodies, metal frames etc ...) accounted for the bulk of steel production. All these commodity materials have a broad field of application. A specific material like PVC can occur in a host of technical objects.

Consequently, any increase in the diversity or variety of materials over the period under consideration was more confined to advanced applications. The diversity of commodity materials was usually increased within a family by altering a few characteristics rather than by creating a new family. Furthermore, this diversity had to measure up to an additional characteristic of materials as commodities, namely the predominant drive, over this period, to achieve higher yields and standardization, coupled with a certain rigidity of production processes.

The drive to achieve <u>standardization</u> and <u>scale-conditioned efficiency</u> gave rise to two phenomena. First there was the <u>rush to build giant-sized factories</u>, as in the case of steel plants whose blast furnace capacity increased five fold between 1960 and 1970 (from 2,000 to 10,000 tonnes per day). The same thing happened in commodity plastics ; the replacement of coal-chemicals by petrochemical derivatives, larger production units and finally automated production processes were the principal methods used by producers of synthetic plastic resins in the race to achieve price reductions. Also, the pursuit of increasing scale-conditioned efficiency resulted in <u>incremental innovations</u> in the <u>quality</u> of materials, though with a far from insignificant effect in volume or quantity terms. The introduction of continuous casting, for instance, made it possible to improve the steels produced, but above all to save energy and increase manufacturing capacity for finished steel products by about 10 to 12 % on average (12) . Thus new processes for the manufacture of materials were introduced where they made for a reduction in costs with a

---

(11) C. WIPPLER, "L'évolution des matières plastiques", ("Development of plastics"), in : "La Technique Moderne", April 1969.

(12) Another example is continuous annealing, first utilized by Nippon Steel in 1972 for reasons of production economies. Only afterwards was it seem to offer very substantial improvements in the manufactured product.

concurrent increase in quantities produced. Their qualitative potential was exploited unsystematically if at all, with no attempt to create a new family or to bring the microscopic complexity of the material under industrial control.

Thus commodity materials were produced in factories where the imperatives of higher scale-conditioned efficiency and standardization prevailed. These materials, of limited economic variety but with multiple uses, were usually "distributed" rather than "sold" in the sense that the marketing aspect was unimportant. The relationship between the materials supplier and the materials customer was in a sense linear, implicitly or explicitly based on a balance of power in which the production end was under no compulsion to diverge from its standards. Furthermore, the customer's processing equipment was hard to adapt and left him little if any choice even where he was in a position to specify his needs in the way of materials. Supplier-customer dialogue therefore did not need to be systematic. The party in a position of relative weakness - usually the customer - would be regarded as compelled to accept whatever was offered, in many cases the sole choice. On the whole, the supplier was in a strong position by virtue of his large capabilities (know-how, research, finance, etc ...) and ability to meet the order cheaply by possibly adjusting some specifications to anticipate or meet the user's requirements.

As commodities in an expanding economy featuring straightforward relations between suppliers and customers, materials showed little variety. Variety only existed at the fringes of the growth economy, for special applications requiring microscopic mastery of the material for severe conditions of use. The dominant materials of the growth economy did not reflect the same economic logic as those which emerged as new materials over this period. The limitation of variety can thus be perceived as inherent in the economic logic which then prevailed.

I.2.1.3. Limitation of variety

The growth economy was reflected by the limits on the variety of materials ; among the materials available for a given application, only one commanded a specific function or group of functions within the application concerned on a lasting basis, and did so at fixed prices, maintenance and recycling costs included. It was, so to speak, a "Hobson's choice" economy.

In the 50s and 60s most of today's new materials were known, particularly technical plastics and composites. So were certain steels, engineering ceramics and non-ferrous alloys. However, irrespective of industrialization problems, the prevailing economic conditions precluded their wide distribution. Technical variety, that sometimes potential overchoice, did not necessarily have an economic equivalent, in view of the economic characteristics of the growth economy (standardization and increasing scale-conditioned efficiency). For a specific material to prevail, it would of necessity have had to conform to the principles of such growth, ie to become a multipurpose commodity material instead of following its logical pattern by splitting into a variety of materials with specific applications. The suppliers of materials can of course be held responsible. The trend in demand for materials was itself a restriction on their economic variety. It was more quantitative than qualitative, geared to standardized mass production and increasing scale-conditioned efficiency. This was still consistent with demand by the end-consumer. It is by no means gratuitous that these years more than any others were the era of mass consumption.

Limitation of the variety of materials is inherent in the features of the growth economy which produced, processed and used materials representing a technical dictate at the lower end of industry. However, this also depends on raw materials, cheap energy sources in particular. The disappearence of such "godsends", in particular one of the chief energy sources, upset this economy and hastened certain developments which were foreshadowed during the years of growth. At that time the advance of new materials was beginning, and the trend of emergence and dissemination was to be speeded up by the recession.

I.2.2. NEW MATERIALS : A RESPONSE TO THE APPEARANCE OF NEW CONSTRAINTS (1970-1980)

The growth economy began to run out of steam towards the end of the 60s. The first oil shock in fact marks the end of sustained growth. The 1974 crisis pinpointed and aggravated the tensions which had arisen in the 60s. Recession was considered conjonctural until the second oil shock (1979), which sealed the failure of policies designed to boost demand. From then on, the recession was regarded as structural and irreversible, and at the same time the expectations of economic agents changed.

As regards materials, the immediate reaction after the 1974 crisis was to relax the energy restriction at all levels, ie production as well as processing and use. Furthermore, materials were also being asked to meet new trends in demand. Thus new materials made it possible to revert to a type of growth where quality takes precedence over quantity, where productivity gains once more begin to show.

Well before the first oil shock, there was a weakening of the growth economy system founded on the pursuit of growing scale-conditioned efficiency, standardization and a number of cheap resources. A falling-off in the growth of labour productivity appeared at the end of the 60s in the USA and at the end of the 70s in western Europe, together with over-capacity and over-production. This is what happened in the steel industry early in the 60s : direct competition affecting steel products appeared in markets outside the EEC and prompted certain European firms, particularly the major German firms, to integrate downwards (13) .

The first oil shock revealed tensions of this kind and abruptly aggravated them. It is indeed striking that a materials industry like steels, should illustrate most vividly how the growth-oriented scheme came under challenge. The steel industry was in difficulties, mainly over a significant drop in demand (14) . Production capacity utilization rates were going down, in both the steel and the plastics industries. The higher price of petroleum, the raw material of the plastics industry, hit the cost structure of European producers while new producing countries favoured by the availability of the raw material began to appear. The European producers'response has been to rationalize production by closing down unprofitable capacity, improving the productivity of that still in use

---

(13) This integration, achieved before the recession, enabled the large German firms to keep out of the red somewhat longer than their counterparts in other European countries ; F. KERN, Thesis, Strasbourg, 1982.

(14) For a more detailed discussion of the causes and factors aggravating the recession in the steel industry, see F. KERN, op.cit.

and striving for a more efficient division of the market. Concurrently, there was a shift in activity towards products with higher added value incorporating more know-how and therefore harder to imitate. (For instance, fresh impetus was given to polymer research and to the production of high-performance plastics).

Moreover, up to the second oil shock, which also confirmed the failure of policies intended, to get the economy moving again, certain industries continued to regard the crisis as momentary and reversible. One of them was the steel industry, which between the 50s and the 70s experienced recurring difficulties on a number of occasions and some setbacks to expansion. The fact that the European steel industry consists of a set of national or nationalized steel industries, compounded by the technological backwardness of some firms (which must be made up at all costs), did nothing to correct this "short-sightedness", which after all was coherent. However, with every year that passed hopes were dashed, financial resources dwindled and outlets withered away. The second oil shock tolled the knell of a sometimes excessive optimism and made these industries revert to a more reasonable or even cautious realism.

For all those involved during the years of recession, <u>energy saving</u> became an overriding concern or even a <u>strategic approach</u> ; almost regardless of cost, this constraint had to be eased in order to increase competitiveness and recover significant gains in productivity. Thus materials received considerable stimulus in so far as they made it possible to influence the energy constraint. Various institutions, though not for the same reasons, mobilized themselves to put a speedy end to this constraint, which was found exorbitant. Certain states, confronted with a sudden deterioration in their trade balance, adopted measures to limit national energy consumption. This is how energy-saving norms and building programmes came to be adopted, encouraging building material producers to develop suitable insulation materials (15) .

On the other hand, certain producers of mass market goods, like the vehicle manufacturers, confronted with keener competition in dwindling markets, are now prepared to accept a premium cost on materials provided that they yield significant gains in vehicle fuel consumption. Lower energy consumption depends on <u>weight reduction</u>,

---

(15) Cf. Chapter III.3.

better engine performance or a different aerodynamic design (16) . The same applies to the aircraft industry. There is a need for lighter materials more resistant to extreme temperatures, tough and malleable enough for shaping and inclusion in the technical object and suited to its uses.

In other words, considerable demands are made on materials producers whether chemists, ceramics specialists or metallurgists, and their responses increase the variety of materials. New materials, backed by information technologies whose advances affect all stages in the manufacture of technical objects, are primarily geared to this requirement.

Furthermore, gains of other kinds are increasingly being taken into consideration ; productivity gain arising from the choice of a material or potential gains in use, maintenance or recycling may mean that of two materials with equivalent technical characteristics the one considered more expensive may be preferred to the arguably cheaper one. The introduction and functioning of new materials nevertheless raises problems as to <u>skills</u> (17) and <u>survival of old equipment</u>. The competitiveness of a new material in fact depends or converting production and working-up processes, if not developing and setting up new ones. The strategy of firms producing materials thus becomes a vital element where these questions are concerned. Perpetuating old equipment and expertise while at the same time facing the future through diversification of materials is one means of making the transition between two very dissimilar technical cultures, ie from one based on macroscopically mastered materials to one based on materials microscopically mastered on an industrial scale. New materials thus ease the energy constraint and offer advantages of many kinds at all stages of production. While they also appear to be a means of restoring productivity, there nevertheless remain <u>human problems</u> which are perhaps the <u>hardest to deal with</u>.

Finally, new materials provide a response to more qualitative and specific <u>new trends in final demand</u>. The recession is not purely economic, nor is it simply a time of tactical retreat, be it individualist or, at the national level, protectionist. It also has a cultural dimension. From this point of view, the social changes between

---

(16) Cf. Chapter III.1.

(17) For a more detailed discussion of skills see Part V.

1965 and 1970 mark the end of an era and the start of another in western economies - the end of an era of consumption where quantity took precedence, and the beginning of another where quality at the lowest cost dominates the concerns of consumers. Thus the years 1965 to 1970 saw the emergence of a culture geared to quality of life in a very broad sense. What is more, the recession of the 70s and 80s has merely served to accentuate the demands of final consumers. This, then, is the challenge taken up by new materials whether in transport, housing, health, sport or recreation.

The boom in new materials is therefore in harmony with the needs of final consumers and the need for renewed productivity and profitability in the industries producing and using materials. This has made the 70s and 80s a springboard for the emergence and dissemination of an "overchoice" of materials (18) . While initially designed to overcome a constraint whose absence was one of the foundations of the growth economy, they are also helping to relax others. New materials are therefore not only responses geared to the constraints generated or disclosed by the economic recession, but are also the instrument which, in termes of <u>overchoice</u>, makes a new technological system (19) coherent and which, in terms of <u>variety</u>, makes a new economic situation coherent.

From this point of view, the sudden rise in the price of oil tended to speed up certain developments already perceptible between the 50s and 70s. Its equally sudden drop may in some cases slow down these developments and occasionally halt attempts to substitute certain new materials (20) . In general, though, there would seem to be no turning back, irrespective of oil price cuts. In other words, regardless of variations in the price of energy, the <u>irreversibility of change</u> is

---

(18) The term overchoice, apparently coined by Alvin TÖFFLER, has been employed to describe the situation created by the emergence of new materials.

(19) The technology system is the sum of the technological coherencies responsible for the production of all goods consumed by man. The system is closely interrelated with the economic system, the social system, etc ... The mainstays of the technology system are materials, energy, information and life. For further details see B. GILLE (1978), M. DAUMAS (1979, CPE-Sciences et Techniques (1985).

(20) This applies, for instance, to maritime transport ; cf. Chapter III.1.

established. As a result, an economic logic founded on new materials is likely to predominate.

I.2.3. MICROSCOPIC MASTERY OF MATERIALS AND ITS CONSEQUENCE

In order to specify what is meant here by new materials, we must place ourselves at the centre of their manufacturing technology. For centuries human effort was devoted to transforming matter by physically reshaping it. Forging, founding and pressing techniques are the best examples in metalwork of this age-old approach from the "outside" (macroscopic level) of matter. With new materials, one seeks to organize the internal (microscopic) structure of matter from the "inside" (21) . The technologies applied to the new materials are complex, reflecting mastery of the microstructure of matter. This presupposes knowledge of the laws governing the molecular, atomic or even subatomic arrangement of matter. However, this mastery would have no technical or economic meaning without the ability to reproduce microstructures industrially. One of the major implications of this second necessity is the need for new procedures for testing the quality of materials. The inherent microscopic complexity precludes a quality check based on the testing of a sample, whose representativeness is not adequately assured since minor variations in the production conditions of materials may adversely affect their microstructure. As they are integrated in the production process, testing procedures must be continuous, which means that they preserve intact the structure of the materials (non-destructive testing).

Mastery of the microstructure of matter explains in particular how certain traditional materials which evolved precisely owing to the use of complex technologies became "new". This applies particularly to steels whose new formulations (high yield-point and dispersion steels) stem from the ability to control the dispersion and geometry of inclusions under industrial conditions . The same applies

---

(21) M. REYNE, quoted by R. THALLER, Thesis, Lyons, p.167, 1986.

to the new alloys and, albeit in different forms, to technical plastics and ceramics (22) . Thus we have a sub-group of materials for which the control of reproductibility operates at a microscopic level but whose preparation depends on the same raw materials as those used in the conventional formulae, viz. iron and carbon for steels and special cast irons, iron and other ores for the new alloys, organic raw materials (petroleum, coal, natural gas) for technical plastics, inorganic and non-metallic raw materials (silicon, alumina, zirconia, boron, etc ...) for engineering ceramics and information technology materials.

What is more, with certain new materials reproductibility must indeed be mastered at the fabrication level, ie at the level of a material derived from a homogeneous family of raw materials, but also at the level where two or more materials belonging to different families are combined or variously shaped. The materials concerned are composites. Fibre-reinforced composites rely on different substances, one for the matrix and the other for the reinforcement. Where the two are identical, the difference lies in their primary shaping as in carbon-carbon or ceramic-ceramic composites. Composite materials with a sandwich structure, as in the case of cellular materials or metal-plastic laminates, follow the same rules as fibre-structure composites except that the various materials or primary shaping processes are interfaced by a bonding process . By analogy with the distinction between primary and secondary energy, the first sub-group of materials will be referred to as the primary materials group and the second sub-group, composite materials, as the secondary materials group.

The first conclusion to be drawn from the foregoing distinction is that it highlights the emergence of a new class of materials. Apart from reinforced concrete and the natural composites bone and wood (23) , not to mention cobwork, pisé and plywood, the secondary materials formed an insignificant branch of materials before the arrival of composites. The peculiarity of composite materials is

---

(22) Mastery of the phenomenon of cross-linking is especially important with polymers. Ceramics manufacturing methods entail the use of high-temperature processes. The static behaviour of ceramic materials further necessitates extreme precision in processing to avert the risk of microcraks.

(23) Bone is a composite of collagen and spatite ; wood is a composite of cellulose and lignin.

that they offer properties exceeding the sum total of the properties of their constituent primary materials, and that they associate properties which engineers formerly considered incompatible (lightness and strength, for instance). This constitutes one source of <u>overchoice</u>, as it becomes technically possible to obtain identical, superior or novel properties from different combinations by manipulating only the nature of the components. In this respect, however, new primary materials are close behind and constitute the second source of materials overchoice.

The most significant example is undoubtedly that of technical plastics, which present a continuously increasing range of properties. They have all the properties common to plastics as a whole, namely low density, good impact and corrosion resistance, electrical and acoustic insulation, etc ... In addition to these traditional properties there are others : good performance at extreme temperatures, high strength (24) , hardness and resistance to aging associated with ease of processing (25) and electrical (26) , biological, optical and other properties.

With composites or technical plastics, overchoice means that for a given product not one but <u>several</u> materials can be adopted. In other words, the variety offered by new materials increases the number of degrees of <u>freedom of economic choice</u> regarding materials. An example illustrating this overchoice is the rivaly among aluminium, high yield point steels, reinforced plastics and steel-plastic sandwiches in car body parts.

The mastery of microstructure which is gradually spreading to all materials, in conjunction with the possibility of testing them non-destructively, thus constitues the decisive technical factor permitting the current transformations in materials. On that basis, it is easier to understand why <u>today any material can be considered "new"</u>

---

(24) For the new RIM plastics, oriented polymers or Inter-Penetrating Network (IPN) type polymers. The strength of these new polymers is comparable to that of high-performance composites.

(25) This combination of properties, formerly considered.

(26) There are intrinsically conductive polymers obtained by doping pyrol, acetylene and aniline polymers in the absence of metallic agents. Extrinsically conductive polymers, on the other hand, are polymers with conductive or semiconductive additives or reinforcements.

if microscopic mastery has led to a revolution in its design, production or use.

What now remains to be considered is how the new technical possibilities combine with the economic factors. To identify the current technological trajectories it is necessary to draw a final general distinction between structural materials and functional materials.

CHAPTER I.3

Structural Materials and Functional Materials

The distinction between structural materials and functional materials arises from a fundamental difference in the manner in which they are used.

I.3.1. **STRUCTURAL MATERIALS** are used primarily to define the form of industrial objects. By giving form to a specific article, the structural material acquires the remarkable property of acting as an interface between production (which has "formed" the article in question) and consumption (where consumer taste tends to favour a certain form over others). Steels, aluminium or plastics in transport, wood, glass and concrete in buildings and paper, cardboard and complex materials in packaging are examples of structural materials which in their respective sectors (the three sectors mentioned are the chief users of structural materials) define the form of industrial articles (car bodies, walls, house doors and windows, cartons, boxes and bags, etc ...).

The importance of structural materials is considerable both in the industries which make them and in those which incorporate them into industrial objects. The very nature of structural materials gives rise to specific industrial phases which make up a very large proportion of industrial activity. A structural material "regulates", so to speak, a great many production sequences (and consequently, the skills as well) which extend from its manufacture to its incorporation into an industrial object.

A metal like aluminium, for instance, once its primary production has been completed, is firstly worked up into semi-processed products (billets, ingots, bars, etc ...), then undergoes a long series of operations in the user industries to acquire its final form : stamping, moulding, trimming, finishing operations, etc ...

In a way, the structural material is inseparable from the sequence and nature of the manufacturing process for industrial objects. Its importance reappears moreover in the proportion represented by structural materials in the end value of industrial objects : in a car, for instance, the bodywork accounts for some 40-50 % of the vehicle's value.

The other specific aspect of structural materials is that they relate directly to the environment. They must consequently take into account not only consumer taste (the province of materials design) but also safety and environmental requirements.

This peculiarity implies that structural materials will often be chosen according to local life-styles. In the case of aluminium, it can be observed that the place occupied by this material in its three main applications - transport, building and packaging - varies distinctly among the main consumer countries, as the figure below indicates.

Owing to this feature, plus their generally high weight, they are comparatively seldom (directly) exported. The markets are confined to major consumer areas (Europe, United States, Japan, Asia). The need to allow for a number of environmental constraints also means that structural materials are by nature dependent on norms laid down by the authorities regarding safety, anti-pollution measures etc ... The study in Part III of structural materials in the main fields of application (transport, building, packaging) accordingly highlights the decisiveness for the evolution of these materials of official measures in the shape of norms.

Above and beyond environmental constraints, a material must of course incorporate society's chief goals. For instance, as has already been pointed out, the energy crisis has placed a very firm weight-saving requirement on materials and has been one of the main vectors of their development in recent years. It is likely that within the next decade other social goals will need to be taken into account.

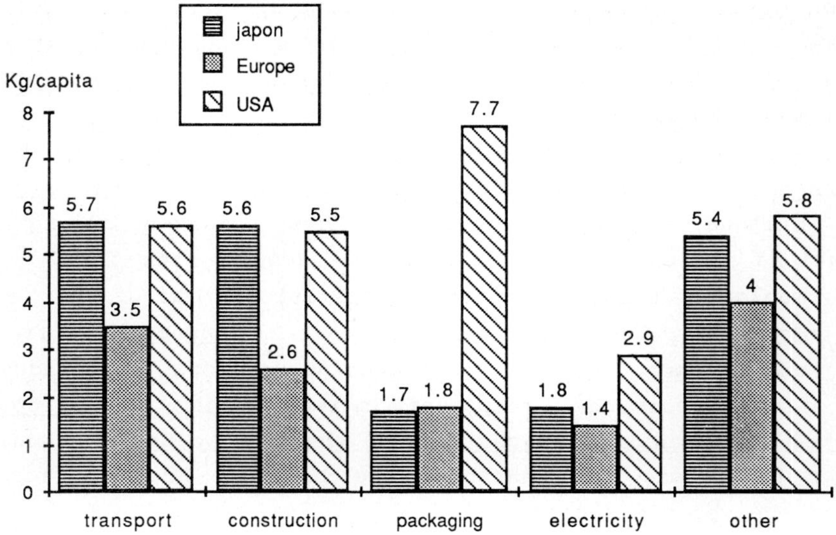

Figure I.3.a : Uses of aluminium. Source : Péchiney - Euroconsult

Among these, the imperative of safety (particularly following recent disasters in which certain materials have been proved responsible) should assume growing importance.

In brief, structural materials are characterized by intense interaction with the designing of production processes and by marked dependence on the general conditions of demand.

These two links with production process design and demand-related constraints pinpoint the expediency of mastering the microstructure of materials. Microscopic mastery leads to major gains in the design of production processes and affords a more satisfactory response to demand through the possibility of manufacturing new materials capable of adapting practically at will to their intended use. Over and above the expansion of the range of properties offered by the new materials, almost every one can provide tailor-made properties.

In this respect, two examples deserve special attention. The development of tough new RIM plastics is founded on the existence of a real CAD process for chemicals (27) . These new polyoleofins of the polycyclopentadiene type make it possible to optimize the organization of the basic chemical groupings in order to obtain certain materials some of whose properties can be improved selectively according to certain lines of constraint. For instance, if the required properties are rigidity, hardness, or resistance to high temperatures, the grouping for the original monomer will be chosen from the esters, hydroxyls or acrylics. If, on the other hand, all these properties are desired, it will be possible to achieve and optimum combination of the corresponding groupings.

The evolution of composite materials towards a functional approach furnishes our second example (28) . The functional approach involves designing the reinforcement of the composite material to suit the actual shape of the object. In this case, the material is adapted to the object not by means of an optimum selection of its technical properties but rather on the basis of the architecture which comes closest to the design of the object to be produced. The ultimate stage in this approach seems to have been attained by the wire winding technique which simultaneously processes the material and shapes the object. The wire winding technique has become established for complex components such as high-pressure tanks, and research is under way at General Motors to apply it to the production of a lattice structure for cars.

The ability of the new structural materials to furnish tailor-made properties and customized shapes is in sharp contrast to the habitual attitude once taken to materials marketing. It is now no longer economical merely to sell them by weight or just to distribute them. The once occasional cooperation with the customer has become necessary in most if not all cases. Thus new materials are high added value products and not commodities. Table I.3.C vividly illustrates this for plastics. The high added value is accounted for by the incorporation into the material or its application technology of advanced skills connected with mastery of the material's micro-structure. However, its actual making - the addition of economic value to the material - usually requires the provision of quality service. It

---

(27) The evolution is described in Chapter II.2.

(28) The evolution is described in Chapter II.2.

is just as important to brief the customer on the material's performance and conditions of use as to participate actively in the design of new technical objects with scope for the use of new structural materials. In such a context, materials marketing tends to become increasingly imperative.

Cost distribution \ Materials	Commodities	Technical plastics
Material	82%	48%
Machine	11%	26%
Tooling	5%	21%
Labour	2%	5%

Table I.3.C : Percentage breakdown of processing costs for everyday (commodity) plastics and technical plastics. Source : G. MENGES 1985, 25 .

Analysis of structural materials is the subject of Part II, an Part III deals with the main fields of application of structural materials (transport, building, packaging).

I.3.2. **FUNCTIONAL MATERIALS** are chiefly intended to accommodate a nuclear or electronic physical property, as opposed to structural materials which must accommodate a mechanical physical property (usually originating in the nature of the chemical bonds between atoms). Thus functional materials are required to conduct electricity, convey information, convert electricity into photons, photons into energy, catalyze chemical reactions, etc ... Functional materials as a whole therefore define they limits to the technology system at a given time and are consequently strategic materials, the object of keen international competition. The market for them is usually worldwide (most are comparatively easily exportable components) and does not normally depends on environmental characteristics or local lifestyles.

They are nevertheless heavily dependent on the progress of basic research.

Apart from their intrinsic value in establishing "the outlines of the possible", functional materials are fundamental to the structuring of major technological branches; the telecommunications, computer, electrical equipment and oil refining industries, for example, are founded on the existence of "dominant" functional materials (semiconductors, optical fibres, electrical cables, catalysts, etc) which structure an entire industry. On the other hand, despite their high added value, these materials generally represent only a small proportion of the value of the articles in which they may be incorporated (a few cents for a silicon chip in a computer).

The structuring of the major branches of technology requires, in addition to the basic functional materials, the development of other functional materials (for connections, carriers, display, etc) in which the essential property required is compatibility. As a result, there gradually develops a coherent group of mutually compatible functional materials which, starting from a small number of basic functional materials, structure the major technological branches. Consequently, when technological progress achieved in seeking higher and higher performance makes dominant new functional materials available, there may be considerable repercussions on all branches of technology.

Mastery of microstructure is the driving force in the evolution of functional materials. As knowledge of the infinitesimal improves, the availability of high-performance functional materials exceeding the physical capabilities of old materials increases. This command of the microscopic aspect is particulary well illustrated by semiconductors in micro-electronics, where the rate of integration is increasing in terms of higher density of information processed over an elemental surface of material. It is now conceivable to build a semiconductor atomic layer by atomic layer. There are even plans to construct complex molecular assemblies capable of storing information.
The paradox, furthermore, is that this would bring us to the very limit of the concept of material ! This constant progress in miniaturization has the major advantage of speeding up the processing of information. Thus the data processing and telecommunications industries are the ones directly concerned by progress made with these basic functional materials.

Functional materials are studied in Part IV.

I.3.3 The distinction between structural materials and functional materials is of course somewhat arbitrary. Any material has a "functional" component and a "structural" component. For instance, it is hard to dissociate the two aspects in the case of certain materials designed to resist high temperatures and used in engine turbines, and the same applies to heat and sound insulation in building. It will even be observed that one of the major tendencies of structural materials is to incorporate ever more "functions" (if only to overcome one of the major problems of structural materials, ie corrosion).

The above distinction between materials nevertheless has the advantage of clearly demonstrating that there are different technological trajectories in the materials field : structural materials contribute primarily to the redesign of production processes inside companies and at the same time strive to meet demand-related and environmental requirements. Functional materials determine the availability of technical possibilities at a given time and to a large extent the development of the major fields of technology. Each group of materials has its own distinct pattern of strategic operation : the main problems of structural materials relate firstly to the dissemination of knowledge and know-how among firms (particulary small and medium enterprises) and secondly to the laying down of norms (safety, environmental protection, etc) by the authorities. Functional materials, on the other hand, primarily depend on basic research, and their development gives the countries or groups of countries controlling them considerable influence over every relevant technology.

PART II

Structural Materials

CHAPTER II.1

Introduction : Variety, Complexity, Integration. The Role of Structural Materials in the Transition to a New Information-rich Production System

Since they demarcate the shape of industrial objects and are located exactly at the point of interaction between industry and its environment, structural materials play a major part in reorganizing production processes. In many respects, upheavals in production processes as a result of the new fabrication methods for materials foreshadow transformations of at least equal importance to those arising from the use of information technologies. Moreover, it is plain that simultaneous mastery of the evolution of structural materials and information technologies respectively is the sole means of managing effectively the transition from an old industrial order founded on standardization to a new one founded on variety.

II.1.1 THE GROWTH OF VARIETY

The gradual introduction of a new variety-oriented system means that firms must come to terms with a dual constraint :

- Firstly, allowance must be made for an increasing variety of customer and environmental requirements. Changing consumer norms, the need for higher product quality and compliance with environment regulations, or in more general terms the response to society's major goals (energy saving, for instance) are all arguments against the standardized production system. The lower end of the production system is increasingly conveying its demands in terms of functions to be fulfilled, quality requirements to be observed, performance standards to be achieved and services to be rendered. The implications for structural materials are considerable. To take just one example, the need to save energy has led to the inevitable lightening of structural materials in vehicles. This factor has revolutionized bodywork design and brought thin sheet steel, aluminium, plastics and composites into competition with each other.
- Secondly, allowance must be made for an ever-increasing variety of technical options afforded by the availability of an ever-wider range of new materials and working-up processes. This great variety of materials heralds in a way the possibility of upending the logic of their application; so far the initial stage (the material) has been a determinant for the later stage (the products embodying the material). Under the conventional industrial order, materials were in fact a typical example of a technical constraint imposed from the outside. The raw material was adapted to its applications by gradually perfecting its properties, which nevertheless remained dependent on the initial "commodity material". This involved "pseudospecialities", so to speak. With the attainment of microscopic mastery, dependence on the commodity is removed and optimization is achieved item by item : the material becomes an intrinsic variable of design. The modular nature of matter establishes itself as one of the key aspects of the new technology. Another important point is that the modular aspect necessarily brings into play several branches of knowledge formerly separated and now needing to be integrated or interfaced. This inevitably has a number of effects on industrial structures, namely development of new trades, a new division of labour and the emergence of new companies.

II.1.2 THE GROWTH OF COMPLEXITY

The need to achieve compatibility between growth in the variety of customer demands and growth in the variety of technical solutions offered by materials will confront businesses more and more with a major phenomenon of growing complexity in the management of production processes. How indeed are viable industrial responses to be found to cater for growth in the variety of demand and in the ever-changing possibilities afforded by materials supply ? How is production to be geared to the dynamics of change ?

The variety factor entails not only altering the rules of management and production, but also rethinking marketing and R&D, functions which must in addition be closely interconnected. Inter-company relations will also be altered as a result; as the variety and complexity of products increase, more skills have to be incorporated and made to interact - hence the difficulty which a firm, however large, has in dominating such a field single-handed. Such issues typically arise in the automotive sector. In managing the complexity resulting from these new variety-oriented production methods, the traditional relations of sub-contracting are supressed by the more fertile and less hierarchical relationship of industrial partnership as a means of combining the various proficiencies needed.

The solution commonly relied on to master this growing complexity is the use of information technologies. Indeed, information technologies (CAD, robotics, flexible workshop, office automation) permit the redesign and subsequent production of multimaterial articles and/or articles made of complex materials. The introduction of computerized forms of automation aids the processing of materials which are to acquire their final form in the product itself. Furthermore, whenever a specific function of a firm is computerized this generates extra information which helps to organize the other functions better, ie makes them easier to express as algorithms and hence to computerize in turn. Nor is there any apparent limit to the scope of computerization, in that once the more routine processes are automated the decisional capabilities of the firm refocus on the problems at the next highest logical level, and these increase in complexity as progress is made in the direction indicated above (systems approach, management of variety, etc). Decision-making processes may themselves be assisted by computer techniques such as artificial intelligence when the experience acquired ultimately makes it possible to systematize them and treat them as a logical abstraction. A cumulative process is plainly at work, resembling the normal capital accumulation process except that here there are successive crystallizations of strata of knowledge, information being linked with meta-information (information about information).

But whatever important information technologies may be, it is fair to say that the implementation of new processes for manufacturing and applying materials plays just as vital a part as micro-electronics in suggesting viable industrial approaches to the mastery of complexity and hence to the mechanisms of the new variety-based system. In this sense, it is decidedly worthwhile for industrialists to control the evolution of information technologies as well as the development of methods for processing new materials.

II.1.3 NEW METHODS AND THE INTEGRATION PROCESS

Before attempting to interpret the function of the processes affecting materials in the remodelling of the production system, attention must be drawn to the main categories of processes which may undergo significant development in the next decade.
Talks with industrialitsts in the course of this study have made it possible to identify the following processes :
- Powder metallurgy (or sintering) is one of the processes on which metallurgists are relying most to enhance their competitiveness. It involves compacting and then heating powders whose particles coalesce by interdiffusion. This approach has the advantage of using nearly all the metal applied to the manufacture of a required object (cf Table II.1.A).

> The term "powder metallurgy" or "PM" may cause confusion since this technique might be thought of as confined to the production of metal powders.
>
> In fact the term applies not only to the powder-producing operations but also to the operations of aggregation and sintering long used in ceramics manufacture. Not until the last century was the technique applied to metals and alloys.
>
> PM has a number of firmly recognized advantages :
> - wastage of material below 2% for all operations;
> - absence of internal defects (flaws, inclusions);
> - easy automation of production-line manufacture, recently supplemented by a characteristic not often exploited in the early days of the technique, namely flexibility in the mean composition of the sintered material and in the macro-microscopic distribution of its components.

The preparation of particulate composites (cutting tools - abrasives) or even filamentary composites is also both an old and a highly promising application.

Also, and more specifically in the field of metal materials, PM has partly been directed at high-quality products in which sintering is aided by hot deforming.

In connection with ceramics manufacture, there are a number of processes with a great future. The manufacturer will in all cases need to arrive at a compromise between the price, the powder (and additives) and the cost of densification in order to obtain an object with the requisite properties. Good-quality powders must be procured. While the conventional methods based on reduction of silica or nitridation of silicon have been widely used, different processes, eg sol-gel processes, need to be developed if better quality powders are to be obtained. This raises the price of the powder, but its reactivity brings down the sintering temperature. Other conceivable approaches are the pyrolysis of certain polymers or compounds, where an essentially superficial diffusion creates "welds" between the powder particles on contact. Next, deep diffusion can improve compactness by reducing residual organometallic pores, or use can be made of plasma-assisted decomposition of liquid or gaseous compounds, inductive plasma vaporization, etc (all soft chemistry applications yielding ceramics but also fibres, membranes, etc from initial materials. For a bulky object there are methods of sintering, vapour deposition and polymer pyrolysis. An identical trend to the one described above in relation to metal materials is also taking shape. Alongside natural sintering, which is economical but leaves high residual porosity, increasingly widespread techniques will be pressure sintering, reaction sintering or hot isostatic pressing, the importance of which is growing in spite of a high cost price.

Table II.1.A : Powder metallurgy (Source : François GOBIN)

- Precision casting, especially for complex engine parts, obviates the use of simpler components and also allows savings on materials and complex milling processes.
- Superplastic forming of metals is achieved by distortion under stress of the material heated to a temperature at which it becomes superplastic, making it possible to give fine sheets of steel the form of a complex mould.
- Diffusion welding is a means of bonding together pieces of metal at a high enough temperature to allow diffusion through the surface of the two pieces.
- Magnetic forming involves charging a coil with a brief but powerful electrical impulse which causes a violent repulsion lasting a few

millionths of a second between the variable magnetic field of the coil and the charge in the adjacent metal sheet. The sheet is repelled at over 700 km/h and clings to a matrix behind it, all the more easily because at the temperatures applied the metal becomes viscous.

- Ion implantation and laser treatments are destined to develop. It should nevertheless be pointed out that the use of powerful lasers is not confined to surface treatment but also has wide prospective applications in welding, drilling and etching. It is making rapid progress in the automotive industry.
- Bonding techniques are assuming considerable importance. They meet the needs of users, who are looking for simple and effective ways of joining materials unsuitable for welding (plastics, ceramics, composites, etc). The increasing variety of materials, and consequently the growing need to associate them, guarantee the value of these techniques.
- Non-destructive testing methods are bound to develop, and their potential should expand since the need for reliability and safety is growing. The use of materials with superior characteristics, ceramics in particular, requires the discovery of more and more sensitive testing methods relying on ultra-sound techniques in particular, and here the introduction of signal processing techniques should allow conspicuous progress. Mention should also be made of research into the use of optical and tomographic techniques in non-destructive tests by extrapolating and supplementing all the knowledge already acquired in medicine.

All these processes naturally present the classical constants of technological progress in industrial processing (savings in material, energy and labour). In addition to these well-known phenomena, however, <u>these processes generally share the feature of offering a solution to the problem of growing complexity in production</u>. The solution takes the form of <u>integration principles</u>. Integration basically represents a simplification of the production process. In practice, it can take the following forms (integration phenomena are examined in Chapter II.5).

<u>Integration of production steps</u> is the outcome of a reduction in the number of steps constituting a given process. For instance, precision casting makes it possible to cut out several production steps in engine manufacture by obviating the use of intermediate components which would need welding. Another significant example is the shaping of titanium. Titanium can be made superplastic at about 950°, where titanium parts can also quite feasibly be diffusion-welded, so that

superplastic forming and diffusion welding can be performed in a single operation. Titanium is therefore indicated in certain applications owing to the reduction in overall cost (about 40%) made possible by the integration of steps. The advantages of reductions of this kind stems from the fact that management of the production process is significantly simplified (smaller stocks, less work in hand, less maintenance, etc).

Integration of the number of parts is particulary well illustrated by the rear hatch of the Citroën BX, as already reported. This part has been made since 1982 by injection of glass fibre reinforced polyester. The adoption of plastic in place of steel sheet has made it possible to integrate in three parts only, in the hatch of the BX, the functions previously served in the Citroën Visa by 27 parts, again appreciably reducing the cost of assembling a complex unit.

Integration of fabrication time also plays an important part in the choice of material. To reiterate an example given earlier on, a thermoplastic polyester is preferred to a phenolic one for making an electric light bulb socket; although it costs more per kilogram, it turns out to be 25% more economical by cutting the moulding time from 2.5 minutes to 16 seconds. The machinery investment for the same output drops from six injection presses to one.

The integration process therefore corresponds to a simplification of the overall process. Conversely, it can be observed that the material itself tends to become more complex by "integrating" several functions at once (eg new packaging films can combine the heat-sealability of polyethylene, the rigidity of cardboard and the gasproofness of aluminium). In certain applications, this integration of functions may have a considerable impact on the production method : futuristic though it may seem, the idea of developing a building sector material integrating load-bearing and insulating functions would probably revolutionize the construction of buildings. Functional integration is indubitably one of the dominant trends in the evolution of structural materials. It coincides with the resolve to cater for the complexity of the technical environments to which the items and structures produced by industry will be subjected.

II.1.4 STUDY PLAN FOR STRUCTURAL MATERIALS

In the light of the foregoing considerations, it is evident that the oft-mentioned competition between materials is also very much a competition between processes. The outcome of the struggle is just as likely to be the establishment of a type of complementary association of materials as a radical substitution trend (in any case, this change very often results in the item being completely redesigned; straight forward part-for-part substitution is only an interim stage in innovation). The innovative uses of materials are as diverses as the items involved. For the technology watcher, the most stricking overall feature is the inter-materials and intra-materials variety : (i) each type of material is adapted in a very large number of specific ways and (ii) the number of types of materials used concurrently in each technical complex is on the increase, so that the producer needs to master several types of materials know-how.

The division between inter-materials and intra-materials variety will form one of the key concepts in this part, as will the other equally important questions raised in the introduction. The coherence of the overchoice of materials, the transitional or permanent nature of variety and the management of complexity will be the three principal themes enlarged upon. Certain characteristics are specific to materials as the source of the issues, while others are merely outlined. But, whether of external or internal origin (spontaneous or induced), all have their effects and help to shape the evolution of structural materials.

Let us begin by considering the case of composite materials as an illustration of the "intra" (internal) variety of materials and by apprehending the nature of the competition aimed at taking over strategic points of their respective manufacturing branches (Chapter II.2). Next we shall discuss "inter-materials" variety and more generally the problem of materials overchoice in terms of competition processes, coherence and management (Chapter II.3). As variety is a phenomenon typical of technological transition periods, we shall seek to demonstrate that today's emergent variety bears the seeds of a new operative principle for the entire system, and accordingly to isolate the transitional features of variety which can be summed up as a problem of managing change, in order to pinpoint the new features of the system associated with informational characteristics and complexity (Chapter II.4). This chapter will close with an analysis of the forms of integration, aimed precisely at reducing complexity and at the same time modifying the industrial structures concerned (Chapter II.5).

CHAPTER II.2

Intra-materials Variety : Competition and Know-how as Exemplified by Composites

During our first involvement in the European Communities' FAST programme, while studying the prospects for chemicals in Europe it became increasingly clear to us that the development of composite materials (CMs) was a possible avenue of diversification for chemicals producers. It revealed a new tendency vital to technological development as a whole; beyond the traditional processes of dissemination of technology and industrial learning, there are signs of a genuine revolution in the method of designing industrial objects. The variety of ordinary materials available to designers is not only being expanded by the newcomers but is even being bypassed by the new conception of customized design which accompanies them. The availability of materials "to order" offers a decisive advantage in an information intensive production set-up where the combinative aspect of product specifications is the main field of economic competition.

This chapter is in two sections. The first will introduce the issues of competition with a brief historical account, a technical rundown and a short description of CM diffusion mechanisms. The second section will analyze strategies proper.

II.2.1 COMPOSITE MATERIALS : THE GOALS OF COMPETITION

II.2.1.1. Historical background

Simple forms of composites existed as early as the 40s, namely thermosetting plastics filled with glass fibre. However, only with the development of the space industry did there really arise a composites technology capable of providing solutions to technical problems which were otherwise virtually insoluble. A typical instance was when lightweight structures capable of withstanding high temperatures were needed. Until the end of the 70s these materials remained rarities set aside for high-performance applications; the lack of any alternative technical solution and the very high stakes (eg defence) made the production cost unimportant. The situation was basically unchanged when these materials began to find a few applications in civil aircraft construction, notwithstanding the very substantial gains made by laying in this way a "sub-stratum" for the replacement of conventional materials. Throughout the period the production methods thus remained at the craft level (the use of advanced methods to produce the materials in no way detracts from the craft aspect of their manufacture, featuring limited runs or commissioned work, lack of organization of the production process to achieve significant cost reduction, etc.

It was in the early 80s that CM applications began to spread rapidly in the aircraft industry to the extent of seriously threatening aluminium alloys. The ability of CMs to combine strength and lightness became especially valuable as the fuel cost content of air transport fares was increasing. They also began to break into consumer products via the top of certain ranges of sportsgoods. Their applications are, of course, still confined for the most part to very demanding uses; they will probably never replace tinplate, for example. Note however that so far we have only been discussing high-performances CMs, the only ones which existed at the outset (unless we extend the notion of a composite material to concrete). Yet a new development has occured with the gradual extension of the "composite" principle to more reasonably priced materials of more average performance. This trend, coinciding with advances in standardized designing and processing methods, makes for the gradual opening up of far wider markets.

II.2.1.2 Technical description

The essential virtue of CMs, their ability to offer tailor-made properties, stems from the fibre-matrix combination. Each of the two components contributes its own characteristics. In addition to the combination of physical and chemical characteristics, however, the exact way in which the combination is made greatly influence the properties of the end-material, and this adds considerable extra variety. Lastly, the properties of the constituents do not necessarily combine in a linear fashion, which for a given characteristics (strength, hardness, etc) occasionally makes it possible to achieve a result exceeding the corresponding values of the separate constituents.

The properties sought in CMs are very diverse : lightness, strength, biocompatibility, etc. The combinations of properties are of course what make them so valuable for technically demanding applications : rigidity + strength + lightness or toughness + endurance, etc.

The extreme properties sometimes possessed by CMs are due to the microscopic nature of their structure. In the form of superfine fibres, component materials can attain the theoretical limits of the substance's physical properties provided absolute mastery of their production and processing is achieved. In this way, it is possible to push substances like carbon, glass, boron and metals far beyond their normal macroscopic performances (as these are substantially modified by the presence of microscopic flaws not usually found in the fibres of CMs). The other side of the coin is of course that the preparation of CMs requires extreme care. More so than with any other form of material, the final qualities of the product depend on the CM fabrication process. The industrial significance of this fact is already obvious : some form of precise control by the user over the manufacture of the material becomes indispensable (assimilation of know-how, partnership, mergers, etc).

For each application, a specific form of the material can (and must) be envisaged. The obvious example here is the direction(s) of fibre weave, to be chosen according to the shape of the object to be produced and the stresses it must take in the system for which it is intended. However, a major advance in CMs is already to be noted in this connection, namely the invention of pre-impregnated fabrics whose final shaping is left to the user. Generally speaking, the fact that thermoplastic matrices can now be used in CMs is a major step in that it allows some degree of freedom from the constraint of integration of the production and use of the material.

II.2.1.3 Learning and dissemination of innovation

Starting from high-performance space-grade CMs, innovation has spread via aeronautics, deep-sea drilling, prestige cars, etc. The distance covered seems impressive considering the framework in which the first applications came into being : with one kilogram of any substance in orbit worth more than its weight in gold, no corners were cut with weight reduction regardless of the cost of the material. In this "heroic" age of technology, small-scale production techniques are quite appropriate, but once the range of outlets increases the demands of industrial production become more stringent. The narrowing of margins calls for consideration of the problems of managing mass production rationally, and so the manufacture of CMs becomes the subject of a gradual intersectoral learning process which grows with their diffusion from high-tech sectors to mass production.

In the user industries, learning from experience usually follows the pattern described by N. ROSENBERG (learning by using) (1) rather than that of K. ARROW (learning by doing) (2) .That is, the explanatory variable is not constituted by learning acquisitions made on a predetermined production apparatus (like a production line whose synchronization is progressively improved), but rather by a process of imitation with redesign and redefinition of objects. The dissemination and economic enhancement of CMs is absolutely typical of such processes of permanent redesign (3) .

In the aircraft industry, for example, composites only broke through stage by stage. They were initially introduced to lighten

---

(1) N. ROSENBERG, "Inside the black box", Cambridge University Press (1982).

(2) K.J. ARROW "The Economic implications of learning by doing", Review of Economic Studies (1962).

(3) E. ZUSCOVITCH and P. BRENDLE "Informatisation - l'impact sur l'organisation des entreprises", La Revue Française de Gestion, n°51 (1985).

certain unstressed <u>ancillary</u> parts such as doors and then certain <u>secondary structures</u> (control surfaces). The final target is of course the primary structures (aerofoils and airframe).

Penetration tends to proceed part by part up to a certain critical point where the technical object is redesigned to suit the new material. As long as the step-by-step substitution phase lasts, it is comparatively easy to calculate that a new material is required to perform slightly better than the former one at the precise point in question. Yet the new material is not necesserily used to full extent of its capabilities in such a context. A move to <u>redesign the object</u> finally becomes unavoidable, a far more complex step akin to a "functional" analysis as meant above in relation to chemicals.

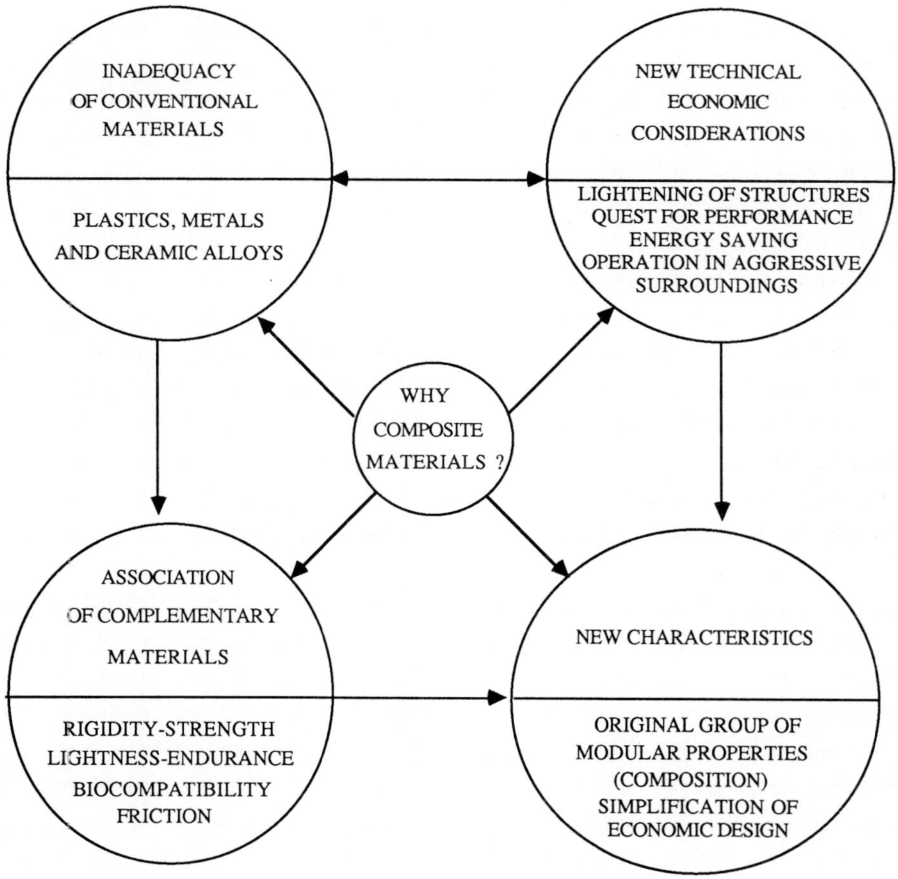

<u>Figure II.2.a</u> : Reasons for introducing composite materials.
Source : Patrick COHENDET ed. (1984), p. 188, according to R. NASLAIN

II.2.2 STRATEGY AND COMPETITION IN COMPOSITE MATERIALS

If the new composite lies at the convergence of several currents of industrial knowledge, it is hard to say which of them holds or may assimilate the strategic knowledge. In contrast with the world of haute couture where the art of bespoke tailoring is mastered, the person who tailors composite materials may be either the producer, or the user, or the manufacturer of intermediate products. Proceeding from a representation of the composites industry, we shall be able to identify a number of knowledge acquisition patterns, each corresponding to a specific conception of the material.

II.2.2.1 Integration of bodies of knowledge and functional redesign

The way in which composite materials are conceived has proved decisive in extending their fields of application. The redesign of parts only became possible through this conceptual revolution.

The evolution of the conceptual approach to material

There is more than one basis for conceiving or designing a composite material : either from the matrix (material - property), or from the reinforcement (material - structure) or directly from the end-product (material - function). Each of these alternative approaches to the composites option has crystallized in a developmental logic connected with the assimilation of a specific know-how.

- Material - property is the outcome of matrix-type design of the material as in plastics technology. Plastics manufacturers have always blended in various granular fillers (silicon, graphite, mica, etc) to enhance given properties of the substance (frequently to reduce costs, hence the term "fillers"). The incorporation of vegetable and later glass fibres only a few millimeters long enabled the plastics engineer to improve considerably the mechanical performance of plastic .Fibre, unlike other fillers, serves as strengthening, which makes it possible to obtain a reinforced or strengthened plastic analogous to concrete. Moreover, the configuration of the fibre filler within the material, unlike other types of filler, plays a crucial part where the mechanical properties of the material are to be modified. It was the systematic exploitation of this idea which engendered the structural approach to material.

- Mastery of a mainly textile fibre technology made it possible to fabricate various kinds of mats from glass, then carbon and aramide, fibre. A number of fibre weaving and orientation techniques (eg pultrusion) are now used to define the structure of the material. A transition has thus been made from "material - property" to "material - structure", thanks in particular to the emergence of new generations of higher-performance fibres. The material has an architecture of its own which can adapt to a wide variety of articles to be produced. This structural conception of the material nevertheless continues to be subtended by an evident desire for standardization at the level of the intermediate products (fabrics). In fact a standardized structure allows no more than an approximation of the "framework" required in the object to be manufactured.

- Fabric makers have therefore begun to offer unconventional forms of fabric produced on the basis of new weaving techniques. The textile industry is able to provide woven preforms (cones, cylinders, etc). These "high-tech" fabrics have applications which for the time being are confined to the aerospace sector (4) . They are nevertheless the practical expression of a new approach to material which might be called functionnal. The idea is to produce the reinforcement to suit the object - hence the definition of preforms whose architecture corresponds to that of the end-product. It is only a short step from the idea of preform to that of form, and technology is already prepared to make that step with filament winding. Initially intended for the production of lightweight tubes with outstanding mechanical properties, this technique has gradually expanded to more complex articles such as high-pressure tanks. Trials are now under way at General Motors to manufacture a lattice structure for cars by computer-guided multiaxial filament winding (5) . The spread of such a fabrication procedure depends on the user's design potential. The expertise of the materials

---

(4) J.M. GIRAUD (BROCHIER S.A.) "Transposition à l'industrie des développements des pré-imprégnés pour l'aéronautique" ("transposition of developments in pre-impregnated products for aircraft to industry"), Composites n°3, May/June 1984.

(5) "Filament wound car frames and other pratical miracles", Modern Plastics International (1984).

user, in other words the designer of parts (for cars, aircraft, etc) is the decisive factor in defining the material.

The development outlined above indicates the conceptual about-turn which is now taking shape. There is an imperceptible transition from the designing of parts on the basis of the material available to designing the material on the basis of the component to be produced. This change in perspective leads to redesign of the components themselves.

Redesign of components by integration

The redesign of components thanks to composites is marked by simplification due to the integration of several components. This apparent simplification is nevertheless the outcome of increased complexity of the knowledge incorporated into the material.

- <u>Simplification and integration of functions</u>

Turning a complex of parts into a single part has in many cases depended on integrating several characteristics into the one material. Composites offer just this sort of combination of properties which often conflict where they occur in a traditional material. The availability of composites leads to two kinds of integration : integration of components and integration of functions. These forms of redesign can best be illustrated by means of examples taken from different fields of application.

- <u>Integration of parts</u>

The most eloquent example from the automotive sector, which has been repeatedly mentioned, is the rear hatch of the Citroën BX. Made of glass/polyester composite by transfer injection (ZMC), it consists of only three parts compared to the 27 in its metal equivalent. More generally, bumpers made of composite material, as in the Renault R5, reduce the number of parts required.

In aircraft construction, a particulary stricking example is the redesign of helicopter parts using composites. The hub of the main rotor of the SA 365 C helicopter built by Aérospatiale now requires only 92 parts compared to the former 293. The same example of integration was presented in connection with the Bo 105 at the MRS colloquy (November 1985) by the MBB representatives (6) .

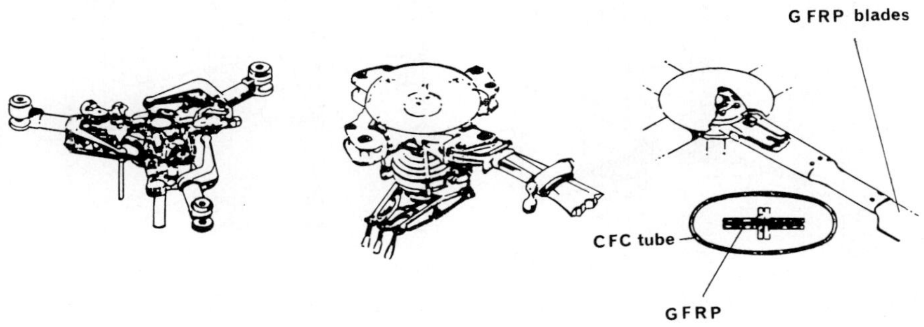

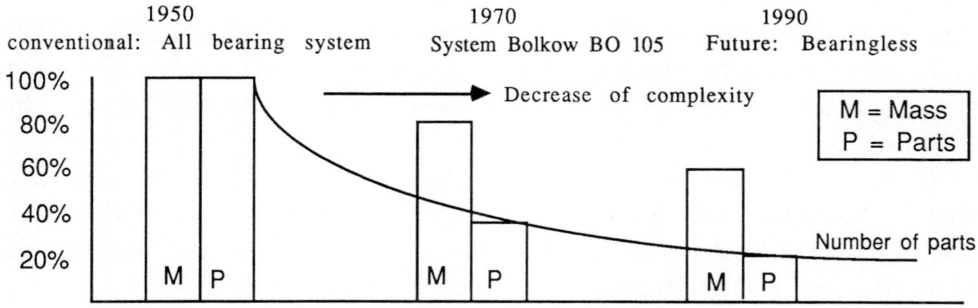

Figure II.2.b : Evolution of the propulsion system for helicopters, according to MBB. Source : W. HARTMAN and H. KELLERER (1985).

---

(6) W. HARTMAN and H. KELLERER, MRS Strasbourg, Nov. 1985.

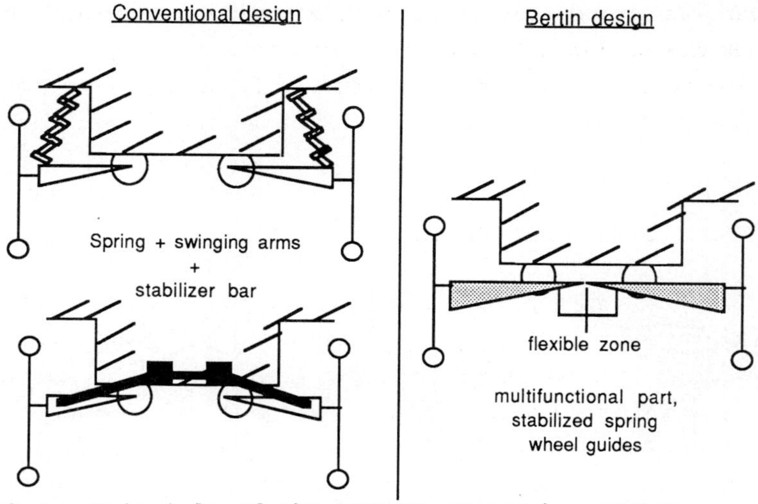

Figure II.2.c : Principle of the BERTIN suspension system.

- Integration of functions

In the automotive industry, one of the most astounding achievements is the suspension designed by BERTIN (7) . A unique arrangement of parts performs the functions of suspension, wheel guide and roll correction which were previously assigned to three distinct groups of metal parts.

Similary, in the aeronautics sector (and shortly in overland transport) carbon-carbon for braking (8) incorporates the dual function of a friction material and a structural component in contrast to the conventional brake which works solely by friction.

Finally, let us mention a case of functional integration in the railway industry - the bogie recently perfected by MBB (9) in which

---

(7) M. de GONCOURT (BETIN & Co) "Suspension par éléments composites" ("Suspension using composite parts"), Dossiers du CODEMAC n°14, "Bourse des affaires offres et demandes de technologies composites" ("Composites technologiy supply and demand exchange").

(8) L. HERAUD and B. BROQUERE "Les carbone-carbone pour le freinage" ("Carbon-carbon for braking") in "Developments in the science and technology of materials", EACM-ECCM, Bordeaux (1985).

(9) C. GUNTHER, R. LEO and R.J. SCHULTZ (MBB) "New Technologies for rail vehicles and car body structures", MRS Coll. Strasbourg, Nov. (1985).

the undercarriage structure also acts as the suspension, considerably simplifying the system as a whole.

In most cases, as the examples show, the integration of components and functions results in simplification of the parts and their design.

- Growing complexity of knowledge

The simplification of components leads to increasing complexity in the knowledge "embodied" in the material. It becomes more "information-rich" when devised from the downstream end. It is located at the frontier between the skills of the designer, the processor and the producer of raw materials (fibres, resins, aluminium, etc). The design stage is therefore likely to assume growing importance in future among users of materials, a trend which is furthermore encouraged by the development of many different methods relying on CAD, CAM/CAD etc. According to a representative of the motor industry, the design phase now accounts for 5% of the total cost of developing a vehicle part but determines 85% of its characteristics. If the designer is required to design the part at the same time as the material, the cost of design will probably increase owing to the standard of skill called for by such an approach. It is not beyond the bounds of possibility that a design industry capable of managing an ever wider range of materials may develop independently in order to solve all the user's problems.

Although the downstream side is more closely involved in defining composite materials, it would be a mistake to conclude that all the knowledge relating to the material will eventually be fully mastered by the user. Some firms upstream, particulary in the chemical industry, possess accumulated know-how which is difficult for those further down the line to assimilate. Indeed, it often happens that these basic industries acquire know-how from further down, the better to impose their norms (guaranteeing quality) on their customers. One should therefore avoid jumping to the conclusion that the lower end dictates the definition of materials. "Composite design" gives rise to various vertical and horizontal interactions and assimilation of know-how. Each of the three approaches which have been outlined corresponds to a specific form of acquisition of strategic knowledge, which will be described after the pattern of the composites industry has been presented.

II.2.2.2 Manufacturing processes and acquisition of knowledge

Analysis of the composites "route"

This is hardly an appropriate term for composites in general; a material goes through very different stages depending on its family (carbon-carbon, aluminium matrix and organic matrix composites). On the other hand, a sequence is vaguely beginning to take shape in the isolated case of organic matrix composites, which have reached a stage of near maturity. Very schematically, it is arranged in two sub-sequences, the first starting with the reinforcing fibre (glass, carbon, etc) and the second with various matrices at the monomer stage.

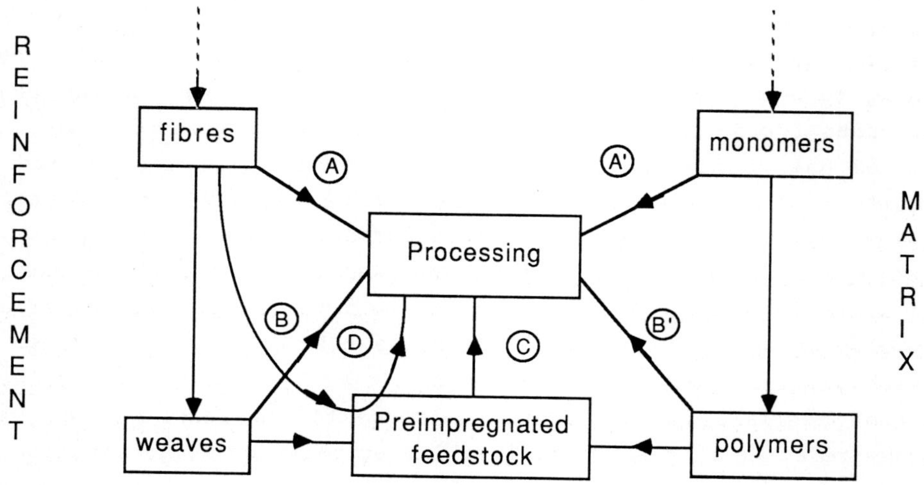

AA' : reinforced reaction injection moulding (RRIM);

AB' : conventional injection moulding or transfer injection (ZMC). Basic Mix Compacting (BMC) and sheet moulding (SMC);

BB' : contact and projection moulding; SMC compacting (using a fibre mat or fabric); mould-to-mould injection;

C : drape forming with pressure polymerization, sheet stamping;

D : filament winding.

Figure II.2.d : Processing techniques for composites.

The various modes of processing organic matrix composites are defined for almost every combination of reinforcement phase and matrix phase.

These various fabrication techniques can be grouped according to whether a matrix-related, structural or functional conception of the composite material predominates.

Alternative knowledge acquisition patterns

Each of the conceptual approaches to the composite material has "crystallized" around a group of processing techniques necessitating mastery of one or more types of know-how. In each instance the acquisition of knowledge about the material is made either by the equipment manufacturer, or by the systems designer, or by the supplier of raw feedstock (fibres and matrices).

<u>The "properties" pattern : educated technology</u>. The "material - property" concept stems from the idea of a (short fibre) filler added to the base material. The processing techniques chiefly use injection (simultaneous injection, transfer injection, reaction injection, etc) and more occasionally BMC. There is recent evidence of development in this matrix-based conception in the spread of a new technology (mainly among vehicle manufacturers in the USA) for producing the polymer, the composite and the part in one operation. Reinforced reaction injection moulding (RRIM) is in fact a means of producing shaped parts directly from monomers and fibres. Mastery of the material is basically a matter of mastering the relevant technology. In other words, it is the machine builder who must assimilate the know-how relevant to the material (polymerization in situ, fibre-resin blending, etc). The equipment maker is learning chemistry, a situation which could be described as technology being educated by the material, resulting in the integration of the intermediate manufacturing steps. At this stage the material only exists in two extreme forms, as a precursor of the material as such and as the completed product. It no longer really exists except inside the mould. Strategic knowledge is fully incorporated into the equipment, and there can be a simplified transfer of technology in which little learning is required of the user.

The "structure" pattern : educated material. In contrast with the pattern described above, the "structure" conception has become organized around a wide variety of processing techniques which can be placed in three categories :

- impregnation of a fabric or mat in a mould (contact, projection, mould-to-mould);
- compacting or pressing of sheets or films (SMC, TPE, etc);
- drape forming and pressure polymerization.

It will be observed that all three families of techniques involve the use of semi-processed products presupposing mastery of specific know-how. Moving down the line, we observe a sequence of processing steps each of which requires a specific type of expertise. The production of weaves or fabrics devised on the basis of textile know-how is followed by the fabrication of sheets, films or sandwiches using specific techniques such as thermo-calendering and pultrusion. The current trend in organic matrix composites is towards wider use of thermoplastics such as PEEK. One of the chief reasons for the trend is that thermoplastics can be reshaped, which makes it possible to produce intermediate standard forms (films, sheets, tubes, etc) which can be shaped by using techniques borrowed from metal processing (pressing, stamping, etc) which have a long learning curve and accumulated know-how behind them.

Knowledge about the material in this case pervades the entire value addition sequence from the chemicals firm to the final processor. Several trades (resin production, weaving, pre-impregnation, etc) successively incorporate their knowledge into the material. There is nevertheless a wide-ranging tendancy for all these trades to assemble upstream so that all the knowledge gravitates towards industrial chemists. They are paying close attention to the textile industry and its weaving methods, judging by the recent takeover of Brochier by Ciba-Geigy, of Stevens-Genin by Hexcel and the cooperation agreements reached between Fothergill-Harvey and Cyanamid. The chemical industry is expanding downstream into pre-impregnation and fabric production. This applies in particular to ICI, which recently took over Fiberide, the largest composites processer in the USA, and to BASF which has acquired the composites division of Celanese.

Thus there is a clear trend towards upstream domination of all materials know-how. This strategy of the European chemical industry can be construed as determination to control the entire composites industry in order to meet increasingly specialized demand from the end-user, and perhaps even to anticipate this individualized demand. At all events, the huge design potential based on the structural approach is causing great uncertainty among traditional suppliers of materials as to the demand which will come from lower down. Two utterly contradictory types of motive can therefore be discerned in this integration trend. The first concerns increasing capacity to meet a more and more diversified demand for materials, which is something of a wait-and-see attitude to the selection of materials eventually made down the line. The other involves making the selection directly at the top, so that the entire sequence must be known, in order to give the user the best material. This activist approach is apparently more characteristics of the chemical industry's strategy. But whichever the dominant motive may be, there is room to speculate on the emergence of a focal point in the composite materials sector supplying a wide range of products to suit user's requirements.

The functional pattern. In contrast with the preceeding pattern, the functional conception of the material developed from specific (ad hoc) technologies and not from borrowed ones. Reference has already been made to filament winding as a means of producing an architecture perfectly suited to the intended part. The non-linear pultrusion method developed by Renault is another instance of a technique yielding fully-formed parts (bumpers) directly. Strategic knowledge here is designer and processor knowledge, as the material is defined according to the intended product. The major revolution in materials can be situated at this precise point : the material, formerly a limitation, has become an optimizable variable which can be adapted to the most stringent requirements in a set of specifications. Hybrid composites can also satisfy technical constraints at the same time as economic cost constraints. The combination of fibres in question does indeed offer particulary valuable properties and substantial economic advantages compared with composites with all-carbon or all-kevlar reinforcement.

Although it is not yet really possible to speak of a strategy in connection with this still embryonic pattern, it may nevertheless be expected to coincide with greater domination nearer to the marketplace in the definition of a material - unless the chemical industry's tendency to absorb all the later stages attains the ultimate link in the value added chain, which would be process of integration in the opposite direction. The strategic knowledge is in fact the designer's and on it will depend the choice of material, stimulating acquisition of knowledge at the upper end. There is therefore a pronounced trend towards concentration of knowledge down the line. The evolution of the functional pattern points to the emergence of a new generation of products in the form of materials, in which each component "material" will be used exactly where its properties are wanted. It is in fact pointless (and perhaps impossible) to manufacture parts entirely from "noble materials" if the real need for the latter is very localized. The future seems to belong to what are already being called "multi-material" parts in which the constituents are optimally distributed.

CHAPTER II.3

"Inter-materials" Variety and the Overchoice Phenomenon

In the last chapter, it was argued that the strategic redeployment and diffusion of activities arose from the complexity inherent in the manufacture of articles from composite materials, and so economic management and the related assimilation of knowledge were geared to mastery of the "intra-materials" aspect. The production of articles which are technical systems in themselves, for example cars and aircraft, involves another form of variety, that existing <u>between</u> materials. With this "inter-materials" type of variety, the problems of technical interface are attended by relentless competition in which each material tries to become dominant in as many technical applications as possible and, through the redesign of industrial objects, to attain a sufficient critical mass to oust its rival. The problem of substitution, whether reversible or irreversible, will therefore be a central issue (first section). A formulation of the "inter-materials" problem will round off the preceding chapter, after which we can raise the question of the actual coherence of the variety or overchoice problem (section two). Having dealt with the technological and competitive issues, we shall go on to discuss the various styles of management of variety (third section) as an introduction to the next two chapters.

II.3.1 THE LIMITS TO OVERCHOICE : TOWARDS ALLIANCES OF MATERIALS

Substitutions of materials have traditionally been governed by a relatively simple mechanism whereby a given "new" material superseded an "old" one on the strength of superior manufacturing or utilization characteristics linked with economic advantage. The examples illustrating this type of mechanism are too numerous and familiar to detain us. With the emergence of new materials, however, the economy has significantly expanded the variety of practical possibilities for responding to needs. The problem underlying this variety is how substitutions are to occur in an overchoice context, and there are two different approaches to the problem. The first would be to identify the technical and economic criteria which determine the choice of materials. The writers find this a narrow approach unlikely to yield any general conclusions, as too many different criteria would have to be considered and weighted differently according to the applications envisaged. It would be preferable to adopt a more general approach which, although involving a certain amount of simplification, should give us a better grasp of the logic behind substitutions of materials and allow us to make appraisals regarding the evolution of the materials pattern. The working hypothesis is as follows : the logic governing substitutions of materials is one of adaptation to the evolution of the production system. One factor in its evolution is in fact the availability of a huge variety of materials. The second factor governing the current evolution of the production system is its tendency to become increasingly rich in information. The increase in the amount of information compatible with the production system is to some extent linked with the new materials. For industry to master their microscopic complexity and their various utilization possibilities, it must increase the amount of information it possesses concerning them. Above all, however, the factor of development under consideration is linked with the existence of a universal, high-performance information processing facility, namely the computer. New materials and computer systems are indeed closely interdependent. To give just one example, without computerized techniques for the three-dimensional calculation of stresses, the replacement of steel by engineering plastics in car bodies would have been greatly handicapped. The evolution of the production system towards an information-intensive order based on variety, and the hypothesis stated above, suggest that the basis for present and future substitutions of materials is the information "embodied" in them. Incorporated information means all the characteristics inherent in the material and all the uses which it acquires through its processing technology, but rules out all instances of value addition where the material is combined in a single "package" with additional information in the form of an impersonal medium or a service.

In a context typified by keen competition among materials, a look at the various instances of substitution reveals two different substitution mechanisms. The first is related to the higher information content of the material or its processing technology along <u>one</u> line of constraint. The second mechanism, which in practice often originates from the experience acquired in substitutions of the former type, yields a <u>combination</u> of characteristics which removes a whole <u>set</u> of constraints. There are specific examples illustrating these two different mechanisms and indicating the resultant pattern of substitution. The first example considered here concerns the evolution of car engine blocks. The material traditionally used in this application long remained cast iron, but the 70s witnessed the appearance of more and more aluminium engine blocks. This substitution reflected the car manufacturer's intention to reduce the running costs of vehicles by lowering fuel consumption through a reduction in the weight of the engine unit. The scale of the substitution and the risk of eventually losing a bread-and-butter market prompted steelmakers to speed up the development of fine cast irons, particulary those containing spheroidal graphite. These have now reduced the weight gap between a cast iron block and an aluminium one to a mere 10 kg. This near equivalence from the angle of the weight requirement has enabled cast irons to begin their comeback in engine block design. Indeed, they enjoy some superiorities over aluminium. The price of cast iron has remained stable while that of aluminium is still speculative. The use of aluminium also makes it necessary to design the engine block separately from the cylinders. With cast iron, by contrast, it suffices to provide apertures of a suitable size and shape to house the pistons, hence a considerable saving on materials and processing. This is an example, then, of a <u>reversible</u> substitution mechanism in which the improvement of the material initially supplanted follows a line of constraint where the replacement material afforded a superior advantage. The strictly economic intention of making the substitution reversible operates by a selective input of information into the material in order to reveal, after equating the critical characteristics, the properties which enable the material with the higher information content to prevail.

Another instructive example along the same lines is worth discussing although it does not represent a straightforward case reversible substitution. In the 70s, tinplate commanded 90% of the drink can market in the USA. In 1985, again in the USA, it accounted for a bare 10%. The scale of its decline on the market suggests an irreversible substitution. However, consideration of the facts qualifies this first impression. The collapse of tinplate on the market in question is due to the advent of aluminium and its potential for the manufacture of drink cans in two seamless, unstamped parts (drawn base plus top) instead of three stamped, soldered parts as with the tinplate technique. The smaller number of parts and the savings thereby achieved in terms of manufacture, storage, handling and stock management are the chief factors which made it possible for aluminium to supplant tinplate in this application. It is to be noted, however, that the offensive waged by aluminium also relied on the creation of a favourable brand image (10) and on a certain strategic conception of producer-user relations. The aluminium manufacturers contacted the drinks manufacturers directly, bypassing the can producers so as to circumvent the objections which they would certainly have raised in view of the differences between aluminium and tinplate processing. In Europe the French firm Sollac, for instance, was able to make a timely counter-attack by developing a manufacturing process for two-piece drink cans from tinplate. Thus the refurbished image of tinplate and the lifting of American import tariffs on European steel products might allow tinplate to stage a comeback in American drink cans, as in the previous example, with a mechanism of reversible substitution operating thanks

---

(10) The aluminium makers emphasized recovery and cleanliness in a direct advertising attack on tinplate. The question of recycling is symptomatic of the passiveness of American steelmakers as tinplate is recyclable and even has an advantage on this count; cans made of tinplate can be recycled into new cans whereas aluminium can only be recycled as wire.

to the greater information input into the material's processing technology (11) .

The reason for expatiating on this example is that it highlights the narrowness of allowing choices to depend on technical and economic criteria alone (we shall come back to this example in Chapter 2 of Part III). Comparing the drink can scene in France and the USA, the obvious conclusion is that the decisive criteria governing choice are strategic : brand image, protectionism and producer-user relations. If such criteria are to be taken into account, the analysis put forward has the advantage of showing why a given criterion intervenes, while at the same time showing how to evade it when an economic intention mobilizes resources which make for an increase in the complexity of the material or its processing. These considerations will be further developed in Part III in connection with packaging. Other examples illustrate a still greater degree of complexity.

In the case of the helicopter rotor mentioned in the previous chapter, the introduction of a carbon-carbon composite into the hub led to a considerable increase in the system's fatigue resistance and strength. The safety criterion alone is decisive in the aircraft industry, while not without economic implications, since improved safety leads to savings on maintenance and parts replacement costs. Other features of the composite material have nevertheless generated major economic advantages. The density and surface characteristics of carbon-carbon have saved 50 kg on the weight of the rotor and improved its aerodynamics, with marked gains in fuel economy. The substitution of materials has also been attended in this case by complete <u>redesign</u> of the rotor tending towards simplification. The savings which result from such a reduction in numbers of components need no emphasizing, and examples of this abound in aircraft construction.

A similar example is provided by the trends in European car bumpers. The energy crisis set car manufactures searching for increased competitiveness through lower running costs thanks to vehicle weight reduction. This economic goal was achieved by replacing steel bumpers with glass fibre reinforced plastics. Once again, overcoming the weight constraint was accompanied by the removal or relaxation of other constraints thanks to the experience gained when the first plastic

---

(11) The increase in information content is revealed by the Sollac company's lack of competitors owing to deficient expertise concerning the dual reduction process for the two-piece tin can.

bumpers were produced : elimination of the corrosion problem and of expensive chrome-plating shops; redesign of bumpers to incorporate front or rear aprons ("shield" design); improved safety; use of the "intrinsic" colour of the plastic. Both examples point in the same direction, indicating the existence of an irreversible substitution mechanism.

The supplanted materials are unlikely to react, since they have to contend not only with superior characteristics but also with a full range of characteristics which have the advantage of stimulating component redesign. In the automotive sector, redesigning can be taken even further as new materials can be expected to cause a total remodelling of bodywork and reorganization of the assembly processes (12) . The irreversibility of the substitutions is therefore directly linked with the multidimensional complexity inherent in new materials, especially where the redesign of technical objects and processes is concerned. The price difference, usually unfavourable to new materials, reflects the incorporation of sophisticated know-how into the material or its processing technology, but is amply offset by the resultant complexity which is a source of large savings in the manufacturing application or final use of the material in the system incorporating it. As increasing complexity is usually attended by simplification of technical objects (integration of components) or of processes (smaller number of production steps; elimination of plant and operations), a compensatory process can be said to exist. It means that the economic yiels of information embodied in the material or its processing technology is on the increase. This can be regarded as a kind of intelligence bonus incentive; the more instrumental theoretical and practical knowledge becomes in advanced mastery of the material, the more adaptable to a theoretical form the material becomes and consequently the more degrees of freedom creative imagination will have in rethinking the uses of the material and discovering new sources of economic advantage. The yield of the incorporated information is likely to achieve only limited growth and, beyond a certain threshold, the law of diminishing returns may apply. One of these threshold can be explicitly stated; that of the overall redesign of technical objects, the best example of which is the motor car. This type of redesign is in fact burdensome to manage; it entails acquiring new and reforming old

---

(12) For further details, see the chapter on transport in Part III.

know-how, creating demand, investing in new capital equipment and rethinking relations with suppliers and plant designers. It is a real <u>industrial changeover</u> involving an effort which is both creative and destructive. The importance of the destructive aspect explains why those involved prefer to reduce the threshold effect as far as possible by taking microdecisions as the degree of substitution rises in order to prepare for the changeover and spread out the associated poor returns. Thus the yield of incorporated information keeps on increasing, though with localized areas of decrease or only slight increase, which become more frequent as the changeover approaches. Past that point, if the changeover has been adequatly prepared for, the overall redesign of technical objects should afford major economic advantages provided it corresponds to a need. Very rapidly, however, the yield of the incorporated information begins to fall because, apart from a few minor improvements, the technical object, its materials and the related processing techniques have attained technological equilibrium. This results in the establishment of an economic standard. Thus the economic intentions at this point are almost exclusively directed towards questions of reproducibility and are no longer concerned with the pursuit of complexity.

Increasing complexity of materials or their processing technology due to the emergence of the new materials has serious implications for the producers of traditional materials. In certain cases they have some hope of reversing or preventing substitutions by also gambling on the incorporation of more information, as did steel-makers by perfecting fine cast irons and high tensile steels. This response nevertheless has limitations where the multidirectional complexity of substitute materials yields advantages connected with redesign. However, substitutions of this kind are gradual and so do not leave the producers of traditional materials totally disarmed. So much is clear in the evolution of materials towards a secondary design. Steelmakers accordingly intend to counter the competition of organic matrix materials in automotive applications by forming an alliance with these materials. Soundproof sheet metal and multilayer laminates combining steel and plastic or steel, plastic and aluminium are evidence of such intentions. If the development of thermomechanical ceramics challenges the development of steel and cast iron, the brittleness and short useful lifespan of the former may enable the latter to play a trump card with reinforcement fibres. These various possibilities point to the existence of <u>dependent</u> substitution mechanisms. In aeronautics, for instance, to avoid the sudden transition which large-scale use of composite materials would entail, certain producers are following the example of Boeing and comtemplating the use of aluminium-lithium alloys. This type of alloy possesses the same properties of lightness as composite materials but does not require departures from existing aircraft construction technologies. Similary, aluminium producers such as Péchiney are developing skills in the aera of reinforcement fibres. The intention underlying this type of strategy is to make the known material evolve towards a secondary form, or to enter the market in the primary materials which compose the secondary materials so as to avoid or exploit future substitutions. In other words, apart from the intrinsic dynamism of composite materials in the strict sense, the tendency of materials to evolve towards a secondary design is fuelled by the options taken on the irreversibility of substitutions.

In the past, the predominant substitution mechanism was that operating between primary materials (13) . The emergence of the new materials demonstrates the role of complexity as a determinant in substitutions, since in simplifying technical objects and processes it generates the economic gains which ultimately justify the substitutions. Thanks to composites, new materials are already giving rise to alliances between materials and helping to restrict overchoice. The evolution of conventional materials towards greater complexity, together with economic intentions resulting in dependent substitutions, accentuates the trend towards the "secondarization" of materials. Reversible or irreversible substitutions will continue to occur locally between primary materials. At present, substitutions are mainly occuring between primary materials and secondary materials, at the expense of the former. Mastery of the higher information content along several lines of constraint makes it possible for secondary materials to escape the reversibility of substitutions. It is likely that in future all the developments recorded will ultimately limit overchoice still further, with the substitution process chiefly affecting secondary materials in relation to each other. This would be a consequence of the substitution pattern favouring the more information-rich materials.

---

(13) Reference is made here to the distinction drawn in Part I between primary and secondary materials according to whether they are controlled macroscopically or microscopically.

II.3.2 THE COHERENCE OF OVERCHOICE

Now that the different origins of the variety phenomenon, and some aspects of the influence of competition on the inherent issues, have been considered, the underlying unity of what is to be called the overchoice phenomenon (a term denoting both variety and its management) can be seen in its proper perspective. The writers regard it as a phenomenon which comes into the evolution of new materials in so far as their evolution establishes a new technological line (14) .

The concept of the technological line can be defined as the existence of a prior technical coherence whose independent development generates a series of technical objects. The development of new materials is still too recent for a full analysis to be made of the technological intentions behind the evolution of the technological line. In the case of plastics, it is nevertheless possible to distinguish some full developed aspects of the progression of the technological line towards overchoice. The analysis presented in this connection is therefore to be regarded as an attempt to understand the overchoice phenomenon rather than as a theory or history of technological change.

The study of any technological line must begin by defining the elementary principle or principles which make its constituent technical objects a coherent whole. In this connection, we are already aware that new materials comprise a definite set of materials all typified by mastery of their microstructure. We have also seen that this definition has three aspects; knowledge of the microscopic arrangement and behaviour of matter, development of techniques and tools for making industrial replicas of the microstructure so defined, and the existence of non-destructive testing methods. Thus defined in its elementary structure, the technological line of new materials has an

---

(14) On this concept and its analysis, see J.L. MAUNOURY "La génèse des innovations" ("Origin of innovations") PUF (1968). Other theories relating to the coherence of technology development have been devised more recently. Principal references : R.R. NELSON and S.G. WINTER "Towards a useful theory of innovation", Research Policy (1977), G. DOSI "Technological paradigms and technological trajectories - a suggested interpretation of the determinants and directions of technical change", Research Policy (1982) and E. ZUSCOVITCH "The economic dynamics of technologies development", Research Policy (1986).

overall coherence, and what is needed is a demonstration of the forms of logic which drive this coherent whole.

The first driving logic to be distinguished is scientific : it is realized that the development of new materials gives rise to new fields of scientific investigation. Materials science, multidisciplinary by definition, tends to establish new connections between different research sectors and makes disciplines converge with the result that "horizontal" themes are defined (15) . This applies to tribology or the science of interaction between surfaces in contact, particularly important where ceramics and composites are concerned. The advance of rheology is due to its ability to describe how materials as dissimilar as steels and plastics behave under conditions of flow and distortion. As to the "horizontal" themes generated by the emergence of new materials, the importance of research into "structure-properties" or "materials-energy interaction" can be noted. These new realms of investigation are being activated not only by scientific considerations such as better understanding of precipitation processes in aluminium-lithium alloys, but also by considerations regarding the problems facing new materials technicians and users, for instance fatigue resistance and shatter proof properties of composite materials, resistance of engineering ceramics to wear and friction, plastics recycling, etc.

The second logical pattern identified is technical and relates to processes in the manufacture and application of materials. The significant event from this point of view is the emergence of a real materials <u>engineering discipline</u> akin to chemical or genetic engineering. All the major European chemicals groups are agreed that industrial chemistry must evolve towards other kinds of expertise and assimilate other forms of know-how if it is to solve the problems facing it as a materials science and industry and achieve downstream expansion of its activity sectors and outlets. The chemical giants, including Hoechst in the Federal Republic of Germany, Rhône-Poulenc in France and Montedison in Italy, have accordingly developed research and industrialization programmes on engineering ceramics. In general, the constitution of a materials engineering discipline depends on the

---

(15) See R. THALER "Pour une économie de la diffusion des innovations technologiques ("Towards an economy geared to the dissemination of technological innovations"), Doctor of Economic Sciences Thesis, ECT-LYONS (1986).

introduction of new techniques for computing and designing structures and parts, specific shaping techniques such as filament winding, suitable instrumentation (laser cutting units) and non-destructive testing systems (ultra-sound, optical and tomographical techniques, etc). It is significant in this connection that certain processing techniques are applied to new materials as dissimilar as metals and their alloys, ceramics and polymers (sintering). Materials engineering naturally also presupposes the formation of new forms of know-how. The shaping of composite materials thus draws on skills originating from textiles (reinforcement), chemicals (matrix) and mechanical engineering (machinery). It is also stricking that chemical companies are increasingly on the lookout for engineers with training in both chemistry and computer techniques. Looking beyond this particular instance and the pursuit of very advanced types of know-how, materials engineering requires the training of real specialists in general applications able to structure and possibly manipulate the complex dimensions of the problems raised by new materials.

This technical logis is not abstract but is motivated by technological aims which help to keep the development of the technological line moving. The role of technological intentions cannot always be clearly discerned where new materials are conerned. The overchoice phenomenon in fact tends to create confusion by multiplying development potential. Besides, we lack the historical detachment needed for this type of study. However, the example of the development of organic materials provides useful pointers for appraisal. As their name implies, technical plastics were born of polymer engineers determination to adapt their products to more stringent requirements. This required absolute mastery of polymer manufacturing methods. It became clear fairly early on that efforts in this area should concentrate <u>more</u> on <u>polymer blending methods</u> than on the discovery of new monomers, the bulk of whose very wide range was already known in the 50s. Polymer blending methods were accordingly developed either in the direction of combined polymerization of two or more monomers (static copolymerization in sequenced or grafted blocks) or in the direction of alloying polymers obtained through the partial or total blending capability of various polymers. Concurrently, process machinery designers, after devoting their efforts to the extrusion process, turned to the development of injection methods. Improvements in knowledge of the organizational and structural mechanisms of polymers and development of injection presses both <u>converged</u> towards the development of RIM materials and machines.

The immense advantage of the RIM process is that it works from polymers instead of monomers; that is, the three steps constituted by the chemical reaction of polymerization, polymer formulation and shaping are all combined in a single machine. This integration of production steps nevertheless encounters two obstacles; firstly it slows down the production cycle in respect of the machines which work from formulated polymers or processed products, and secondly it uses only a comparatively limited range of monomers (liquid-phase thermosetting monomers). The former handicap is now beginning to be overcome thanks to the development of self-cleaning high-pressure heads which help to increase the pumping and blending capacity of RIM machines. As to the second handicap, apart from the still far-off hope of eventually adapting the process to thermoplastics, we have already had occasion to mention the innovation represented by high-strength RIM plastics.

These new technical plastics are the first sign that a real chemical assembly kit has been achieved for polymers. Indeed, it is the first instance where industrial mastery of the molecular array of polymer chains has progressed to a point where chemical structure blocks and different functions can be obtained to measure. The direction of this trend is reminiscent of the route taken by the heavy chemical industry towards mastery of unit operations.

Likewise, a second part of development can be retraced. The performances of polymers were enhanced by blending in foreign matter. Polymer technicians thus adopted the practice of doping the plastic base material with vaious granular fillers in order to improve load-bearing properties (and cost). This in turn engendered the idea of improving dynamic stress properties by incorporating vegetable and then short glass fibres. Systematic development of this idea and command of textile know-how led to polymers being reinforcement with long braided fibres of glass and subsequently of carbon and aramide. Substending this composite design of technical plastics there lingered the overt desire to standardize reinforcing fibres, which impeded the perfect adaptation of the architecture of the material to the object to be produced. This explains how fibre weavers have increasingly tended to offer unconventional fabrics. Finally, the idea of suitable preforms gave way to that of forms, thanks to filament winding which enables long fibres to assume the net shape of the object to be produced.

As this trend in design and technology progressed, so the composite material principle spread to other materials like ceramics, carbon and metals. Through their specific fabrication techniques and their extension to other materials, composites have demonstrated that they are no longer technically dependent on their original field. Thus with technical plastics proper, the technological trend has been towards the discovery of simple fundamental mechanisms. With composite materials, however, the technological tend has been to extend a single principle, that of reinforcement, to the materials to which it was initially applied, and then to others.

While differing in their respective technical aims, these two forms of development of the technological line of new materials nevertheless converge on the same ultimate objective. Controlling the basic organizational and structural mechanisms of polymers should make it possible to arrive at an interplay of properties leading to rationalization of overchoice from the chemistry angle. Each successive stage in the extension of the composite material principle creates new ways of matching materials with the shape of the objects to be produced, while endowing materials of different origins with a common unity enabling them to compete with each other and increasing the degrees of freedom in materials choice.

Having demonstrated that the overchoice phenomenon in materials corresponds to the tendency of new materials to trace out a technological line, we should now consider the economic intentions which pervade and guide its development.

II.3.3 MANAGEMENT OF VARIETY

The hitherto predominant materials economy has fostered the linear organization of companies around the R&D - production - sale sequence on the inside, and within and around the primary producer - user sequence on the outside. This order is radically contested by the emergence of the variety associated with new materials. As previously suggested, the problem of managing such variety, and hence its impact on company organization, lies outside the strict context of new materials. The real issue is the firm's adaptation to a new data-rich technological and economic systems. Thus the real dimension of the problem of managing variety is at the nexus of variety and information.

This section will simply circumscribe the various possible forms taken by the management of variety. What follows can therefore be read as an introduction to the next two chapters, which may be referred to for a closer analysis.

II.3.3.1 Scientific and technological management of variety

The main changes affecting science and technology as a result of the variety of materials have already been mentioned : emergence of new fields of scientific investigation and constitution of a materials engineering technology. These changes need not be reiterated except to indicate their internal and external repercussions on firms.

Such changes firstly entail an alteration in the behaviour pattern of those involved in R&D. The breaking down of the former boundaries between traditional sectors calls for co-operation among physicists, chemists, metallurgists, engineers and many more in order to find common denominators for the description and comprehension of materials. This requires integration of research work by the various specialists so as to cover the problem of materials in its entirety and master it from the working-up stage to the utilization and testing stages. Pure and applied research are both unavoidably implicated, as materials science is a continual alternation between the theorist's prototype material and the engineer's real one. Any firm having its own R&D activity in the materials field must therefore reorganize its laboratories by decompartmentation in order to achieve new combinations and parallels and to acquire and build up multidisciplinary qualifications and new expertise. In most applications relating to secondary materials, engineering, ceramic and polymer science specialists can no longer afford to work in isolation. Correspondingly, the need for familiarity with the microstructure of materials commands new interest in research outside the firm. The scientific and technological management of variety also requires appreciably more inter-company co-operation, which at present is scant, and the continued expansion of cooperative relations between firms and official bodies.

The development of this <u>network</u> of scientific and technological cooperation is essential because the problem increasingly faced by firms is no longer how to secure a procedure associated with a material but rather how to manage the variety of procedures relating to materials and thus respond promptly and effectively to demand. The new materials in fact constitute just one of many illustrations of the revolution in intelligence. Knowledge does not follow the same logic as goods; in particular, it cannot be appropriated.

In a data-rich system, what becomes vital is not so much the information or knowledge as their controlling network, since this alone can be appropriated (16) . The formation of scientific and technological cooperation networks should also enable firms to cope with diminishing returns on research by contracting out their R&D work. Technically speaking, the acquisition and formation of know-how are the controlling factors in the management of variety. However, as each new material often goes with a specific working-up technique, these forms of know-how will be able to rely on the development of <u>data banks</u> incorporating all the scientific and technical information associated with each material and related process. The growth of cooperation among firms in different sectors but nevertheless interested in the same materials makes the need for such data banks all the more acute.

II.3.3.2 Industrial management of variety

Industrial management of variety applies essentially to the purchase of new equipment, which has in any case been developed and used by firms irrespective of the emergence of new materials. Reference has already been made to the need for non-destructive test intruments. In addition, there must be equipment which can make variety of materials and information complementary. In most cases, the application of new materials would not have been possible without computerized design or calculation facilities. Furthermore, the advantage of the variety of new materials is the capacity for a flexible response to demand. Computer-aided manufacturing and flexible workshops are therefore indispensable adjuncts to the industrialization of new materials. Firms that are already highly automated should take a further step and set up computerized networks to interlink the various workshops and departments by means of a uniform language. While this approach may not be unique to new materials, they make it all the more necessary by reason of their information content and their potential for adaptation to customized demand. The first attempts along these lines are being made in the American automotive industry (MAP networks)

---

(16) Even the patent, which is the most extreme case of ownership of knowledge, is a means of disseminating or divulging it.

in anticipation of the likely organization of scientific and technological cooperation networks.

II.3.3.3 Commercial management of variety

The emergence of new materials makes materials marketing ever more necessary. In the past, marketing was often regarded as a secondary link in the firm's organizational chain. New materials require a marketing approach to capitalize on their greater incorporated information content and to supply information on the situation and trend of materials demand. The establishment of organic links between marketing and R&D department is the best way of turning the variety offered by new materials to account. A general distinction can be drawn between two types of marketing strategy which are compatible with the objective of capitalizing on the included added value of new materials. The first strategy relates to customized demand. It presupposes accurate knowledge of customer requirements and monitoring of their development. Ultimately, it calls for close cooperation between the materials supplier and the materials user. One of the key areas of this strategy is in the industries where technical requirements are set very high. The performances that can be attained using new materials are often such that the buyer can more or less afford to disregard their price in the light of the gains accruing, particulary from redesign of articles and processes at the user's end.

Materials marketing, in the context of the demand for custom-made goods, sets out to demonstrate to customers that the material is suited to their needs, and above all to collect all possible information on the advantages secured by using the material. This enables the producer to capitalize more effectively on his products and pay off his high R&D costs despite the low sales volume achieved in transactions which are compatible with the customized demand.

Another marketing strategy to ensure that new materials are used to full advantage is functionalization, ie dealing with demand under a systems approach taking account of all aspects contributing to the definition of a need. Thus the point is no longer to satisfy the random specificities of customized demand but to solve a concrete problem (Problemlösung). The favourite medium of this strategy is a "package deal" consisting of the material and back-up services. The

Rhône-Poulenc group, to take but one example, has established a specialized centre for technical plastics, where technicians and engineers concentrate on solving the problems encountered by users of these plastics, from the design to the pre-production moulding stage. Software for designing articles and tools is offered, as well as colour software kits. These are of course built especially to suit the plastics sold by the French firm. They are an excellent means of assessing the trend in demand, responding to its diversity and controlling it to extent that it conforms to standards. In the final analysis, the service offered is what becomes vital, and far more important in terms of the profit made than the actual material.

II.3.3.4 Strategic management of variety

One of the major repercussions of the appearance of new materials has been the fragmentation of the sector and the corresponding strategies. The variety of materials accentuates the need to have strategies minimizing the risk of unfavourable substitution. When for instance synthetic materials compete with metal ones, it becomes irrational to base a strategy on control of the metals branch. As we have seen, an intelligent strategy in this case is to form an alliance of materials, ie to establish connections between different branches. Here the most ambitious strategy is undoubtedly to put together an industrial portfolio of different materials. The aim pursued is to guard against unfavourable substitutions by acquiring a group of activities and skills related to the competitor materials. In this way Péchiney, specializing in aluminium, has acquired expertise in reinforcing fibres and ceramics so as to meet the competition put up by composites and thermomechanical ceramics in technical applications. This portfolio strategy is supplemented by an integration strategy whose advantage is to speed up the firm's progress along the learning curve of the competitor materials through takeovers of firms with suitable expertise. Such a strategy is favoured by the comparative "youth" of the new materials and the fact that most of them are developed by SMEs. For the major European chemical groups in particular, it is a considerable boon.

Finally, a less ambitious and expensive strategy is that of industrial partnership which helps to combine the skills needed to put new materials into production in a specific field of application. It is a fitting response to the desire of a firm producing a primary material to enter the market for a secondary material. In this way, the Japanese building and public works firm has entered into a cooperative agreement with the fibre producer Teijin for the industrial production of concrete goods reinforced with aramide fibre. Their two main competitors, Mitsui in building and public works and Toray in fibres, swiftly reacted by signing a similar agreement.

II.3.3.5 Organizational management of variety

The science and technology, production, business and strategic aspects of the management of variety all converge and combine in the direction of company reorganization. The type of coherence which will structure the firm is probably no longer linear but reticular. The new organization of the firm has selling as its fixed point; demand has become more and more qualitative, complex and changeable. The standardization of products allowing scope for scale-oriented efficiency and limitation of sales efforts is no longer applicable in a context of variety. In conjunction with R&D, marketing assists in the definition of projects in a constant give-and-take relationship. At design and production level, all the possibilities offered by new materials can be manipulated to increase the availability of products and related services. The firm paradoxically becomes less dependent on materials in the sense that the materials enable it to adapt more fully to variations in demand. It becomes correspondingly less integrated since the all-item production tendency and the sectoral concept give way to the search for high added value products. Traditional inter-company relations are modified because, whereas the shortage of information formerly obliged firms to establish rigid mutual links, information can now circulate via flexible links thanks to the organization of scientific, technological and industrial cooperation networks. If a firm is to master this new approach to its surroundings, its internal organization must be founded on the network principle.

Precipitated by the variety of new materials and amplified by the new information technologies, the firm's evolution towards an organizational form modelled on the network and on the management of information is the first pointer to the emergence of a new economic system.

CHAPTER II.4

Transient and Permanent Variety. From Management of Uncertainty to Management of Complexity

Having analyzed the emergence of variety (intra-material and inter-materials) in the previous chapters and discussed the various aspects of the management of both forms, the time has come to analyze the choice available to firms by distinguishing the transient from the permanent. Facets of emergent variety genuinely constituting the basis of a new technological system should be distinguished from those which merely represent a cyclical adaptation. Any period of change is characterized by a search for wide variety; in a sense this is a technological projection of a general behaviour pattern in response to uncertainty which entails the diversification of portfolios and in financing circles, is termed a "preference for liquidity". We shall firstly isolate this cyclical trend for specific analysis and in a subsequent section go on to deal with the variety phenomenon as a new structural factor. Before doing so, however, let us look briefly at the peculiarities of each of these two adaptations.

Transient variety is a form of reaction to the accumulation of uncertainties typifying a crisis period and resulting largely from the emergence of the new options which all recessions generate by devaluing existing technologies. In the writers opinions, the emergence of these new options raises two sets of questions. The first concerns what stance to adopt in relation to new options, and the second concerns the process and consequences of learning about a new material.

The growing variety of available materials poses a problem of management and implies in particular the need to master the relations of substitution and complementarity existing between some materials and others. The usual response is to build up a <u>portfolio of materials</u>, a strategy nevertheless having multiple aims which can be classed in two categories. The first comprises passive behaviour patterns involving either self-preservation from a risk of major substitution or, on the other hand, extension of the range of future options by immediately embarking on the necessary learning process. This approach is nevertheless observed to have been generally subordinate to portfolio strategies of a far more activist kind. Within this second class of objectives, certain producers of traditional materials seek to dominate the substitution processes in order to check the spread of technological progress and thereby avoid undesirable obsolescence of equipment and accumulated knowledge. Other protagonists, however, aim to speed up the penetration of new materials. In either case, the ambition is systematic exploitation of the synergies which may result from alliance of the different materials, in order to take part in the spread of technological progress.

The second series of questions relates to the way in which new materials are introduced into existing technology; they seek to prove their capabilities by developing affinities with traditional materials occuring in existing technical objects. Composite materials are a significant example of this process, having been introduced "step by step" into the aerospace sector. The specificity of the substitution process further complicates the accumulated knowledge making it possible to devise new ways of solving technical problems whenever the process meets with constraints. Once the percolation process reaches saturation point, systems are completely remodelled thanks to a simplification of design which is possible once a complex body of knowledge is mastered.

Mastery of complex knowledge is the main thing at stake in a technology system where variety is the operative principle. This problem will be analyzed in a later part dealing with the variety phenomenon as a new dimension of the system. The main input into such a system no longer has a strictly material basis but consists chiefly of information. Rationalization of production used to be almost entirely a matter of larger-scale production and therefore standardization. It is henceforth possible to reconcile variety and efficiency and to alter at reasonable cost the arrangement or the actual <u>coordination</u> of designing and manufacturing phases. This is achieved by accumulating and constantly processing information on the characteristics of demand, information on new scientific and technological knowledge and its evolution, information on preferences, and so on. To exploit the variety potential offered by flexible manufacturing technologies, the technological system must in fact be constantly fed with new information. This "data constraint" raises two further series of questions, firstly about the acquisition of information and secondly on mastering it.

We shall begin by looking at the acquisition of information, an expensive process in terms of demand and a complex one in terms of technical opportunities, since the proliferation of micromarkets in a variety-based system inevitably pushes up transaction costs which in some cases may act as a brake on the emergence of added variety. It will be seen that the existence of variety-derived economies can largely compensate for higher information costs. Furthermore, the proportion of information costs incurred in seeking the new opportunities offered by the materials, can be covered to some extent by introducing organizational flexibility, ie increasing the firm's ability to open up to its environment.

Next we shall address the question of the simplification of certain functions, a procedure which stands out as the prime instrument for mastering complexity. The gradual introduction of automatic decision-making procedures into the various corporate functions allows a greater proportion of resources (particularly "human" ones) to be allocated to information functions. Moreover, the computerization of one specific corporate function makes the others more amenable to computerization. If this incremental process is rapid, the firm will inevitably devote a large amount of resources to its strategic dimension, that is to determining which aspects of its environment to incorporate into its operation. By releasing resources from activities which can be controlled by routing procedures, ie simplifying its internal operation, the firm can release enough resources for allocation to the work of mastering external complexity. It will be seen that by freeing itself in this way from its internal constraints the firm can pay more attention to coordination activities, plainly the essential functions in a variety-based system.

II.4.1 MANAGEMENT OF TRANSITION

This first section will analyze the strategies of firms in respect of new materials and then the process by which new materials are disseminated.

The company strategies to be discussed in the first paragraph are chiefly strategies linked to the substitution of materials. Two types of strategy can be identified; those reacting passively to variety and those playing an active part in the shaping of technological progress.

The second paragraph will examine the process by which new materials are disseminated in a given technology system, using composites as an example and demonstrating that new materials become established by acquiring affinities with materials and articles of a conventional pattern. The subsequent increase in the complexity of knowledge in the course of this "step-by-step" substitution process ultimately results in a quest for new technical solutions.

II.4.1.1 Management of substitutions

A problem familiar to finance specialists is that of deciding which eggs to put in the one basket, or how to distribute a given capital over a range of possible investments. For a satisfactory reply to this question several criteria must be considered, the main ones being risk, yield and the degree of liquidity of available options. These features are common to all problems of speculative decision-making. They apply in particular to the materials field and more specifically to problems of selecting materials for given applications. From this point of view, two classes of objectives can be distinguished in passive multimaterial strategies. At the first level, a response is made to the risks of substitution presented by new materials. At the other level, producers of materials take up options on materials which now have a fair chance of predominating in the near future, earmarking part of their resources for acquisition of knowledge about such materials.

These passive types of behaviour constrast with active strategies transcending the analogy with finance. Some firms build up an active portfolio out of determination to achieve the synergic effects to be gained by mastering several materials at once. This entails active participation in the shaping of technological progress. For other firms, however, technological command of several materials is a means of checking the spread of new materials which constitute a threat to their viability. These various points will be enlarged on below.

Constitution of passive portfolios : the value of options in an uncertain world

The factors of <u>risk</u> and <u>yield</u> explain to a large extent why industrial chemists are taking an interest in ceramics and acquiring textile firms, or why steelmakers are exploring the possibilities of plastics. It is primarily a matter of taking precautions against a risk. The intrusion of new materials brings uncertainty for firms traditionally established in materials markets. Part of the uncertainty concerns the future distribution of markets over a wide range of materials, and it is enough to introduce significant unknowns into the producer's long-term profitability calculations. This upheaval must not be put down to the fact that new materials have a necessarily higher future profit potential than traiditional ones. Although they place a question-mark over the profit potential of traditional materials, their own profitability is at least as uncertain. It is precisely by hedging investments against rising uncertainties that producers of materials can secure minimum cover against hazards to their operational results.

At the second level the <u>liquidity</u> factor intervenes. Here again, the aim is not to reduce the variability of operational yield but to secure possibilities for varying activities in the future, or in other words to hedge one's bets. The acquisition of a new technology or know-how takes time (formation of bodies of knowledge, access to an industrial network). If it is anticipated that a large part of the present uncertainty will be eliminated in the fairly near future, it may be advantageous to invest forthwith in materials or technologies regarded as valid candidates for a new technological system.If <u>in addition</u> it is anticipated that the new system will function predominantly on a variety principle, this points to an ongoing need for liquidity and thus boosts the incentive to devote a major part of resources forthwith to the diversification of the firm's internal knowledge about materials.

The interest taken by European chemicals firms in engineering ceramics is largely guided by the foregoing considerations. The <u>large-scale</u> use of ceramics in thermomechanical applications (engines, turbines, etc) and as industrial cutting tools is only a fairly remote prospect according to most futurists. Alloys and steels will hold their own in these areas for the time being. This deferred issue is certainly the reason why ceramics are on everybody's lips (17) . Industrial mastery of ceramics is nevertheless an immediate issue despite the uncertainty surrounding some applications (18) . At all events, if ceramics spread rapidly, eventually those who today have not made the effort of investment needed to acquire knowledge are in great danger of being outclassed by other competitors. Japanese manufacturers are already at the stage of setting up engineering ceramics production on industrial lines despite the lack of a significant market. The reason for this venture, which would be considered catastrophically unviable by any financial analyst, is precisely to bring ceramics under industrial as well as technological control. If they ultimately become widespread, it will have paid the Japanese manufacturers to simulate the future technology system on a vast scale. Lastly, let us note that these knowledge acquisition strategies aid the emergence of the new materials by enhancing their credibility and speeding up their introduction.

Anticipation and forward appraisal of the future may thus cause firms to opt for strategies favouring the breakthrough of these new materials. Ceramics will of course ultimately gain in importance, but that importance will be heightened if all recognize it and adjust their approach accordingly. A consistent effort to dominate this movement stimulated partly by expectations (19) of technological progress is the way to attain a higher strategic level.

Constitution of active portfolios : accelerating or impeding the dissemination of new materials

An alternative to being guided by uncertainty or by available information is to participate actively in the transition by looking for

---

(17) The enhancement of high-temperature engine performance through ceramics is an activity partly depending on the more or less unfavourable trend in energy prices.

(18) The preparation of a genuine industrial option calls for preliminary work already gambling on a fairly large scale of production. Nippon Carbon is applying this industrial infrastructure strategy to silicon nitride fibres.

(19) On this subject, see N. ROSENBERG "On technological expectations", Economic Journal, September 1976.

synergies or trying to dominate substitution processes. New materials are obtained mainly by combining traditional ones thanks to mastery of their microstructure, which in industrial terms means that the future, and indeed the present, belong to alliances of materials. The old utilization sequences of materials must therefore be discarded and a new industrial fabric created. Strategic knowledge lies at the interface of the traditional materials, and here the strategy adopted by European chemical firms in respect of composites is once more revealing. Traditionally matrix producers, they are acquiring controlling interests in the textile industries. Ciba-Geigy has bought up Brochier, the leader in high-performance weaves. Stevens-Genin has been taken over by Hexcel. However, to master the interface it is not enough to master fibre orientation and fibre-matrix compatibility. A great many of the properties of composites are determined at the time of shaping, and this calls for mastery of processing technologies. The purpose of ICI's takeover of Fiberide, the largest American composites processor, and of the controlling interest acquired by BASF in the composites division of Celanese, is of course to secure access to the industrial network of composites and the American market, but above all to control an essential phase in the definition of a composite material, namely its processing.

While the formation of a materials portfolio is one means of coming to terms with transition (more or less actively, depending on the research approach), it is occasionally also a means of slowing down technological progress, as we shall see.

Materials portfolios are in fact usually built up in order to cope with or control the future, so that the strategies involved are resolutely forward-looking. In numerous instances, however, possession of a materials portfolio also makes it possible to curb the substitutions foreshadowed by new products. While such an approach is only one aspect of a wider strategy within which eventual large-scale substitutions are anticipated, it is frequent among producers of traditional materials. Aluminium producers are a very informative example . In order to limit a sweeping tendency for aluminium alloys to be replaced by composites, Péchiney is diversifying its activities by incorporating new materials such as carbon fibre. The best strategy may indeed be active involvement in adverse competition with a view to controlling it more effectively. Péchiney is clearly trying to go along with the dissemination of high-performance composites by becoming involved at the earliest stages of their production (carbon fibre). By concurrently developing new alloys such as aluminium-lithium (Al-Li), whose strength and density come close to equalling those of composites, the firm partially controls the range of substitution possibilities. In the process, aluminium makers are turning to account the long years spend in researching and developing Al-Li in response to the upsurge in rival composites (20) . Steelmakers have similar motives for turning to plastics. This class of problems also occurs further down the line among the users of materials. For instance, bumpers are still being made of steel so as to avoid unduly rapid detooling. However, what goes for materials and plant also goes for trades whose skills are liable to become obsolete. Here the problem is more acute. The all-plastics car, for instance, threatens to downgrade traditional automotive trades. While there may be an economic advantage, undeniable in this case, in curbing substitutions, incentive to slow down the conversion trend is heightened by human and social considerations. Yet the management of human resources and the relaying of generations of knowledge go far beyond the company framework, so that questions regarding the profitability of investments in human capital must be placed in the macroeconomic context.

To return to the microeconomic level, however, while an effort is made to check substitutions in the short term, the longer term objective is alliance with the new materials which will spread in any case. Producers of traditional materials are trying to develop affinities with the new materials in order to keep in the running. Aluminium makers, for example, may form a partnership with a fibre producer and a plastics producer to develop new sandwich sheets for aeronautics. A case in point is the cooperation by Alcoa (aluminium), Enka (kevlar) and 3M (resins) under the leadership of Fokker (aircraft). Similarly, tinplate makers are combining with resin makers. By producing evidence that they remain indispensable, traditional material producers hope to retain the bulk of their markets by joining forces if necessary with the new materials. As we shall see, the strategy of new materials producers exactly coincides. They begin by seeking to prove that they have affinities with existing materials and are therefore indispensable in the fields where those materials are applied, and then try to gain ground gradually.

---

(20) This point is further discussed in the Chapter on "New materials and transport" in Part III.

II.4.1.2 Pursuit of complementarity

The field of application of materials can at any time be seen as divided into two kinds of area. There are areas in which only a very specific material can solve the technical problem arising, and others in which several materials compete to serve the same function. The following diagram illustrates this situation.

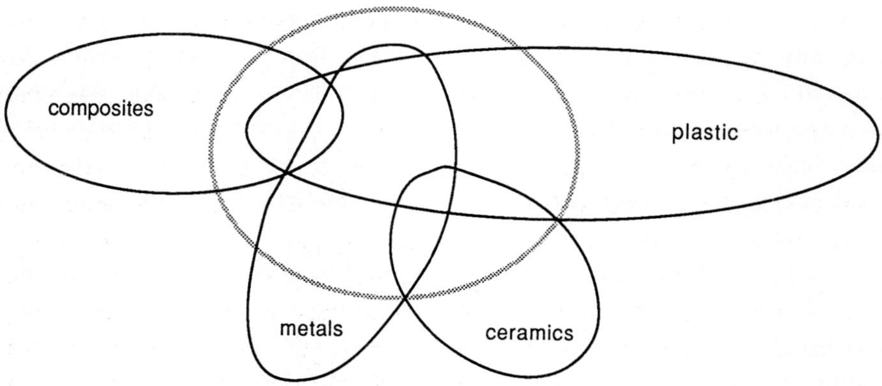

In a area where a single material dominates there is not really any economic problem as such, whereas areas of overlap are competed for according to the comparative resource prices of the factors and technologies involved, these being elements which determine the comparative costs of the available alternatives.

The foregoing simplistic description is intended as an aid to understanding the dynamics which apply to the development of a new material. The technical needs which promote the appearance or emergence of new materials are quite obviously those for which no technically or economically satisfactory answer is forthcoming from existing materials which are firmly founded in well-established technologies. It is the outer limit of the field of existing technical paradigms, where they are not very effective and therefore extremely expensive (areas of decreasing performance of the usual materials). In response to this stimulus, some firms are developing new materials which might meet insoluble or economically unviable technical challenges. New materials are understandably expensive to develop, but may potentially compete with standard materials in their own specific area because there the latter are also very expensive to apply. The technological research conducted at that emergent stage of the new material is of a particular kind. The properties sought are ones which complement the other materials and production techniques. In the incremental R&D approach, an effort is made to move in where the others are ineffective, but this is done in a climate of complementary and mutual tolerance because at this stage one cannot hope to redesign the whole technical object with the new material. For instance, an attempt will be made to include certain ceramic components rather than to produce all-ceramic engines at the outset.

When the new material passes the point of technological and industrial mastery and is at the centre of a tripartite network of bodies of knowledge, skills and industrial relations, disseminations passes a threshold beyond which the new material endeavours to gain on materials already confirmed in their applications. In addition to the complementary features which were pursued prior to the emergent phase, the search is now for substitution properties in order to compete with the other materials, even on their own preserves. This war of materials is all the more relentless because there is a high degree of irreversibility in the choice of materials. The mere fact that at a given point in time a specific material may be the right one to meet a (presumably haphazard) specific need will tend to make it progress further than others; its greater standardization and the resultant cost reduction will enable it to take a larger slice of the general cake. A firm building up a portfolio of materials in this framework conserves options which are at least to some extent mutually exclusive. Their mutual exclusiveness is of economic origin, being founded on a momentary need for standardization.

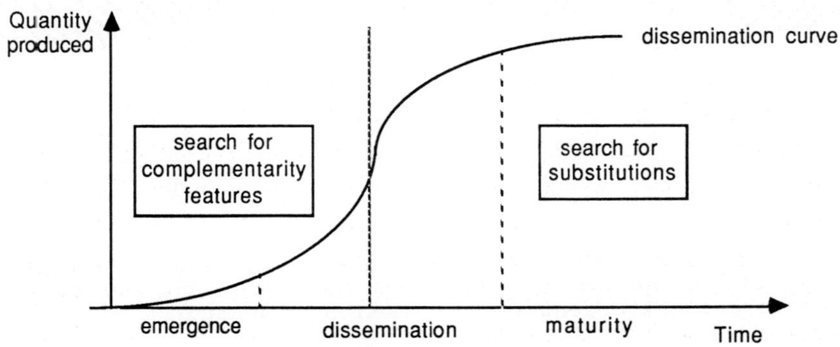

The possibility of reaching the threshold of redesign largely determines dissemination potential and consequently standardization potential, as we saw in the preceding chapter when comparing reversible and irreversible substitutions. We shall now investigate the properties of this new logic.

Economic features of the new technological system

The keystone of the old technological system, still dominant today, is standardization of production founded on a process of technical and organizational changes needed to increase the scale of operations in response to growing demand. This style of production, however, does not provide scope for exploiting scientific potential to full capacity or for offering a qualitative response to consumer needs. As a result, economic regulation as a whole used to be centred more on production requirements than on the requirement of satisfying the variety of needs (21) . In the postwar period, the marked rise in standard of living in the industrialized countries therefore had its price. The standardization rationale took its toll in limited communication between science, technology and production since only an insignificant proportion of knowledge could be put to use. For the same reason, the satisfaction of consumers' needs was also standardized to afford consumers the benefits of mass production.

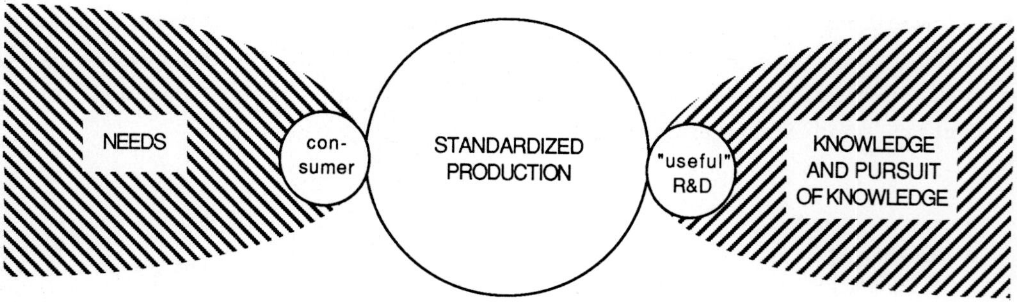

---

(21) In this respect, it is hardly surprising that the Keynesian concept of "effective demand" does not refer to the market but rather to the entrepreneur's perception of demand.

In the emergent scheme of production, the pattern described above is disrupted. The resultant features can be summed up by the flexibility concept.

A flexible production apparatus means, to use an asymptotic reasoning process, that the firm's internal constraints in terms of standardization are far slighter, so that variety can increase significantly. Under the new economic system which is gradually taking shape, the firm's role is one of <u>information regulation</u>, in an endeavour to combine the variety of properties demanded by users with the variety of its database, which enables it to cater for these characteristics. A function of this kind is likely to gain increasing importance and calls for unceasing research into the characteristics demanded by users, and also research concerning the scientific and technical knowledge able to meet these requirements. In other words, a system operating on a flexible equipment basis requires a constant input of information, not only on the characteristics of demand but also on the potential of the technological properties of supply.

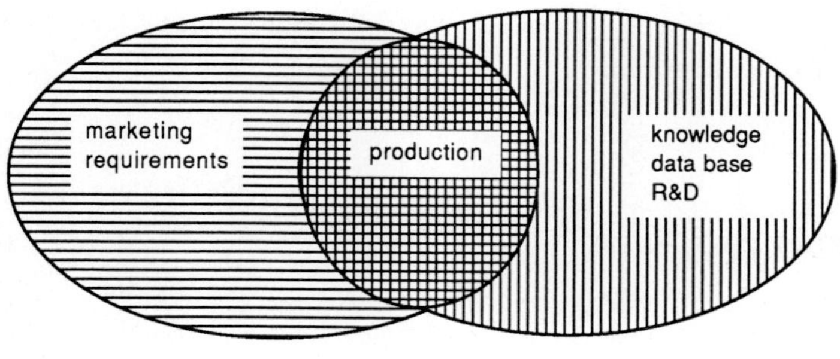

Two remarks should be made at this point. Firstly, growing variety implies increasingly complex articles bringing more types of skills and expertise into play. Secondly, the product cannot even be defined without the user's active cooperation since he must disclose his needs and preferences and specify the required solution in consultation with the firm. This raises interesting theoretical problems regarding the definition of the product and the division of the user's surplus which cannot be discussed in this limited framework.

For present purposes we need to consider how the new system affects the quest for properties in the new material.

Variety and complementary

Firms are henceforth characterized by a constant process of research and incorporation of new knowledge into production. This occurs without potential standardization conspiring to restrict variety or eliminate research sectors. In these conditions of variety, if the knowledge to be incorporated grows more complex, options will have to be kept open continuously, and not only until one of them should predominate. The groups of materials portfolios must therefore continually maintain the options on materials, ie the attention paid to technical properties should diverge widely from the axis of substitution and concentrate increasingly on complementarity since the less standardization there is, the less mutually exclusive materials are. It naturally follows that the dissemination curves flatten out, and it will not be easy for each material to predominate in the economic sense. Consequently there is a likelihood that scale effects will originate more and more from technological and economic control of the interfaces; in economic terms, this means that the coordination function becomes paramount.

II.4.2 MASTERING COMPLEXITY

The emergence of a new technology system operating on a variety principle raises many queries. We have chosen two that seem to be of capital importance and relate to the complexity of the informational environment which the firm must control.

The first question concerns the acquisition of information. A variety-based system must be fed constantly with information if it is to function efficiently. The evolution of demand characteristics must therefore be very accurately registered, new technical solutions must constantly be sought and new forms of knowledge assimilated. These acquisition processes are long and costly and use up an increasing proportion of resources traditionally earmarked for the production phase. Section II.4.2.1 will demonstrate how the existence of "variety-derived economies" in a flexible production system can partially offset higher costs of information about the characteristics of demand. We shall then discuss the process by which complex know-how is acquired, taking composites as an example. Emphasis will be laid on the need for firms to acquire a capability for access to the skills network.

The second question, to be dealt with in II.4.2.2, concerns the management of that information. It is already apparent that in order to master the informational variety of an evolutive environment, firms are increasingly freeing themselves from their internal constraints by introducing routine procedures. More and more of their internal functions are subjected to automatic control processes making it possible to release human resources from an increasing number of internal activities. The simplification of the in-house environment means that firms have more effective control over their activities and can devote more effort to mastering the complexity of their external environment. As a result they can develop new functions, including coordination, which is plainly a crucial activity in a variety-based system.

II.4.2.1 Information acquisition

We shall begin by discussing the question of acquiring information on the characteristics of demand and on the feasibility of variety-derived economies to meet the transaction costs arising from the multiplicity of micromarkets. Having done so, we shall state the principle of viability in a system where it is clearly vital to keep on acquiring new skills in order to furnish an increasingly specific and evolutive demand with technical solutions.

Transaction costs and variety-derived economies

The production system's evolution towards a more flexible mode of operation has two major implications.

The first is the possibility of producing limited runs of varied goods under conditions of efficiency and profitability equivalent to those prevailing in a standardized production setup. This change, particularly evident in the automotive industry, is due to the evolution of manufacturing plant making for greater production flexibility. In the past, the production system operated on the principle of standardization so as to exploit the economies of scale derived from the indivisibility of production capital. The new generations of plant which have absorbed advances in microcomputing and robotics have little in common with earlier generations. The effects of size achieved by this type of equipment are negligible compared with their "effects of variety". These are obtained by easing the "horizontal constraint" afflicting conventional equipment. A horizontal constraint denotes the "specificity of production capital to a definite production" or particular objective (English uses the term "scope"). This constraint is sharpest in crisis periods. Crisis is in fact associated less with losses of efficiency or diminishing yields of scale owing to recession than with the physical constraint represented by a capital designed for a group of products of doubtful economic viability.

With flexible equipment this constraint is felt far less because it is possible to produce a comparatively wide range of products efficiently. However, efficiency does not mean "achieving economies of scale thanks to the effects of size", but rather "achieving economies of variety due to the non-specificity of the capital". These economies exist where a firm producing n separate products with the same plant has a total production cost lower than the sum of the individual production costs for each of the n products (22) . The aggregate output is thus more efficient than the sum of the specialized productions. However, in losing specificity, flexible capital generates new constraints for the firm, and this is the second major implication.

The second implication relates to production process inputs. In order to exploit the advantages of plant flexibility and so achieve economies of variety, it is necessary first to have adequately varied inputs and second to be more accurately informed about the state of demand. The first of these conditions is one of efficiency and the second is one of profitability.

The condition of efficiency amounts to a principle of generality. For a flexible production system to operate efficiently the potential and variety of the processes must not be restricted by a

---

(22) The argument will be developed in the next chapter.

limited range of inputs. This is why the materials revolution is a prerequisite for the introduction of a more flexible production system. The ability of new materials in general to offer tailor-made properties is, we believe, the main factor in disrupting the old production system. Under the standardized production system, materials constituted the constraints on which the whole edifice of the production system was founded. Now that they have ceased to be constraints and become variables, materials have permitted the transition from a rigid to a flexible mode of operation.

While the efficiency of the new order depends on the emergence of the new materials, its profitability depends on information. Under a flexible system, which by definition can meet a far more varied demand, much more information must be generated. The proportion of data inputs thus tends to grow by comparison with other inputs; it is not enough to respond to the existing variety of demand at a given point in time; there must be constant adjustement to variety. Thus it is vital for the production system to be fed with flows of new information. This constant pursuit of information is vital in exploiting the variety-derived economies afforded by the flexible production system. Consequently, one must seek information on the specific characteristics of demand and no longer reason according to global quantitative statistical indicators. This being so, the reliability of sampling techniques declines with the effort to determine the needs or preferences of a microgroup of clients. Where initially the existence of a large aggregate market was presumed, a cluster of specialized micromarkets is found to exist.

To meet the specific demand raised by these micromarkets, a larger part of production resources has to be allocated to information-related activities : selling, informing, communicating, researching and so on. The production phase thus becomes a secondary feature company activities; transaction costs (contract prospection, negotiations, advertising, market research, etc) tend to account for a larger proportion of overheads. The viability of such a production system thus depends largely on these costs and more generally on information costs, and so the question is whether or not variety-derived economies suffice to counterbalance the information costs, which are likely to be mounting.

Analysis of the viability of a variety-based system

Two products only will be examined here.

Let q_1, q_2 be the unit quantities of products 1 and 2 respectively, and let it be assumed that these quantities are measurable with the same units so that $q = q_1 + q_2$. The cost of production is represented by $C(q_1 , q_2)$.

Variety-derived economies exist when :

$$C(q_1 , q_2) \leqslant C(q_1, 0) + C(0 , q_2).$$

The production of product 1 generates economies of scale if :

$$C(2q_1 , 0) \leqslant 2(Cq_1 , 0).$$

(a similar definition applies to product 2).

If economies of scale exist for one of the two products and if they both meet the same need on average, the industry will specialize in making a single product. In this case, the information costs will represent no more than the cost of a quantitative estimate of market scope and will be classed as fixed overheads.

For product 1 the total cost is :

$$C(n . q_1 , 0) + CI(n)$$

where n is the number of customers and CI(n) is the cost of information on the size of the market (n). Where economies of variety exceed economies of scale, we have :

$$C(n_1 . q_1 . n_2 . q_2) \leqslant C(n . q_1 , 0)$$

$$\text{with } n_1 + n_2 = n.$$

The total cost of production is then :

$$C(n_1 . q_2 , n_2 . q_2) + CI(n_1 , n_2).$$

It is no longer enough to know the total number of customers; their number per category must be known. Thus it can be expected that :

$$CI(n_1 , n_2) \geqslant CI(n)$$

More generally, the cost of information tends to increase with the number of demand characteristics, chiefly because of more complex negotiation and transaction procedures.

The question is then whether :

$$C(n_1 . q_1 , n_2 . q_2) + CI(n_1 , n_2) \lesseqgtr C(n . q_1 , 0) + CI(n)$$

The two graphs below give a more general formulation of the problem where the number of characteristics can vary continuously. Assuming rising information costs, there emerges an area of viability for a variety-based system where the production level $q = \sum_{i=1}^{m} q_i$ is given, over $m \in [0, M]$.

The extent of the viability area depends on the relative directions of information costs and variety-derived economies.

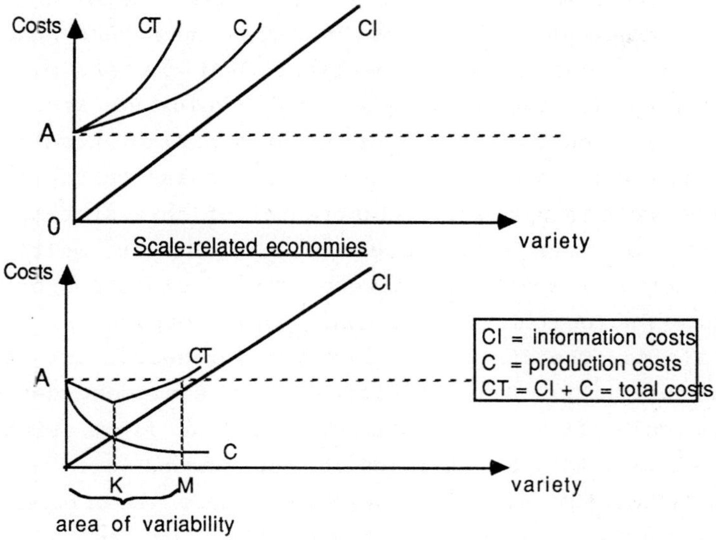

Variety derived economies

This problem is examined very schematically above. The analytical example given reveals an area of viability for a variety-based system. The example also highlights the existence of a point (K) of optimum variety corresponding to the minimization of total overheads (cost of generating variety; information costs). The existence of such a point means that it is advantageous to <u>standardize variety</u> by reducing it to small number of characteristics. However, this point is not guaranteed to be unique or even to exist, and the trend of demand characteristics over a period of time alters its position on the axis of variety. It is therefore necessary to continue seeking new ways of meeting demand. This problem is considered in the next paragraph.

Acquisition of technical knowledge : organizational flexibility

While it is vital to become more receptive to information from below, one must also look up the line from the production units because to meet specific and changing demands it is necessary to be constantly on the lookout for new technical solutions. These, however, are more prone to fluctuation than was formerly the case and are therefore liable to become obsolete earlier. What is at stake is therefore the ability to master complex and evolutive knowledge; complex because it is central to the overlap areas of traditional branches of expertise and evolutive because demand characteristics are unstable. Composite materials are an illustration of this statement.

Such products as the rear hatch of the BX or multilayer materials for aeronautical applications were born of close cooperation between firms whose scientific and technological expertise lay in vastly different fields. The first case involved a research project in which a vehicle manufacturer (Peugeot-Citroën), an equipment designer (Billion), a chemicals firm (Rhône-Poulenc) and a fibre producer (Vetrotex) were associated. An example of the second case is the production of an aluminium alloy-resin-kevlar fibre multilayer for aeronautical applications, involving cooperation by an aircraft builder (Fokker) with a resin producer (3M), an aluminium maker (Alcoa) and a fibre producer (Enka) under the sponsorship of the University of Delft aeronautical engineering department. These examples, many more of which could be given, clearly indicate the diversity of the fields of expertise brought together for a specific purpose. The principle is of course not new, but the proliferation of co-operative ventures of this type is.

The underlying objective is to master the aeras of compatibility of the various skills. With composites, it is in fact easier to produce ready-formed objects by manipulating the direction of the reinforcement. Thus it is vital to master the material's microstructure together with the technologies used in producing the primary materials, processing and non-destructive testing methods. The need for more and more industrial partnership agreements as a means of access to know-how reveals a new dimension of technology, as previously stated, but also entails a new principle of economy in activities of research and knowledge acquisition. With the old technological system, the pursuit of knowledge occured mainly within large standardized production units which could be equipped with specialized research centres in contact with outside laboratories thanks to the existence of information-linked economies.

As with variety-derived economies, which emerge when one and the same plant can be used for many different types of product, information-linked economies can be achieved where the same information can be used for a large number of different products or processes.

Consequently, with a standardized system where research costs are spread over a large number of identical products and processes, information-linked economies are very substantial. For instance, one of the reasons which prompted the French electricity board to opt for a standard pattern nuclear power station was the possibility of information savings on the learning process. Any improvement originating in one station is automatically incorporated in all of them, thereby substantially increasing the marginal yield of research.

By contrast, with a variety-based system information-linked economies are more restricted as the capital yield of research projects is confined to a small number of short product runs. This is a familiar problem in the pharmaceutical industry, for example. Under these conditions, research centres have to master increasingly complex and rapidly superseded knowledge. It is therefore found more profitable to conduct research jointly, thus making it possible to bring in small units with very specific expertise. This type of joint research has the advantage of generally preserving the system's ability to adjust to fluctuations in requirements. However, where a production unit incorporates a large part of the expertise which proves necessary at a given time, it is compelled to turn its knowledge to account. It then has to immobilize the internal links between the various skills, activating those links which are crucial to the attainment of its technical goals. This necessary focussing of the in-house network in order to meet profitability requirements also restricts the firm's ability to mobilize new external links owing to the "cross-linking" of its own network of skills.

In a production system requiring the virtually continuous invention of new technical solutions, it is therefore necessary to substain the organizational flexibility of firms, ie their ability to activate latent links in the network of skills which are diffused through the technology system, so as to generate flows of new information. Yet this ability cannot be gained by preselecting a specific subsector of the network and dominating the currently active links by controlling it. Constantly changing demand permanently destabilizes the division between active and non-active links and calls for the exploration of new links. Thus it is vital to develop the firm's receptiveness to its environment and take every step to foster modes of access to the network of outside expertise. The final part of the chapter will show how firms can achieve this objective.

II.4.2.2 The firm in an information-rich production system : management of information

We shall begin by demonstrating that in order to become more receptive firms must simplify their internal environment. It will then be demonstrated that this receptiveness leads to the emergence of new functions. Thus the coordination function appears as a major activity of firms in a variety-based system where an effort is made to master the external environment.

Simplification of the internal environment through algorithmic formulation of functions

Permanent receptiveness to new information, ie simplification of the firm's strategic dimension, entails the corresponding ability to integrate more and more elements in its automated procedures. Problems of decision-making must be continuously transposed into the sphere of standardized responses. Only to the extent that this transposition process proceeds easily and with some measure of autonomy is the process viable. There must be some indication that the new operating principle, founded on informational processing, is capable of generating a surplus. During industrialization the principle of creating surpluses was founded on a chain reaction interlinking standardized organization of labour, downgrading of individual skills through automation and increasing openess of markets as the goods themselves became standardized. The accepted expression in economics is increasing returns. For the new data-intensive production system to be viable, a new chain reaction must take place without, however, depending as before on standardization. It is not possible at present to answer this question satisfactorily in terms of economic regulation, since we are in a phase of transition and the two schemes of development coincide. The answer already seems affirmative, however, since information technologies embody their own chain reaction principle, one needs simply to analyze its developmental phases.

Mainframe computers penetrated the firm's more structured areas as far back as the 60s, since here it was possible to apply batch processing and improve routine operation efficiency (accounting, stock management, invoicing, etc). Then in the late 70s and early 80s the development of more decentralized computer generations with the integration of data processing into production through CAD, robotics and finally flexible workshops allowed the pattern to swing towards more flexible operation which triggered rapid changes. However, the second stage largely depended on the first, as the ability to switch from one production system to another entails the optimization of both, an optimization which is itself the outcome of improvement in methodological knowledge achieved in the preceding phase. Mastery of production organization through large-capability computer techniques is the key to the direction of the variations which need to be introduced when products and methods are undergoing structural change.

In other words, the computerization of standardized functions is a means of acquiring a surplus of information on conditions for the efficient operation of the less organized functions. This makes it possible to establish procedural norms more clearly whereas formerly the process of decision regarding them was much more uncertain. These more "open" functions become standardized in turn, and then algorithmically formulated and thus more suited to information feedback technologies. There is plainly a chain reaction, each computerized phase of which creates the conditions required for the new generation of computers to be installed. At every stage, the approach presupposes the introduction of procedures to identify the internal logic of the various activities concerned. This gradual transformation of the very perception of corporate and organizational structures is thus tending towards increasing automation of decision-making procedures, which are becoming part of the firm's operating routine. This evolution tends to alleviate the uncertainty surrounding more fundamental decisions and in this respect partakes of an environment with increasing qualitative scope. Put another way, the introduction of increasingly sophisticated data storage, processing and transmission systems tends to displace the limit between the formulated and unformulated sides of the company; at each successive opening up of decision-making to other fields and other complexities, yet another open part is encoded in an algorithm.

The description of the cumulative process briefly presented above is naturally simplified (23) . It is in no sense a continuous structural transformation process. Threshold effects appear whenever the algorithm subtending the operation of part (or the whole) of the firm can no longer react to the development of the environment. Yet this is not a conceptual novelty, since it already applied to alterations of scale in conventional industrial processes.

Mastering the external environment : emergence of the coordination function

Whether or not the "growing pains" are cute, the cumulative nature of the assimilation of information through successive opening up and algorithmic formulation is obviously not about to cease. A flexible apparatus is valueless without a constant turnover of information. The constraints born of standardization are gradually pushed back in this process whose asymptotic limit is the wedding of specificities. Whenever the variety of a product increases somewhere in the system, its components tend to become more varied so as to meet more and more differentiated and specific needs. Materials are already taking their place in this universe. The mastery of complexity raises a problem as to the type of complementary features which will receive priority in this process , and the case of materials places the problem in an interesting context. Precisely stated, the problem is that the possibility of designing and producing a growing variety of articles depends crucially on knowing the <u>areas of compatibility</u> of the various properties.

Flexibility does not mean that the property dimension is convex. Combination potential will probably be the future focus of competition in the developed countries because the crucial distinction between two

---

(23) For an in-depth analysis of this process, see "Informatisation - l'impact sur l'organisation des entreprises" ("Computerization - the impact on company organization"), op. cit., E. ZUSCOVITCH and P. BRENDLE.

equally flexible systems will be established by the possibility of extending these areas of compatibility beyond the frontiers marked by standard data banks. This is the avenue where R&D activity is sure to become the major constraint for firms, as their sole guarantee of achieving informational specificity and consequently a relative advantage. This may be a forecast rather than a solid reality, but even now this problem is plainly central to the strategic mastery of new materials. Will process designers (vehicles, aerospace, electronics) best dominate these areas of compatibility, or will traditional suppliers of materials like industrial chemists and steelmakers be better equipped to do so ? The process designer is doubtless aware of the areas of variation in the compatibility of constraints for a given technical object whereas the materials supplier, traditionally associated with several user industries, has a better command of the full spectra of properties. Intersectoral technological partnership already indicate that the matter is not yet settled; indeed there is probably no single solution to it.

If there is no single answer, the question is what are the determinants of competitiveness in such a complex system. Who will master technological progress most effectively ? How will its results be absorbed ? We are of course at the dawn of this transformation and it is too soon to foretell the exact configuration of the new industrial system. One feature is nevertheless plain. The growing complexity of the system will make the <u>coordination</u> function crucially important. As well as firms which define a specific type of information, those in control of networks will perhaps fulfil the essential function. Technological progress in conjunction with such a system has a high organizational content. There is moreover an area where this type of operation used to predominate, namely the space programme. Space technology is not a single body of scientific knowledge giving rise to a technology as in the case of chemicals or electronics, but is essentially an organizational one. A concatenation of technical and organizational constraints (delivery time, specifications, etc) from clients to process and plant designers is what makes space technology progress. Transfer of knowledge to other technical and commercial areas is very obvious in the space programme, in accordance with the nature of participation in this large technological and organizational coordination network (24) . Depending on their role in the network, firms have not developed the same learning profile, in the sense that the categories of induced advantages were qualitatively different. Thus, beyond the traditional arguments of sectoral division, according to the type of activity and the market's structural features, the part played by firms in connection with the organization of the network of skills will also determine their competitiveness.

CHAPTER II.5

Reducing of Complexity through Integration

The preceding analyses have established that the development of firms and their strategies must be interpreted essentially in terms of information. The crucial issue for a firm under a variety-based system becomes the mastery or even reduction of complexity, not only in its market relations but also in its own internal skills and production network.

Before presenting the various methods, which will be classed as integration processes, used to attain the goals described, our first concern will be to state (and sometimes reiterate) the factors of the integration process.

---

(24) See P. COHENDET and E. ZUSCOVITCH "L'apprentissage technico-économique du programme spatial européen" ("Technological and economic learning in the European space programme"), SITEF, October 1985.

II.5.1. KNOW-HOW AND ACQUISITION OF INFORMATION

Mastery of know-how and acquisition of information are clearly the most specific and important factors for the purposes of elucidating the development of materials manufacturing structures, and that applies not only to their production but also to their processing and uses.

We shall firstly recapitulate the phenomena relating to know-how and then go on to discuss the integration mechanisms whereby the effects of environmental risk factors can be curtailed through more rapid acquisition of information.

II.5.1.1. Strategies and know-how

Among the observable strategies which are of significance in the management of know-how, two patterns of logic can be distinguished.

The first pattern is one primarily involving "systematic" development of the firm's pre-existent expertise. Its rationality lies in "economies of scope" associated with the diversified exploitation of a common "production factor" for all the various types of production : here the factor is know-how or the firm's generic technologies.

The second pattern, on the other hand, reveals a will to anticipate potential developments and substitutions of materials. It is displayed in the acquisition or amalgamation of external skills and in the formation of new fields of expertise. The main objective here is to gain a surer strategic foothold in relation to the anticipated developments.

These two patterns are not unrelated, however, and may even form a sequential scheme of development as follows : formation of the new know-how (pattern 2), then exploitation of the economies of scope (pattern 1), or even the reverse : exploitation of economies of scope (pattern 1), then constitution of new know-how (pattern 2) followed by exploitation of economies of scope (pattern 1).

No illustration of the mechanisms pertaining to the first pattern need be given as they are described, elsewhere in this part. Let us simply consider a few interpretations. Reasoning in terms of economies of scope introduces a novel view of the evolution of a sector such as chemicals. This sector has a tradition of applied and basic research on materials which makes its diversification worthwhile and also necessary, since competition by oil-producing countries in respect of "commodities" has become unbearable for European, American or even Japanese firms which can no longer guarantee their viability through mass production, ie economies of scales specific to a given commodity. To growing extend they are developing speciality products fitting particular slots. Rhône-Poulenc is investing in rare earths, biomaterials technologies and ceramics and Monsanto in silicon for electronics while Du Pont is backing the communications market and Hoechst, ceramics. However, owing to special characteristics linked with their end-use, these products have limited outlets in terms of volume. The chemical groups must therefore rely on economies of scope, and to this end they are turning to systematic exploitation of their research potential.

The development of engineering ceramics by chemicals firms is a highly symptomatic example. Each chemical group, in Europe at least, has a ceramics line - Rhône-Poulenc and Ceraver, Hoechst and Ceramtech, Rosenthal and others. More generally, these groups are seeking to expand their production range into electronics, information technology and biomaterials.

The exploitation of economies of scope is thus a means of remedying the lack of scale economies in micromarkets and enjoying relative advantages over the new producers.

The second pattern, which involves the combination and constitution of know-how, seems more instructive as far as the evolution of manufacturing structures is concerned.

The constitution of new know-hows often stems from a combination of old ones. This necessity is eminently characterized by composites or optical fibres, and it is also a far from insignificant factor in the evolution of materials market structures. To illustrate this we shall concentrate on two significant cases : organic matrix composites and optical fibres (note that the latter are functional materials).

It should be reiterated that organic matrix composites stand at the crossroads of chemicals (or petrochemicals) and textiles, the former defining the matrix and the latter providing the woven reinforcement (25) . Obviously enough, the know-hows involved are

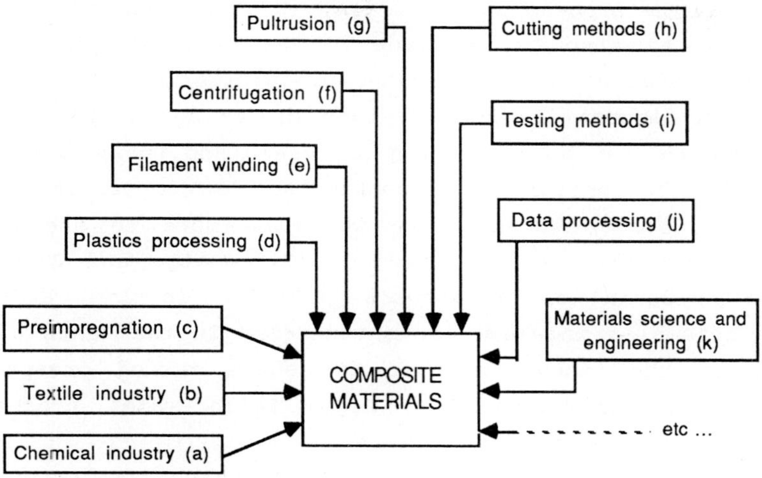

(a) Resins, adjuvants, additives ... eg "low shrink" resins.
(b) Spinning, fibre-forming, weaving.
(c) Example : SMC ("Sheet Moulding Compound") : semi-processed product (sheets of fibres preimpregnated with resin) for pressure moulding ; example : thermoplastic impregnated fibre (Ato-Chem's TIF).
(d) Injection, press and other types of moulding.
(e) For "rotational" parts : tubes, cylinders, spheres, cones, tanks, etc ; analogy with electrical engineering, textile winding or sintering.
(f) For "rotational" parts : analogy with rotational moulding (TMP) and centrifuging of metals.
(g) For profiles (eg fishing rods).
(h) Laser, water jet ...
(i) In particular, non-destructive testing : holography, scanner, ultrasounds, acoustic emissions ...
(j) CAD, CAM ...
(k) Physics, mechanics, tribology, rheology, structural calculatic and modelling, behavioural laws of anisotropic materials ...

Figure II.5.a : Composites processes (26)

specific to these two industrial sectors and must be combined to make up a "composite" know-how. Furthermore, these materials "form unique

---

(25) See the second chapter of this part.

relationships of technological convergence and proximity (eg filament winding and centrifuging) and encourage process combinations (eg weaving + impregnation), techniques (eg cutting and testing), proven methods (compression, injection, etc ...) and scientific disciplines (eg the science of heterogeneous materials, rheology, calculation and dimensioning) to evolve so as to adapt to new probles" (26) . This complexity of knowledge can be represented schematically.

The main figures in this development are again the chemical firms. Their knowledge of manufacturing processes and their industrial strength give them the possibility of commanding the composites branch, where changes in production structure would be more to the matrix producer's advantage than to the weaver's (Ciba-Geigy and Brochier, Hexcel and Stevens-Genin) or processor's (ICI and Fiberide, BASF and Celanese). The assembled know-how is concentrated in the hands of the chemical firms so that they are able to secure varied prospective outlets whose variety stems largely from "tailor-made" production intimately linked with composite materials.

The case of optical fibres presents certains similarities in that their production requires a command of scientific and technical knowledge deriving from various industries (Table II.5.A).

Technical operation	Decisive element	Industry of origin
PREFORMING	- Equipment - Primary material and principle of growing materials on a substrate	GLASSWARE SEMICONDUCTORS
FIBRE PRODUCTION	- Hot drawing of a material into filaments	WIRE DRAWING GLASSWARE
CABLING	- Cladding and cabling of filaments	CABLE MAKING

Table II.5.A (27)

(26) According to THALER, op.cit., p.190.

(27) Excerpt from N. MASSARD : "Recherche technique et mise en oeuvre des connaissances scientifiques dans la production : le cas des fibres optiques" ("Technical research and application of scientific knowledge in production") in "Recherche technique et .../...

The industrial structure corresponding to this technical expertise indicates that there is no specific integration process. No firm incorporates the entire technological branch (Table II.5.B). The degree of integration greatly depends on the background of the firms involved. Their strategy is largely a market - and specifically a telecommunications market-strategy.

SILICA TUBE	Heraeus (F.R.G.) Quartz et Silice (F) General Electric (USA)	
PREFORMS FIBRES CABLES	FOI (F) Corning Glass (USA) Optical Fiber (UK) Siecor (USA) Acome, Fobtica (F) LTT (F) BICC (UK)	ITT (USA) Western Electric (USA) Northern Telecom (Canada) Sumitomo (Japan) Furukawa (Japan) Fujikura (Japan) Philips (Netherlands) Câbles de Lyon (F)

Table II.5.B

Note, however, a special case, the French firm Fibres Optiques Industries (FOI) (28) . This is a financial association of three major groups : Corning (glassware), Saint-Gobain through its subsidiary Quartz et Silice and Thomson through its LTT cables subsidiary, and is a government promoted venture. It clearly illustrates the huge importance of mastering the entire technology branch.

A number of facts can be observed from the examples of composites and optical fibres. Firstly, no set pattern of integration appears to be associated with the need to build up know-hows. That being so, the integration process will depend on the other variables in

---

.../...
innovation industrielle", Vol.3, STS, CNRS, ECT, November 1985 (p.14).

(28) N. MASSARD, op.cit., p.27.

the balance of power between those holding the expertise (chemicals - weaving), company background (optical fibres) and policy direction (FOI).

Moreover, opposite motives apply to these two examples : with composite materials, the object of integration is to form and master a new know-how in order to take advantage of the economies of scope originating from a "tailored" response to the requirements of the various markets (Ciba-Geigy/ Brochier, for instance). With optical fibres, the main aim is to exploit economies of scale. "Within this (industrial) structure, company strategy is concerned mainly with the pursuit of economies of scale" (29) . Note that there is a second, related difference organic matrix composites are structural materials and optical fibres are functional materials.

Lastly, the differing degree of integration in the optical fibres branch would seem to originate from the high assimilability of the relevant expertise, at least for certain segments of the branch, as opposed to the composites branch. To conclude this paragraph, then, a form of "allocation of know-how" can be observed. Its mechanism might the defined as follows :

- Firstly, <u>where the specific expertise does not yet exist</u>, but :

. If it depends on existing expertise and knowledge, the industry most conversant with them will carry out the diversification. The ceramics field is a good illustration. Users like Renault and Fiat have discontinued or scaled down their ceramics activities while the chemicals firms are developing them.

. If the expertise is to derive from a compound of know-hows (eg composite materials), there is no set pattern, as we have seen. Other variables come into the structural development process.

- However, <u>where the expertised already exists</u>, even if it is destined to evolve further, the integration process will depend crucially on the extend to which the expertise can be assimilated and transferred. Thus, where there are considerable initial barriers either because of specific scale economies or because of a long learning curve developped by the holder of the know-how in question, integration will proceed

---

(29) N. MASSARD, op.cit., p.28.

more to the latter's advantage. A typical example of this is provided by Péchiney in the production of Al-Li and aluminium processing.

If, however, the expertise is easy to assimilate, other determinants have to be incorporated. A very fragmented manufacturing structure appears possible. The development of small independent plastics processors up to a few years ago is the stock example, and its development is also significant. The increasing specificity of market requirements in relation to these plastics presupposes more and more complex, and correspondingly less and less transferable, knowledge. This would explain the integration process observed in this sector.

Now that we have identified the structuring effects of know-how on industrial strategies, we shall go on to look at their possible forms in section II.5.2. Before doing so, however, we shall elucidate the possibility of changes in manufacturing structures brought about by a wish to limit market uncertainties.

II.5.1.2. Strategies and processes of information acquisition

The main aim of these strategies is to obtain on the best terms the information needed for manufacturing.

For instance, exploitation of economies of scope requires knowledge of the markets and their special characteristics. A knowledge of customer needs thus become increasingly vital to the suitable design of the material. A process of information acquisition operating downwards is therefore necessary. Analytically, the model is as follows :

Firm B down the line from firm A produces an end-product using a material supplied by A. B's requirements are highly specific to the final demand and are utterly uncertain as far as A is concerned. However, as from the period t-1, B knows the demand specifications. In order to deliver within a time t a material with the requisite features, A must select a equipment at stage t-1. In the circumstances thus described. A will acquire in t-1 a capital such as to maximize the profit expected.

However, let us now assume that A has the opportunity to absorb B. If so, it will be aware as from t-1 of the specifications and amount of material needed by B and will thus be able to select a capital when in possession of all the requisite particulars of its downstream side and of demand. Ex ante, the profit to be made in this case in the expectation of maximum profit, not the expected profit (30) .

On the face of it, the second situation is more advantageous to firm A, which hitherto has had to bear the risk. Thus uncertainty itself acts as an <u>incentive to integration</u> equal to the anticipated value of the information carried by the transmission, with a certain lead, of the characteristics of demand (31) . The same analysis can be applied to upward integration. The anticipated value of the information according to this definition may even indicate the maximum cost which a firm would be prepared to pay to undertake integration.

The risk as such does not necessarily depend on integration for its resolution. An increase in the number of co-contractors might also limit the outlet risks, and would constitute a "financial portfolio" type strategy. However, materials have acquired highly specific characteristics and processing in the course of their evolution. Know-hows are still rare and often not easily transferable, so such a strategy may be difficult to apply. We have here a game with a very limited number of players, which is why integration in the case of materials proves an ideal way of reducing risks. Therefore, whether in order to secure a specific supply (upward integration) or to specify needs (downward integration), materials technologies undergo a highly developed process of vertical integration. These processes can take various forms in practice, the object in all cases being to acquire relevant information as fast as possible. This variable becomes more strategic as the specificity of requirements increases. The forms of the process range from simple agreements to the development of industrial partnerships, and ultimately to full mergers ; they will be analyzed more fully in the next paragraph.

---

(30) Mathematically, the expectation of maximum profit is always equal to or greater than the maximum expected profit.

(31) In other words, the value of the information is the difference between the expectation of maximum profit and the maximum expected profit.

II.5.2. FORMS AND EFFECTS OF INTEGRATION PROCESSES

The factors of the integration processes described above result in different forms of inter-firm integration ; horizontal or cross-integration forms associated with the exploitation of economies of scope or the formation of new know-hows, full vertical forms of integration and all forms of quasi-integration. These various forms of inter-firm integration, summarized in Table II.5.C, will be described before evaluating their implications for the manufacturing process as such.

	Types of integration	Strategies	Motive factors
Inter-firm integration	Horizontal/cross-integration	. obstruction or control of substitutions of materials - directly - through capital goods . possibly : materials portfolio	. know-how
	Vertical quasi-integration	. reduction and control of market uncertainties . reduction of transaction costs	. risks . costs . know-how

Table II.5.C : Inter-company integrations, strategies and factors

II.5.2.1. Forms of inter-firm integration

Theoretically, markets are the natural places for communication between firms. It is through them that firms take in information (prices, quantities, qualities, etc ...) which normally suffices for their decision-making. Yet some of these markets are non-existence or very imperfect.

The transfer of know-how is perhaps the most significant instance. It can take place through skill training of the workforce, through equipment where the know-how is incorporated, or through the dissemination of technological and scientific information. A large proportion, however, especially where incipient know-how is concerned, cannot pass through these channels.

The complexity of the specifications required by customers and the variety of materials available also bring uncertainties and costs into transactions on the existing markets.

The process of integration can thus be interpreted as a response to the imperfections of the markets. Basically it takes three forms : horizontal and/or cross-integration, vertical quasi-integration and full vertical integration.

Horizontal and/or cross-integration

This form of integration involves a firm's becoming interested in other branches or other materials than those routinely produced or used by it. This applies to chemical companies like Ciba-Geigy in buying up Brochier the textiles manufacturer or to Rhône-Poulenc with the ceramics firm Ceraver. The first merger is a case of integration between branches (chemicals-textiles) and the second is an example of expansion of the materials "portfolio".

To effect these integrations, the companies endeavour to involve themselves in the mechanisms governing the substitution and evolution of materials. They want to acquire expertise which may be necessary to avoid losing out on substitution, which they deliberately anticipate.

There are two possible courses of action. Either the industrial groups keep pace with the developmental process by moving into the competitor materials, as Ciba-Geigy is doing with composites thanks to Brochier's weaving know-how, which means that Ciba-Geigy is already taking up a position in the production of a promising material, or they develop materials of their own to prevent (or simply slow down) a substitution favouring a material which is presently foreign to them. For instance, the replacement of metal allows by composites is acknowledged to be an irreversible substitution. Péchiney's effort is developing a capacity to produce Al-Li alloys is specifically aimed at checking the impending substitution, and the effort may succeed because Al-Li is easier to adopt than composites, as its application techniques are akin to those of other alloys and aluminium.

At the core of these strategies there invariably lie the acquisition and formation of new know-how and exploitation of their applicational and evolutive potential. These incursions into the mechanisms of change may nevertheless prove more complex or, on the other hand, simply equate to a more conventional strategy.

One possible complex form is cross-integration of activities targeted on the manufacture of machine tools. German steelmaking traditionally controls part of the machine tools production, or to be more precise the capital goods, intended for it. Thus by controlling one of its outlets it also controls the instruments of the technology which it uses. Similarly, extending the illustration to new materials, firms like Kobe Steel in Japan are found dominating activities related to new materials. In the case of Kobe Steel, machines are being produced for the HIP (Hot Isostatic Pressing) process for use in superalloy powder fabrication or ceramic sintering.

The conventional strategy in question is the choice of a materials "portfolio" ; some of the combinations in fact closely resemble pursuit of "static flexibility" in which the variety of the materials offered by the firm is a response to that of the immediate or more remote substitutes which could fill the demand. This strategy also makes it possible to exploit the synergies (or "economies of scope") which may exist between the various materials. Péchiney's venture into carbon fibre (with Elf-Aquitaine and Toray, in Soficar), engineering ceramics (Criceram for powders, and Demarquest), and the development of special alloys (Hownet Turbine Components Corporations - HTCC) could also be analyzed in these terms.

It should not be forgotten, however, that there are other less drastic ways of benefiting from expertise unrelated to the firm's own specialty. Instead of acquiring outside know-how, it is possible to solve an isolated problem by resorting for a time to a "partnership" arrangement, ie close cooperation among several firms with different specialties, which is far from uncommon. A representative example previously mentioned is the cooperative designing of the BX rear hatch by Peugeot-Citroën, the Saint-Gobain press, tooling and fibre subsidiaries Billion, Seva and Vetrotex.

However, for such cooperation to be feasible, prior possession of a critical mass of knowledge and expertise may be necessary. Aérospatiale had to acquire its own expertise in the chemical structure of materials before the chemical industry agreed to work with it in the production of composites.

Vertical quasi-integration

This integration principle operates within one materials sector ; it is a downward or upward process. Such quasi-integration occurs in the shape of takeovers, by acquisition of holdings in a firm and by sub-contracting. It involves having some of the manufacturing steps performed outside the firm. Control over the products offered by such firms is nevertheless maintained - there is no question of submitting to the laws of the market - in order the reduce uncertainty ("secondary information") and transaction costs. This procedure is particularly significant where manufacture needs standardized knowledge which is not decisive for the dominant firm. The firm can thus control the contingencies of market transactions and reduce or even eliminate certain transaction costs. These transaction or information costs can arise in particular from the need to seek out suppliers producing extractly the desired materials, either raw or processed. The variety of "possible" materials (tailor made composites and so on) favours such special close relations between producers, processors and even users of materials.

Quasi-integration of this kind also makes it possible to appropriate the added value generated by the use of a specific material.

Vertical quasi-integration often occurs in assembly industries such as automotive industries where certain tasks are performed by a group of subsidiaries or subcontractors. The trend is especially meaningful in the processing of plastics : "... well before the century ends, the respective shares of actual processors of plastics and firms having integrated plastics processing will be in inverse ratio" (32) ie integrated groups will presumably represent 60 % instead of the present 40 %.

---

(32) M. REYNE : "Plastiques : la croissance continuera", ("Plastics expansion will continue"), Sciences et Techniques n°9, November 1984, p.37.

The creation or existence of specialist subsidiaries in a given field provides a more accurate idea of market requirements and of processors' or even users'expertise. Some examples : Péchiney has set up an exclusive packaging subsidiary, Cebal, and acquired Demarquest for ceramics ; Hoechst has bought out Rosenthal (ceramics) ; Du Pont took over Berg Electronics in 1972 for connectors.

Lastly, there are also more one-off forms of transfer of information and know-how, for instance joint venture agreements between producers and users of materials : Du Pont with Philips for optical disks ; Hoechst with Alcatel-Thomson ; Du Pont with British Telecom for opto-electronics and optical fibre components.

Full vertical integration

Full integration, unlike the above, means making products wholly within the firm ; it is a more radical form of integration.

Two related reasons for an arrangement of this kind have been encountered ; they are both very revealing and concern know-how.

In the first instance, namely production of plastic automotive parts, the specific know-how already existed in a subsidiary and was fairly readily transferable. However, the plastics "culture" had not yet penetrated the parent firm doing the assembly. This absence of know-how caused a very high and costly rate of rejects and parts sent back as imperfect. So that learning could take place, it was accordingly decided that the parent firm would produce the parts itself. This full integration permits "learning by doing" and is not necessarily final ; it could easily be undone once the "culture" is established.

In the second instance, the materials are of a sophisticated design and require advanced skills as it takes a thorough command of complex knowledge to develop them.

Consequently, production is initially carried out within the firm possessing that knowledge, ie the chemical firms or end-user. It should also be noted that one the theoretical and practical knowledge and expertise are standardized, it becomes more profitable to have the material produced by a less expert subsidiary or outside firm with more attractive production costs.

It should nevertheless be born in mind that full vertical integration also makes for reduction of the uncertainties and costs connected with marketing. The decision by mineral water bottlers switching to plastic bottles to produce them in-house was taken mainly to save transport costs.

These various forms of integration must be analyzed with reference to the explanatory factors and mechanisms described in the preceding section, ie the exploitation of "economies of scope" and the desire to constitute new know-hows.

In general, these horizontal, cross- or vertical integration processes imply an alteration in the very nature of the firm, both in its forms of management and in its conception of the product.

II.5.2.2. Integration processes within the firm

Over and above the factual aspects noted in the preceding sections, and the relevant strategies, the nature of the firm must be reconsidered. If, using a rough but analytically convenient approximation, we regard know-how as a stock of crystallized information, the strategic goals lie chiefly in mastery of know-how and systematic exploitation of its potential.

The strategies mentioned give rise to a sequence of formation and exploitation of know-how, and hence to a stock of information. The firm's adaptations, made possible in particular by more and more extensive use of information technologies, have an effect on the production unit itself and on the management of information within the in-house network of connections between agents in the economic system.

In-house integration

As we have seen, the adoption of new materials proceeds through two phases which are distinguishable at least in analytical terms : one where materials are complementary and another where substitutions occur. It is most important to stress the need to rethink both the production process and the actual product. The remodelling tends towards a higher degree of in-house integration in at least three forms :

- integration of manufacturing steps ;
- integration of the number of parts in the end-product ;
- reduction of production time.

Integration of manufacturing steps

Integration of the number of manufacturing steps involves condensing into a single operation the work which hitherto required two or more consecutive operations.

The advantages of this development lie partly in reduction of the number of contingencies to be controlled during each manufacturing step and partly in improved and simplified management of work in hand, elimination of certain linkage times and reduction of the time taken for any retooling. Thus the central variables are broadly production risks as well as costs.

The plethora of examples of numerical integration of manufacturing steps in symptomatic. Altering processes often causes this type of phenomenon. It is exemplified by precision casting, which obviates using intermediate components which would have to be welded. The use of titanium allows shaping and welding in one step. As we have seen, titanium becomes superplastic at 950°C, hot enough for diffusion welding, so that although titanium is more expensive it can supplant aluminium since the integration of steps reduces the overall cost. The most striking illustration, however, is definitely the RRIM (Reinforced Reaction Injection Moulding) technique whereby a single operation suffices to create both the composite material (matrix by polymerization plus reinforcement) and the finished object.

Integration of the number of parts

Reducing the number of parts in the finished object simplifies the overall management of production. It nevertheless calls for a totally new approach and a new design to suit the specific qualities of the material and the processing methods used.

At the assembly stage at least, it presupposes numerical integration of manufacturing steps. Here again, production cost are the chosen target.

As a change from the standard exemple of the Citroën BX rear hatch, let us take some examples from aeronautics.

	Metal		GFRP	
	Parts	Fasteners	Parts	Fasteners
Tornado CFC taileron	422	5 000	104	2 150
Front fuselage	1 650	8 000	258	1 050
Airbus rudder	600	17 015	335	4 800
Airbus fin	2 072	42 000	96	3 600

Table II.5.D : Examples of numerical integration of parts in aeronautics (33) .

Another less familiar but equally interesting example is Soulé's use of a zinc oxide (ZnO) ceramic for the varistors in its lightning conductors. This material reduces not only the microweld times (important from the data processing angle) but also does away with one part, the spark discharger. ZnO is a good low-tension insulator but at high voltage (lightning) the resistance goes down drastically and the current is transmitted, obviating a spark discharger. Note that in this example there is an integration of two functions (varistor and spark discharger) in the same component.

Reduction of production time

Resuming the example of the lightning conductor, simplification of the end-product also makes it possible to contemplate advanced automation of the manufacturing process, leading to a

---

(33) Source : HARTMANN and KELLERER "Criteria ...", op.cit.

reduction in production time which can also be the specific motive for a substitution of materials. Reduction in production time permits improvement in machine work rates and consequently in plant profitability. Thus a thermoplastic polymer, though dearer than a phenolic one, is preferred for an electrical socket as it reduces assembly time from 2.5 minutes to 16 seconds, and this gain allows a reduction in the number of injection presses (from six to one) and thus in capital cost.

There is a generally observable reduction in the complexity of manufacturing processes (fewer manufacturing steps) and/or articles. This reduction (or mastery) of complexity is a response to the growth in variety demanded by the firm's environment. In-house integration processes can thus be regarded as the operative principles of management of the complexity of manufacturing processes (34) .

Such changes nevertheless raise problems over estimates. Several methodologies are conceivable at the present stage. In fact, the characteristics of systems differentiated basically by the material(s) used must be compared in a multidimensional context which is uncommon in traditional economic analyses. Just two complementary methods will be discussed here ; firstly the multi-attribute utility analysis technique of J. CLARK (35) , a means of evaluating the importance of the characteristics of a given material for a specific application, and secondly the formulation of a global unit cost indicator.

The multi-attribute utility technique is based on data supplied by manufacturers. It is a means of correlating the various combinations of technical and economic characteristics of a material for a given system. This makes it possible, for instance, to plot iso-utility graphs indicating the cost combinations of the system according to the material and the (technical) performance of the product. The diagrams below illustrate two example of the application of this technique : plastic bumpers and cutting tool inserts. For a process to be competitive, its cost and performance must fall below the iso-utility line.

---

(34) Cf. P. COHENDET and P. LLERENA "Productics and new representation of the production process", in P. COHENDET, P. LLERENA and P. PECQUET, Economica, 1987.

(35) Cf. J. CLARK and M. FLEMINGS in "Advanced materials and the economy", American Scientific, October 1986, p.46.

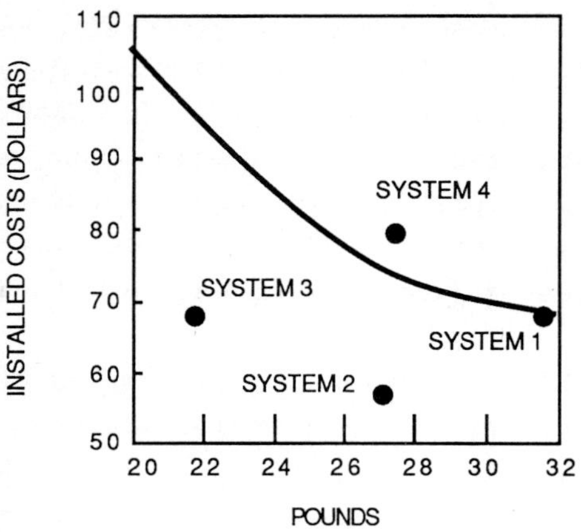

Figure II.5.b
Iso-utility graph for plastic bumpers

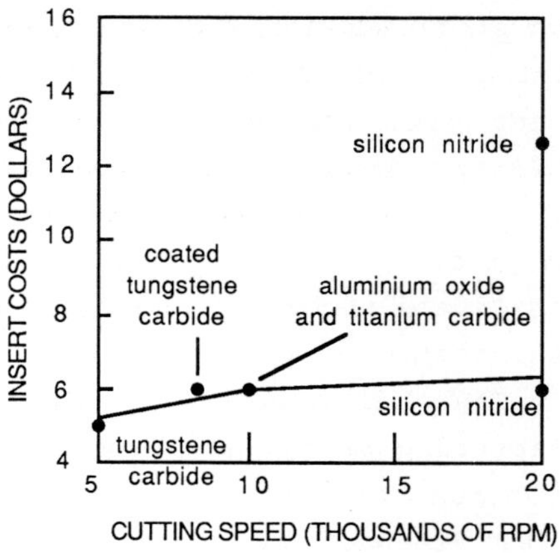

Figure II.5.c
Iso-utility graph for cutting tool inserts

The first graph compares weight and installed product cost for two bumpers presently in production (systems 1 and 2, both consisting of polyurethane on a steel backup beam but differing in their energy-absorbing systems) with another bumper of the same type but used with a third variety of energy absorber (system 4). In the case of cutting tool inserts, the performances are given in terms of the maximum cutting speed. The viability of a silicon nitride ceramic insert not only depends on its superspeed capacity ; it must achieve a price under 6 $.

The second evaluation methodology, a global unit cost indicator, is designed to include manufacturing process transformations in estimates. We find this indicator more suited to the various forms of integration, particularly in-house ones. Indeed, production cost assessment raises at least two problems with no explicit solution in conventional management textbooks : how to take account, indirectly at least, of the organization of manufacturing steps and how to turn the periods making up the fabrication cycle to good account. The organization of manufacturing steps can be represented schematically by the entire network of links between these steps, links which are themselves "quantifiable" in time units, particularly linkage times between manufacturing operations proper. Thus a synoptic expression of fabrication cycle duration might at least partially reflect the corresponding production organization. For the time being we shall confine ourselves to this tentative approximation and concentrate on the question of global unit cost. Thus the second problem above proves to be central.

The proposed formula of the global unit cost indicator is expressed as follows :

$$IC_{gp} = \sum_i [(t^i_E + t^i_C + t^i_L)C^i_{Mh} + C^i_C/L^i] + C_{Ch} \cdot t_{HP} + CM/N$$

in which :

IC_{gp} = global production cost indicator for one product unit ;
t^i_E = working time on machine tool i ;
t^i_C = additional time for handling the workpiece or the moulds or tools required by machine tool i ;
t^i_L = linkage time, ie time taken to supply moulds, workpieces and tools (transport, storage, testing, etc ...) between i and i+1 ;
t_{HP} = non-production time (design, research, organization set-up section, etc ...) ;
C^i_{Mh} = machine tool cost per hour, for the manufacturing stage i
C^i_C = costs additional to the manufacturing step i, specific to each batch (change of mould, tools, etc ...) ;
C_{Ch} = cost of design and conception per hour ;
CM = cost of materials processed ;
L^i = size of batch used by step i ;
N = size of run during reference period.

More precisely, C^i_{Mh} means the global hourly cost for manufacturing step i including machine depreciation, payroll costs, energy, tooling, site costs and interest charges, etc ... This indicator is a form of unit cost as it is the sum of the costs accumulated by the product from start to finish of the production process. Its originality lies in the incorporation of the various manufacturing times, but most of all in suggesting a mode of "capitalization". Indeed, although it seems logical to apply an evaluation of machine cost to a machine's work rate and an hourly cost to the design of a product, a problem arises as regards the incidental times : retooling, quality controls, transport, etc ... Although some of these operations do have specific costs (sometimes negligible and included within C^i_C), here we have elected to use an opportunity cost expressed precisely in terms of the hourly machine cost C^i_{Mh}. The unit cost for step i is therefore the opportunity cost of the product's "dead" times in manufacturing step i. This gives us a measurement of unit production cost which explicitly incorporates the time structure of the process applied and allows <u>comparison of the various alternatives</u> without overlooking the advantages which may accrue from the integration processes linked with certain materials. This general formulation does in fact permit analysis of the advantages to be gained from in-house integration. In this case, costs are chiefly what motivate the process of integration.

In connection with the various elements of production time, it is observed that the integration of manufacturing steps will influence the terms of the sume of IC_{gp} in that the times occupied by the eliminated steps are likewise eliminated. Similarly, integration by way of the number of parts, and by definition the integration of production time, will reduce the unit cost index by $(t^i_E + t^i_C + t^i_L)$.

However, deviations in the different variables in the unit cost indicator do not operate in one direction only. One of the advantages of the proposed formulation lies precisely in the fact that its overall effects on the unit cost indicator can be estimated. For instance any form of integration often entails investment, re-allocation of costs among manufacturing steps (maintenance, stock, work, etc ...), reorganization of process time and different materials costing. All these elements are reflected in the formulation of the indicator, and so it is quite feasible to use a more expensive material (higher CM) which reduces the number of parts, simplifies their production and cuts production times. A picture resembling the following diagram may then materialize (Figure II.5.d).

The key factor nevertheless remains the effects of integration on overall production time - all the more important since its reduction gives the firm the additional advantage that the shorter the time, the more rapidly it can respond to changes in its environment. Here we have the expression of a "dynamic flexibility" ; it is through this variable of "reaction time" in relation to environmental changes that firms will try to enhance their competitiveness, and this variable will become the strategic goal pursued by firms in order to cope with new information on their environment (price of raw materials, characteristics of demand, progress of research, setting of new norms, etc ...).

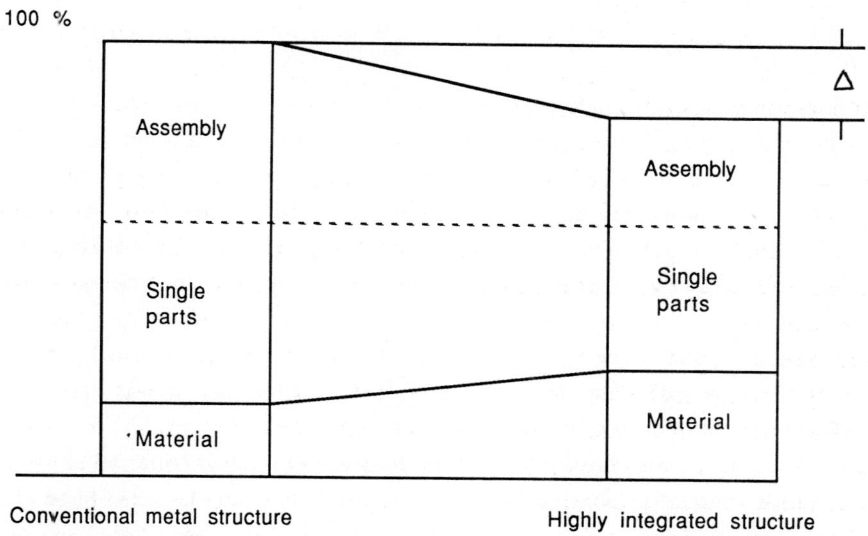

Figure II.5.d : Effects of part and material cost integration on production costs (36) .

In fact, in-house integration processes can also be interpreted in terms of in-house data management. These processes are the extreme forms of simplification of information flows within manufacturing processes. For example, reducing the number of parts significantly reduces the volume of information to be managed regarding inventories, assembly methods, compatibilities, etc ... More generally, in combination with the progress of automation, in-house integration processes associated with substitutions of materials allow a firmer grasp of the complexity of a production process. If such an

(36) HARTMANN and KELLERER, op.cit.

interpretation is accepted, then the extension of the algorithmically formulated sybsystems, (37) in the production process becomes a specific form of "soft" integration process. The aim of this sort of process amounts to the pursuit of faster feedback between production and information which is external to the production. This greater outwardness is a must for a company needing to come to terms with a variety-based system, whether transient or permanent. However, in order to make its pursuit of outside information effective, the firm must redefine its position in the explicit or implicit network of relations between agents of the economic system in which it is immersed.

The firm and outside information

The procedures used to gain privileged access to certain information have nearly all been described above. Our intention here is simply to conclude by placing them in the context of changes in the nature of the firm.

The various forms of vertical integration previously referred to are means of acquiring information, of "internalizing" to varying degrees the sources of the requisite information.

Certain firms have developped special divisions with various titles to adapt their products as far as possible to diversified demand : "Chemical of effects" at ICI, "Problemlösung" at Ciba-Geigy and Bayer, "Centro Sviluppi Impieghi" at Montedison. These divisions deal with quite specific sectors of materials application (housing, household appliances, etc ...) ; in fact they exist to redefine the firm's overall purpose, over and above mere market research. Firms of this kind are engaged in "functionalization" of their production. The concept of functionalization involves a perception of the demand as a system and a response tailored to the demand by supplying a "package"

---

(37) E. ZUSCOVITCH and P. BRENDLE "Informatisation - l'impact ...", op.cit.

consisting of a product but above all a service which is furthermore a means of acquiring fuller qualitative knowledge of demand (38) .

Thus, more complex and explicit networks grow up to make the relevant information available to firms as swiftly as possible. Saint-Gobain, for instance, through the agency of its subsidiary Saint-Gobain-Développement, is setting up a network of innovative small and medium sized firms in allied technologies for life-size "technology watching". But although these networks make information acquisition easier, there is an acquisition cost. As previously pointed out, the cost increases in proportion to the specificity of the information. Consequently the establishment of explicit or implicit information networks outside the firm must be backed by procedures for reducing information costs. Apart from the development of telecommunications, which certainly favours this trend, these procedures may range from the establishment of compatibility rules (General Motors'MAP "network") (39) to the designing of data banks and/or the determination of norms. All such procedures imply a lower information premium, avoid the vicious circle of potential but unavailable scale economies associated with a particular product, and encourage the development of economies of scope suiting the characteristics of a limited and increasingly specific demand.

---

(38) On the concept of functionalization, see B. ANCORI and P. BRENDLE in P. COHENDET (Ed.) "La chimie en Europe", op.cit.

(39) MAP (Manufacturing Automation Protocol) is General Motor's standardization of dialogue protocols between the various element of computerized manufacturing systems. The firm has calculated that the absence of these norms was costing it close on 30 % of its investments in these systems.

PART III

Structural Materials : the Main Areas of Application

There are three main aeras of application for structural materials : transport, packaging and building/public works. A look at the development of structural (dealt with in turn in this part's three chapters) reveals that these areas share a number of common features : keen competition among the various structural materials, the widespread rethinking of production processes, the substantial part played by governments - by setting technical norms and consumer norms and issuing environmental regulations - and the importance of major social objectives such as the search for energy savings.

There is however one major difference between these sectors. Not all of them play the same role in the spread of new materials. It is clearly apparent that the aeronautical and automotive sectors are now operating as catalysts in the application of new materials : aeronautics in the case of "high-performance" materials, and the car industry in the case of more "common" ones. The car industry is particularly closely scrutinized and serves as a test-bed, because it demonstrates the combined effects of a substantial market and the simultaneous problems of reorganizing production, with the help of information technology, and redesigning materials applications.

When compared with the aeronautical and automotive sectors, the other major sectors, and especially building, appear to be lagging somewhat behind. Yet an appropriate industrial policy could have the effect of reversing that situation where certain innovative aspects of the building sector are concerned.

Lastly, the sometimes very dynamic part played by certain "miscellaneous" sectors such as household appliances, leisure or toys, must be acknowledged. There is not doubt, for example, that small household appliances were one of the first sectors in the post-war period to open the door to plastics, enabling new shapes to be created and costs to be brought down. Likewise, the leisure sector now offers highly fertile ground for experimenting with new materials, whether pleasure boating (glass fibre, epoxy resins for surfboards and large sailing boats), skiing (multilayer materials), fishing rods (carbon fibre) or tennis rackets (carbon fibre). However, while a scrutiny of these sectors clearly underlines the importance of the new materials, capable of opening up novel possibilities, a detailed study of them does not seem likely to yield any specific pointers to the development of structural materials compared to the three major sectors discussed in this part.

CHAPTER III.1

New Materials and Transport

The transport sector has been one of the aeras of activity most severely affected by the economic crisis. The substantial rise in the cost of energy following the two oil shocks of 1974 and 1979 greatly affected the cost of using transport and, albeilt to a lesser extent, production costs in the transport industry. The general conditions governing activities in this sector, especially where competition and demand are concerned, have also undergone profound change. The recent favourable trend of oil prices does not obviate the need to adapt to the new economic conditions. In more than one respect the transport industry has managed to respond appropriately to the challenges which have faced it since the crisis began. A two-stage analysis of these responses underlines the preponderant role played by new materials.

To begin with, the sharp rise in the utilization cost/research cost ratio, relative to the production cost/research cost ratio, drove the industries in this sector to develop large-scale materials replacement programmes in order to make transport systems lighter. The trend towards weight reduction benefitted greatly from earlier research done in sectors which were not strictly subject to the requirement of profitability, such as space and defence aeronautics. From there, the new materials were able to spread to civilian applications - first aeronautics, then motor vehicles. The second consequence of the trend towards weight reduction was to reverse the order of the relative cost ratios. Once the first steps taken to make vehicles lighter enabled the user cost of transport to be brought down, there was automatically renewed interest in the lowering of production costs. It was all the more necessary to return to this strictly industrial objective as the use of new materials to reduce weight could only be achieved at a premium.

The experience gained from this trend towards lighter materials has proved decisive. Faced once again with this industrial constraint, which is fundamentally unfavourable to an extension in the use of new materials, the experience acquired revealed the ability of these materials to yield major cost reductions through the redesigning of transport. Thus the new materials, initially adopted as a short-term economic expedient, proved justified in their own right because they offered an outstanding means of structural adaptation. In this latter regard, the analysis which follows is qualifies by differences between the industries which go to make up the transport sector.

Rail transport, for reasons bound up with the importance of the safety factor, and merchant shipping because of the particular manner in which this sector has evolved, have been unable to follow the trend towards weight reduction. The industries involved in these two modes of transport do not at present have sufficient know-how in new materials to devise replacements which would make the redesign of the respective means of transport conceivable. This explains why their development continues to be conditioned by economic fluctuations and is not amenable, for the time being, to an analysis in which the central issues would relate to the innovations brought about by new materials.

Most of what follows is consequently concerned with the aeronautical and automotive industries, both of which have managed to apply the experience acquired from weight reduction to the redesign of their mode of transport. The rate at which material substitution culminates in redesign depends mainly on the constraint represented by industry's mastery of the new materials. There is however one important point on which these two modes of transport differ. The aeronautical industry is an industry producing short runs. By contrast, since the innovations introduced by Henry Ford, the automotive industry has been a typical example of a mass production industry. Yet both adjustment to demand and the prospects for redesign are converging towards a transformation in automotive technology towards the production, certainly of large quantities, but by way of relatively small series.

Our present analysis is in three stages. In the first section we shall examine the way in which the transport industry has responded to the shock of the energy contraint. This leads us to describe in turn the anatomy of the trend towards lighter materials and its characteristic substitution of one material for another. The limitations on this trend provide our transition to the next section, which is concerned with analyzing the aircraft and motor car manufacturers' response to the requirement of industrializing the new materials. Here we shall emphasize substitutions of materials motivated by the prospects of new designs - airframe and engines in aeronautics, engines and shapes in cars. Finally, in the third section, the automotive industry is further analyzed in order to show the changes it has undergone as a result of developments in technology and demand. In this final section, attention is drawn to the role of anticipation in the progression towards the changeover point at which the automotive industry moves to short series production, and to its adaptation to pollution control norms and the demand for customized products.

III.1.1 SUBSTITUTION OF MATERIALS IN THE PURSUIT OF WEIGHT REDUCTION : THE EFFECT OF THE ENERGY CONSTRAINT

The transport industry's response to the shock of the energy crisis has been to make transport vehicles lighter. This trend towards weight reduction stems from a simple economic motive : reducing vehicle weight is a way of restoring competitiveness by offering more fuel-efficient products. In concrete terms, this trend has been made possible by programmes of systematic research which have benefitted from the experience gained in the space field and military aeronautics. Lighter weight has been achieved mainly by substituting composites for aluminium alloys in aeronautics and plastics for metals in the automotive industry. Substitutions of this kind have made little headway in other modes of transport. The cost premium paid when one material is substituted for another, and the possibility of cutting transport utilization costs through aerodynamic and engine changes, are the most important limits to weight reduction. It was when these limits encountered the redesign opportunities afforded by the new materials that the need for the industrial development of these materials gradually emerged.

III.1.1.1 The shape of the trend towards weight reduction

Motives

Weight reduction - technically the most obvious strategy - spread throughout the transport sector following the two oil shocks of 1974 and 1979. Aeronautics was by far the sector hardest hit by the rise in fuel prices. From representing only 30% of operating costs in 1973, fuel costs rose to 60% in 1981, with a direct impact on both freight and passenger tariffs. Faced with a likely fall in demand for air transport, it was becoming necessary to achieve savings in fuel consumption by reducing weight. Lower consumption is indeed known to be largely proportional to weight reduction (cf. Figure III.1.a below). In a Boeing 747, for example, it is possible to achieve an almost 40% saving in fuel consumption by lightening the structure. More precisely, a weight saving of approximately 100 kg on an Airbus A310 corresponds to a drop in consumption somewhere between 8,600 and 19,400 gallons of kerosene each year, giving an annual saving on operating costs of between 10,300 and 23,300 dollars. The same calculations apply, though in different proportions, to road transport. Thus a 100 kg weight saving in a car normally reduces fuel consumption by 0.5 litres per 100 km travelled. In urban driving a 2% saving can be achieved by a weight reduction of 50 kg. Just as in the aeronautical industry, the result can be achieved either by reducing the weight of component parts or by reducing that of the load-bearing structures. The two aspects are generally combined : for example, a saving of 1 kg obtained in a car body part affords an induced weight

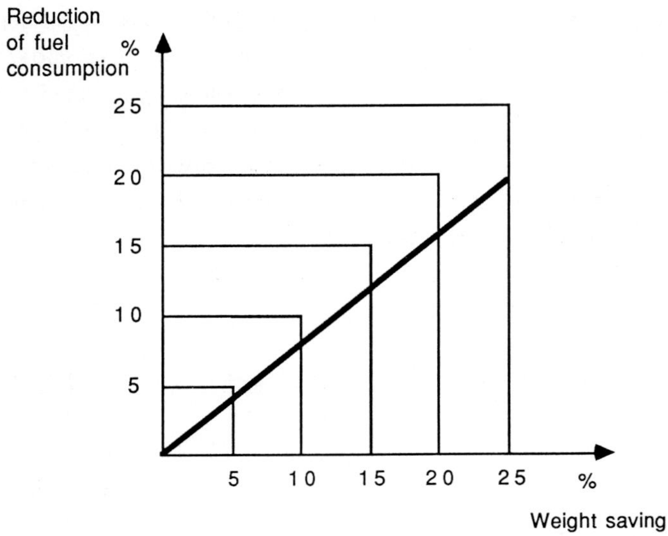

<u>Source</u> : MBB, cited by W. HARTMANN and KELLER (1) .

<u>Figure III.1.a</u> : Effect of weight reduction on aircraft fuel consumption

saving of 0.4 kg in the weight of the load-bearing structure. Even though apparently less important, the effect of weight reduction on the fuel consumption of trucks is not negligible : a reduction of 100 kg yields a fuel saving of 0.07 litres per 100 km. These apparently low figures hide an undeniable relative benefit, since weight saving has enabled commercial road transport to keep up with rail transport at a time when the latter was benefitting from the increase in oil prices in countries where a substantial proportion of electrical power generation does not depend on oil.

---

(1) "Criteria for the material selection for aircraft structure", in "Advanced materials R&D for transport", Fall Meeting of the European Materials Research Society, 26-28 November 1985, Strasbourg, European Materials Research Society.

The background against which western car manufacturers have been operating over the last ten years has been characterized by two major phenomena. On the one hand there has been fiercer competition due to the deregulation of air transport and the strong penetration of Japanese cars in western markets, without any compensation in return (Tables III.1.A and III.1.B). On the other hand, the economic crisis has caused transport demand to fall. In the case of personal road transport, that fall in fact hides the veritable <u>saturation</u> of the market, the proportion of replacement demand in overall demand now being between 85 and 95%. If we add the effect of the energy constraint to these two phenomena it is easy to understand the need for western manufacturers to reduce the weight of their transport products.

	1975	1980	1981	1983	1984	1985
USA (1)	9.4	21.3	21.8	20.9	18.3	18.8 (9 months)
EEC (10 states)	n.av.	9.2	8.3	8.7	9.1	9.3 (6 months)
of which :						
FRG	1.7	10.4	10.0	10.1	12.0	13.1
France (2)	1.5	2.9	2.6	2.7	3.0	2.9
United Kingdom (3)	9.0	11.9	11.0	10.8	11.3	9.4
Italy (4)	0.1	0.3	0.1	0.1	0.1	-
Netherlands	15.5	26.4	26.4	23.5	22.0	21.1
Belgium	16.5	24.7	25.0	22.9	20.2	19.4

(1) Self-limitation on exports to 1,680,000 vehicles per annum from April 1981 to March 1984; 1,850,000 vehicles from April 1984 to March 1985; quota of 2,300,000 vehicles from April 1985 to March 1986 set by the Japanese government.
(2) Unofficial quota : Japanese manufacturers have never exceeded 3% of the French market.
(3) Gentlemen's agreement renewed annually since 1978, under which the Japanese undertake not to take more than 11% of the market.
(4) Strict quota : imports of Japanese cars into Italy have been limited for many years to 1,700,000 units per annum plus 500 off-theroad vehicles.

<u>Table III.1.A</u> : Development of market shares (in %) of Japanese manufacturers in the western world

	1975	1980	1981	1983	1984	1985
All western vehicle manufacturers (mainly VW-Audi, BMW and Daimler-Benz)	1.7	1.7	1.4	1.1	1.4	n.av.

Source : Fabrimetal, cited in BFBBL (2) .

Table III.1.B : Development of market shares (in %) of western manufacturers in Japan

Weight reduction was seen as an effective way of restoring competitiveness. So it is not surprising that it was pursued at all costs, or virtually so, since the aeronautical industry was prepared to pay between 3,000 and 7,000 FF for every kilogram saved, while the motor manufacturers were willing to pay a premium of 10 FF per kilogram saved on the weight of cars and between 20 and 25 FF per kilogram saved on the weight of trucks. At least for a time, the requirement of competitiveness, which meant offering customers less thirsty vehicles, took precedence over the requirement of profitability.

Research programmes

Materials replacement programmes in the civil aircraft industry have benefitted greatly from the spin-off of space research and progress achieved in military aeronautics.

In the space field, the extreme constraints placed on space vehicles and the need to overcome gravity steered research towards new materials with low density and high mechanical, thermal or chemical strength - metal matrix, carbon-carbon, 3D composites, etc. Such achievements as the Ariane rocket or the space shuttle incorporating a good many high-performance composites would certainly have been inconceivable without new materials. The use of lightweight, high-strength materials is also justified on the basis of cost

---

(2) "L'industrie automobile : de nouveaux défis", Bulletin Financier (Financial Bulletin) of Banque Bruxelles Lambert, 15 January 1986.

considerations. For if a kilogram saved in a car is worth 20 to 25 FF, the value of that same kilogram in a satellite can be estimated at 30,000 FF. Composites therefore came into general use very fast in the space field, rising from 20 to 80% between 1960 and 1980 for unmanned craft.

No doubt for reasons of prudence, but mainly because defence programmes were funded by the public authorities, the plane makers began by using composites in military aircraft before considering their use in civil aircraft. The backing of the public authorities enabled manufacturers to learn composite technology, and then develop it for applications in which advanced materials such as alloys were already being used. This tendency in the introduction of composites is particularly apparent in the United States, where composites moved in turn from NASA's laboratories to the US Air Force and then civil aviation, and finally to commercial applications (3) . Carbon-carbon composites for brakes, carbon-epoxy and "honeycomb" sandwiches for primary structures, owe their development to military research programmes. In military aircraft the percentage of composite materials rapidly reached sizeable proportions, as is evidenced by their use in American fighter aircraft : 0.8% in the F14; 5% in the YF-17; 9.5% in the F18; 30% in the AV8B; in the VTX (which is at the development stage) the proportion will lie between 25 and 40%, and future models (AFF, V/STOL) are expected to comprise more than 50% composites. The reasons for this substitution are (a) weight saving and (b) the prospect of building a "black" aircraft rich in carbon fibre and invisible to radar detection. The same trend is emerging in Europe. As long ago as 1972 the Dassault Mirage F1 contained composite components, and the Dornier ALPHA-JET soon followed suit. The Mirage 2000 incorporates several important parts made of carbon-epoxy and boron-carbon-epoxy composites. Similarly, the AMX which is built in Italy and the C101 built by the Spanish firm Caja bear witness to the important developments taking place in military aeronautics. The future EFA European combat aircraft, the Dassault Rafale and the Hermes shuttle are certainly serving to spread the trend, which will soon communicate itself to civil aeronautics for reasons of structural lightness, but also, as we shall see, as a way of reducing production

---

(3) G. SOCCAL : "Les matériaux de demain" ("Materials of the future"), Athena, N° 15, 1985.

costs. Lightweight solutions using new materials, once tested and tried in the aeronautical field, have then been taken up by the automotive industry.

Like every other important vehicle innovation, lightweight solutions were first tested in experimental vehicles. For example, the Vera research vehicle developed by the French manufacturer PSA on the basis of a Peugeot 305 model has been used to test weight reductions afforded by the use of synthetic materials. The weight was brought down to 750 kg as against 925 kg in the all-metal version - an overall gain of about 19%. The experiment was continued using a diesel version (Vera 02), which yielded a weight gain in the region of 15% (4) . In 1979, Ford in America was experimenting with the concept of a lightweight vehicle using large quantities of composite materials, purely for the purposes of technical feasibility studies. The test proved conclusive, giving a weight saving of almost 33% (5) . Other tests of the same kind were carried out by the majority of manufacturers.

These early experiments were followed by systematic research programmes to explore weight reduction, the commercial applications of which should come into effect around the year 2000. They have involved such experimental vehicles as the Vesta (Renault), Eco 2000 (PSA), Auto 2000 (VW), VSS (Fiat), LCP 2000 (Volvo), Probe (Ford), etc. In Europe the objective is a mass-produced car consuming 3 litres per 100 km. (6) Setting that objective is equally important to manufacturers and to the public authorities. The latter are supporting the weight reduction programmes pursued by manufacturers in order to reduce the burden of oil imports on national economies. For example, in conjunction with the Italian government the Fiat company was recently able to demonstrate the validity of such an objective in connection with an experimental vehicle produced by conventional techniques. However, the underlying

---

(4) A. GERARD "Nouveaux matériaux pour l'automobile" ("New materials for motor vehicles") in "Matériaux et énergie" ("Materials and Energy"), Colloque international de Sophia-Antipolis, 20-21 January, Paris, Agence Française pour la Maîtrise de l'Energie (AFME) 1983.

(5) P. BEARDMORE, J.J. HARWOOD and E.J. HORTON ""Design and fabrication of a GrFRP concept automobile", in : "Advances in composite materials", A.R. Bunsell et al. (eds), Vol. I, 11-24, Oxford, Pergamon, 1980.

(6) 4 litres per 100 km for cars produced by Volvo, whose models are famed for their safety and strength.

trend is the use of new techniques, and Italy's biggest manufacturer is no exception. In particular, motor manufacturers are preparing actively for the utilization of new materials, and especially reinforced plastics, the technologies of which have been worked out and perfected by the aeronautical industry. This is also an area of European cooperation, as is evidenced by the preparation (as part of the second stage of Eureka) of the Carma 2000 project for a low-cost vehicle using new materials.

III.1.1.2 The principal substitutions

Aeronautics : composites versus aluminium-lithium

The primary and secondary structures of civil and military aircraft are nowadays made for the most part of lightweight aluminium-based alloys (averaging 70% of the weight, not including engines). Pressed by customers wanting to bring operating costs down, plane makers have been forced over the last ten years to consider introducing order, even lighter, materials. Composite materials based on carbon fibre and kevlar were the most appropriate candidates, mainly because these materials offer substantial weight savings. In the Airbus A310, the use of 14% composites has yielded a weight reduction of one tonne. It is estimated that an average weight saving of 10% can be obtained in the airframe as compared with aluminium alloy. In addition, composites offer properties which are just as interesting as those of alloys, and their fatigue behaviour has proved markedly better. This characteristic is so crucial to the life of an aircraft that it has boosted the use of composites, which have passed certification tests with flying colours. Civil aviation has been able to benefit from the experience gained with military aircraft and adopted composites very rapidly from 1970 onward, so that today their use is envisaged for primary structures such as tail units. At the present time they are being used mainly for certain static components such as doors, fairings and seat frames, but also in moving parts such as airbrakes, control surfaces, helicopter rotor blades, and so on.

In many cases, aluminium alloys were replaced by composites without any change in component design. Thus no effort was made systematically to optimize properties but mainly to take advantage of the opportunity to save weight. Aeronautics also appears to have wanted to keep its options open for a return to traditional materials, by not redesigning parts from the outset. This attitude on the part of manufacturers is explained by the difficulty of foreseeing today the scale of the substitutions which tomorrow may bring. In other words, it is by no means obvious that the move to composites will take place as fast or as massively as is usually said. Boeing is reckoning on 65% composites in aircraft by the end of the decade. Its European rival, the Airbus Consortium, forecasts only 50% by the year 2000 (7) . Leaving aside the disinformation which stems from the fierce competition between these two manufacturers, a great deal of the uncertainty is bound up with the response from the aluminium industry whose market is under threat. To counter the challenges thrown down by composites in aeronautical applications, the aluminium industry announced in 1984 that it had achieved industrial mastery of Al-Li alloys. Lithium is the lightest metal (with the lowest density), so that the addition of 1% of lithium to aluminium makes the alloy 3% lighter and 6% stiffer (modulus of elasticity). These characteristics give Al-Li comparable advantages in terms of weight savings to those of composites. But the most attractive feature of Al-Li is that this material does not require a departure from the industrial processes of rolling, extrusion and forging, which are normally used by the processors of alloys for the aeronautical industry. Al-Li can be processed on existing equipment of which there is substantial installed capacity. Moreover, the processes have already reached an advanced stage in their learning curve and it is not necessary to redesign components. Another argument in this comparison is price : Al-Li is about four times as expensive as a traditional Al alloy, whereas the price of a carbon fibre based preimpregnated composite is 10 to 40 times higher than that of an aluminium sheet. For an identical shape, therefore, Al-Li appears the better bet. On the other hand, the bulk of the gains obtainable with composites derive from redesigning shapes, as we shall see later.

---

(7) R. DEVELAY "L"aluminium : un métal d'avenir, progrès récents, perspectives nouvelles" ("Aluminium : a metal for the future, recent progress, new prospects") in "Matériaux et Energie", op. cit.

While the main contest in aeronautics lies in competition between aluminium alloys and composites, there are also other materials which can offer weight savings - notably, in the longer term, aluminium matrix composites. Moreover, the development of multilayer composites for panels, bulkheads and fuselage elements may well have a major impact on the present competition between the various materials suppliers and convert it into close cooperation. Already, aluminium scientists and chemists are working together to develop new multilayer materials combining aluminium, kevlar and resins. Furthermore, the fact that kevlar is by no means ousting aluminium in "honeycomb" sandwiches is an additional argument in favour of closer cooperation in future.

Motor vehicles : plastics versus metals

The weight reduction techniques tested on experimental vehicles gradually appeared in production vehicles from the early 1980s onwards. The wide range of components and equipment concerned shows how widespread these solutions are.

With regard to engine parts and related equipment, a great deal of research is currently under way to apply composite materials to both moving parts (shafts, suspension wishbones, clutch plates and pedals, brake discs and pedals, etc) and to stationary parts (carburettor and petrol pump housings, steering column mountings, etc (8) . All these applications afford substantial weight savings - in some cases as much as 66% compared with the earlier equipment (9) . Not much of this research has been commercially implemented as yet. One example is the glass fibre/carbon fibre composite double wishbone

---

(8) A. BASSI, L. MERLINI and L. OROFINO "Materiali innovativi nel motore a combustione interna : prospettive a medio termine" ("New Materials in the internal combustion engine : medium-term outlook"), in : "Convegno internazionale materie plastiche" ("International Plastics Congress"), 10-11 October, Turin, Unione Industriale di Torino, 1985.

(9) This is true of the glass fibre reinforced nylon steering column mounting cited by F.J. PRESTON and J. GARDNER "Technical innovation in plastics for automotive interiors", in "Convegno internazionale...", op. cit.

suspension engineered by PSA (Peugeot Société Automobiles) in collaboration with SEP (Société Européenne de Propulsion). Tests have shown that the weight of this safety component was only 50% of its metal equivalent (10) . The epoxy-glass fibre composite leaf springs which have been mass-produced by the British firm GKN since 1985 and are much lighter than the metal design should also be mentioned (11) . In all these applications, composite materials are direct rivals to steel and aluminium.

Plastic fuel tanks have established themselves in Europe since 1982, mainly due to a weight saving of 30 to 40% over metal tanks.

The front bumper of the General Motors 1981 Oldsmobile sports model shows a weight saving of 50% through the use of reinforced polyurethane produced by the RIM technique. Since 1972 and the Renault R5, the trend in bumper design towards a "shield" incorporating all or part of the front or rear skirt and involving composite materials has become routine. The weight of this component has been considerably reduced as a result.

In the bodywork field, the French company Citroën has demonstrated the relevance of weight reduction on a mass-market car (the BX). The bonnet of this model is made of SMC polyester and affords a weight saving of almost 36%. The success of the BX, to which we shall have occasion to return, has encouraged other European car makers (VW in the Federal Republic of Germany and Fiat in Italy) to make plans for cars incorporating similar bodywork parts.

Although they have already made substantial inroads into vehicle interiors, plastics are continuing their market penetration, particularly in seats. The back seat of the Audi 100 achieves a 40% weight saving through the use of a reinforced polypropylene.

The driver's cab of a MACK heavy goods vehicle is 30% lighter when made of reinforced plastic rather than its metal equivalent. The technique is expected to become widespread in all heavy tonnage trucks over the next few years.

---

(10) A. GERARD, op. cit., p. 176.

(11) M.S. STARKEY, P. WOOD and J. DIMMOCK "Glass epoxy composite as a leaf spring material", in "Advanced materials R&D for transport", op. cit.

The applications which we have been looking at are just a few among a great many examples of the technical and economic feasibility of lightweight solutions. The weight reduction afforded in automotive components and equipment, both internal and external, is important, invariably lying between 30 and 50%. Overall, in terms of the complete vehicle, lightweight solutions should make it possible between 1985 and 1992 to save 14.7% of the average weight of a European car . The trend towards weight reduction ought to be even more markeed in the case of American cars, which have traditionally been heavier than their European or Japanese counterparts . The same conclusion is also arrived at if we look at cars in the bottom of the range or vehicles which have to meet high strength requirements, as in the case of vehicles produced by the Swedish Company Volvo .

Most weight savings are achieved by <u>substituting new materials</u> for old ones. A note of caution is called for here, because the expression "old materials" sometimes covers materials which have been successful in evolving and adapting to the requirement of lightness. One example is the new sheet steels, which owe their lightness either to the use of high yield-point steels or, generally speaking, to the structural calculation applied to ordinary steels. These new sheet steels account for 40% of the sheet used in the Peugeot 205 model and between 60 and 70% of that used in the Porsche, Audi and Volvo top-of-the-range models (12) . Cast iron, suffering heavy competition from aluminium as an engine block material, has also evolved and become lighter and finer (spheroidal graphite cast iron). Consequently, to a significant extent the slowly declining share of steels and cast irons in vehicle weight (see Figure III.1b) is not the result of replacing these materials by other ones, but can be explained by their own development. Furthermore, aluminium having become widespread in "first generation" parts (crank-cases, cylinder heads, pistons, suspension arms, wheels, etc), new fabrication, processing and utilization techniques (13) should enable it to move into "second

---

(12) J.J. CHANARON and E. BAINVILLE "Le système automobile français" ("The French Automobile System"), Study by the Centre de Prospective et d'Evaluation (CPE), N° 56, 1985.

(13) The techniques are, respectively, powder metallurgy, plasma or electron bombardment or laser processing, and new foundry techniques (melting-forcing, squeeze-casting and pore-free casting).

generation" parts, that is to say mainly brakes, suspensions, transmissions and steering.

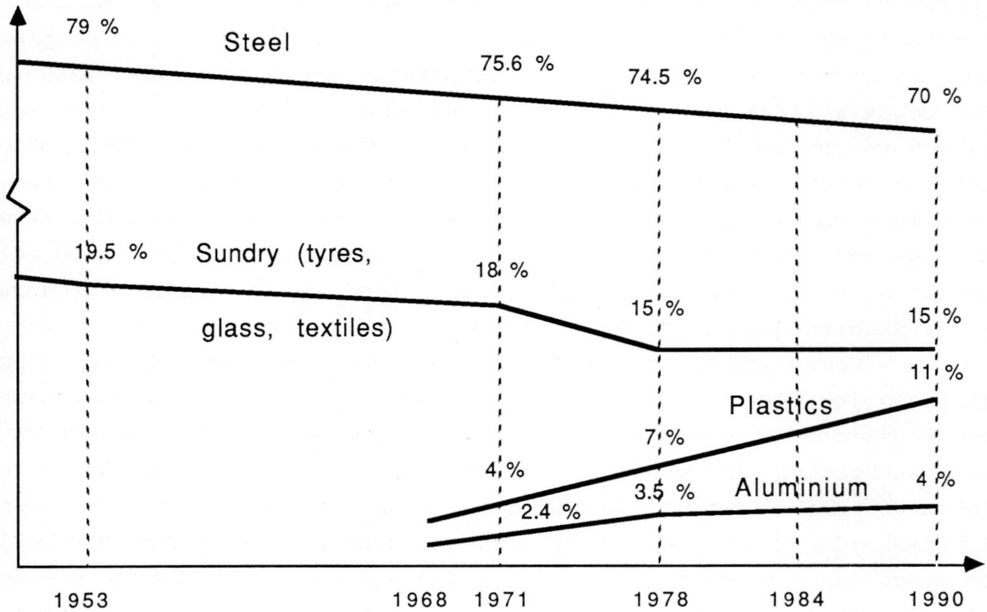

Source : Renault, cited by M. DRIBAULT and M. GIGET (14) .

Figure III.1.b : Relative weight of different materials in a European car

There are also plans to use magnesium alloy, especially for crank-cases, although it is at present impracticable because a number of technical problems remain unsolved . Despite these remarks, the fact remains that the penetration of plastics at the expense of metals

---

(14) "Amélioration et substitution des matériaux dans la problématique de l'allègement des produits industriels : tendances générales" ("Improvement and replacement of materials in the context of lighter industrial products : general trends") in "Programme d'étude et l'allègement des produits industriels", Paris, Euroconsult, 1984.

and alloys is the underlying trend in materials substitution in the automotive industry.

The penetration of the automotive sector by plastics is not a recent phenomenon in itself : the German car maker Audi had explored some of their possibilities long before the last war. Since then, the average amount of plastics in European cars has been growing constantly, and by now has reached about 72 kg. The figure is expected to reach 98 kg by 1990, ie. 11% of average vehicle weight (see Figure III.1.b). There is little difference in this respect between American car makers and their European counterparts . On the other hand, the Japanese car industry appears to be lagging somewhat behind its western rivals in this field (15) .

To begin with, plastics were introduced into the least technically demanding part of the vehicle, that is to say the interior. They now account for two-thirds of present applications and seem to be nearing saturation point. The products used here are common plastics (PVC (Polyvinylchloride), PE (Polyethylene), PS (Polystyrene), etc) whose static stress characteristics and cost are improved by the incorporation of granular fillers of either mineral or

---

(15) According to calculations by Y. SAITO ("The structure of Japanese car industry and up-to-date plastics in Japanese cars" in "Convegno internazionale..." op. cit.), the average amount of plastics in Japanese cars was 67 kg in 1985 and likely to reach 80 kg in 1990. The difference between these figures and those given in the text for European cars is due in part to the different methods of calculation employed. However, while Japanese manufacturers are competing on equal terms with their western rivals with regard to items of external equipment (bumpers, skirts, spoilers) they seem to be less advanced when it comes to major bodywork items and more especially opening parts such as doors, bonnets and rear hatches. A significant example here is the recent Ballade Sport CRX model by Honda. There are several explanations for this relative backwardness of the Japanese automotive industry. Firstly, western manufacturers produce a higher proportion of prestige models than do the Japanese. For reasons of production economics, bodywork innovations have to be applied first of all to vehicles at the top of the range (de luxe and sports models). Secondly, the Japanese industry has focussed most of its energies on the <u>organization</u> of the processes involved in the production of models at the lower end of the range ("sub-compacts") rather than on the search for product innovation. Finally, it should be borne in mind that the main world producers of <u>engineering</u> plastics are either American (General Electric, du Pont) or European (Bayer, Hoechst, Ciba-Geigy, ICI, Rhône-Poulenc, Montedison); none of them are Japanese.

vegetable origin (16) . The remaining third of plastics applications is shared more or less equally between stressed parts - the chassis and bodywork on the one hand, and the engine and related parts on the other . The plastics used in these applications are engineering plastics : thermosets, generally reinforced with glass fibre (plastic composites) for bodywork and chassis; thermoplastics, either reinforced or unreinforced, in the engine and related parts . The former are replacements for steel sheet and the latter for steel and light alloys. Substitutions have been motivated by various criteria besides the low density of plastics. Plastics have very good corrosion resistance without needing to be specially treated like metals. Engineering thermoplastics behave well at high temperatures and are therefore suitable for certain engine parts. The average mechanical strength of plastics can be considerably improved by the incorporation of fibre reinforcement. Where the bodywork is concerned, composite solutions offer a twofold advantage over sheet steel : they absorb kinetic energy better in the event of a collision, and do not present dangerous sharp edges when broken. Their fire resistance is improved by the incorporation of flame retardants. They have the advantage from the safety standpoint that plastic fuel tanks limit the risk of explosion. Seen in a narrow economic perspective, despite their generally higher cost plastics have managed to remain competitive since the crisis began . This relative stability is a factor which has militated in favour of plastics, particularly as compared with aluminium whose price is unpredictable.

One of the major advantages of plastics over metals is their ease of processing. While the speed of these processes is still a handicap to the wholesale introduction of synthetic materials, the difference between metals and plastics in this regard is tending to be blurred by the constant advances made in injection moulding machines. Overall, however, the utilization of plastic materials continues to be less economic for long part runs because of the problem of production rates and the need to paint bodywork parts off-line. This goes some way towards explaining the cost premium which manufacturers pay in

---

(16) One example is provided by the unusual solution devised by the Japanese company Mitsubishi with a 50-50 or 40-60 mixture of polypropylene and paper to serve as a dashboard insulant or interior door trim. Cf. A. PEREZ "Plastiques et automobile : le doute" ("Plastics and the motorcar: uncertainty"), Industries et Techniques, 10 June 1982.

their search for weight reduction. But the use of plastics does enjoy one preponderant relative advantage : it makes it possible to obtain complex parts relatively simply, even to high dimensional tolerances. In conjunction with other factors which we shall examine, the possibilities of component integration underline the fact that there are limits to weight reduction as an autonomous trend. It remains undeniable, however, that had it not been for strong incentives to use low-density materials, the other characteristics of plastics would not alone have established them so rapidly and incrementally as substitutes for metals in motor vehicles. The need to find lighter materials is therefore of interest because it has led to the adoption of materials whose novel technical and economic potential, over and above simply reducing the utilization cost of vehicles, has come gradually to be realized.

Other modes of transport

In rail transport, weight reduction programmes have taken axle loads as their target. Traditional steel structures have been in part replaced by structures made of aluminium alloy or steel with a high limit of elasticity (HLE). For example, the French Railways have achieved weight savings of 15% on their "Corail" coaches, half of them through the replacement of traditional steel structures by HLE steel. Because of reliability and safety requirements, the use of composite remains limited, little being known as yet about their aging characteristics. The German Company MBB, however, has developed a bogie using composites which yields substantial weight savings. But this is an example of a functional substitution in which the system design is considerably modified. Composites are also little used as yet in the TGV (French high-speed trains) though the use of aluminium alloy instead of steel has enabled important weight savings to be made. The introduction of carbon-carbon brakes will mean considerably greater savings. The saving per axle through using Sep-Carb is estimated at 680 kg, representing a weight saving of 15 tonnes over an 11 bogie train.

Sailing (pleasure boats and racing yachts) has benefitted greatly from progress in the materials field, though curiously weight reduction programmes have made little impact on the merchant navy. Yet shipping is a big oil consumer because of the particular resistance offered by the medium in which ships move. This does not appear to be simply a matter of late development. The present downward trend in oil prices is placing a large question-mark over the viability of materials substitution programmes. In addition, shipbuilding is currently in crisis, mainly because of over-capacity which is estimated at more than 25%. Nor does the downturn in shipbuilding favour investment in equipment and techniques calling for new materials. As of now, innovation in materials has mainly affected military vessels, one example being the use of composites to build patrol boats hulls.

III.1.1.3. The limits to weight reduction : towards structural adaptation

There are important limitations on the movement towards weight reduction that was sparked off by the energy crisis. It is obvious, from the technical standpoint, that a continuous reduction in motor vehicle weight must in due course pose a problem of safety through loss of grip. This (absolute) physical limitation on weight reduction makes itself felt all the more strongly as there are a number of factors which magnify its effects. For example, one of the underlying tendencies in the European car industry has been to make cars more powerful by increasing engine capacity. Advances in plastics must also be taken into account. The availability of new high-strength RIM plastics (it has been announced in the United States that these will be in commercial use for the whole of the load-bearing structure by 1995) should generate weight savings of the order of 400 kg (17) ! If we add to this likely innovation the other weight saving solutions already being or about to be implemented, we shall understand more clearly why the designers of tomorrow's cars have already begun to think about the most effective way of solving the problem of grip

---

(17) CPE : "Nouveaux matériaux RIM à haute résistance", ("New high resistance RIM materials"), Bulletin du Centre de prospective et d'évaluation (CPE) 1985, 22, 10.

without forgoing what has been achieved through weight saving. However, the principal limitations on this trend are economic ones.

Opportunities for saving weight through the use of synthetic materials in the aeronautical and automotive industries remain very limited if one thinks in terms of part-for-part replacement. The extra cost to the manufacturer has to be recouped from the customer sooner or later for reasons of economics, so that to the customer the advantage of lower fuel consumption is offset by greater expense. In turn, this affects the extra competitiveness enjoyed by the manufacturer as a result of weight saving. Generally speaking, for a given form, metals and alloys are more attractive to manufacturers and have similar advantages in terms of lightness. So the use of synthetic materials is a rational option only if it is based on other properties, some of which are still not fully developed in certain cases. For example, the composites used in aeronautical applications are mainly hand-built. This results in variations in fibre orientation of as much as 5%, which corresponds to a loss of stiffness of the order of 25%. So the advantage of lightness obtained by simple substitution is partly offset by losses in safety (18) . Because the new materials make it possible to <u>redesign</u> parts and structures, it is possible to achieve substantial fuel savings by varying aerodynamic shape and engine output rather than weight alone. In the case of cars, for example, the weight factor is known to be negligible in relation to wind resistance when speed is incorporated into the calculation for reducing energy consumption (Table III.1.C.).

A 10% improvement obtained in aerodynamic drag (CX) yields a fuel consumption advantage of nearly 5% on the open road. So it is desirable to optimize the body shape, since it accounts for 65% of the CX value. By optimizing the properties of new materials, and ensuring that they are reproducible using automatic processing plant, manufacturers will clearly limit the extra cost involved in part-for-part substitution. So it is this capacity to redesign transport vehicles which makes the new materials interesting. It enables them to fit into the general trend towards lower utilization costs in which initially, for technical and economic reasons, weight reduction was the only factor at work. Such redesign, moreover, is not

---

(18) G. MENGES : "Die Zukunft hat gerade erst begonnen", ("The future has only just begun"), Plast-Verarbeiter, 1985, 36, 9, 23-32.

unconnected with weight reduction, since it generally results in a lighter structure through a mechanical advantage. But redesigning

Speed	Energy consumed
60 kph	44%
90 kph	62%
120 kph	72%

Source : A. GERARD (19) .

Table III.1.C : Energy consumed in overcoming wind resistance.

reverses the perspective : weight reduction ceases to power the trend in view of the goal of economic viability, and tends increasingly to be (merely) induced.

III.1.2. MATERIAL SUBSTITUTIONS RELATED TO PROSPECTIVE REDESIGN : THE PRESSURE OF THE INDUSTRIAL CONSTRAINT

The underlying reason for introducing new materials into aircraft and motor vehicles is that it should afford a substantial improvement in profitability through the redesign of the vehicles in question. This prospective redesign can only be the culmination of a progressive movement. Apart from overcoming the inertia of existing technologies, in the majority of cases the industrial applications of the new materials have to be mastered more fully and consequently substitutions will take place stepwise. In the aeronautical industry, it is for structural components and improved efficiency of propulsion systems that materials will be substituted, being replaced by composites. In the automotive industry, it is probably engines and bodywork that will be redesigned, using engineering ceramics and engineering plastics respectively.

---

(19) Op. cit., p 169.

III.1.2.1. The development of aircraft structures and engines

The aeroplanes built at the turn of the century were made of cellulose "composites" and varnished fabric, mounted on a wooden frame. Until the 1940s, their mode of propulsion consisted of an internal combustion engine driving a propellor. Aluminium alloys and the jet engine later took over from "natural composites" and propellors. Composites and propellors are now making a strong come-back, through not in the form of simply returning to earlier techniques as the result of favourable price developments. The new generations of composites have nothing in common with the natural composites at the turn of the century, and propellor shapes and drive systems owe nothing to the pre-war propellor engines. While the introduction of modern composites was initially motivated by weight reduction factors, it has now led to a process of intelligent substitution, the practical outcome of which is the complete redesign of parts and structures. If energy savings are taken as a criterion, they are far greater when components are redesigned than when materials are substituted part-of-part. Weight reduction and structural redesign have snowballed, leading to gains in engine systems. Materials have also played a decisive part in improving the thermal efficiency of engines. As in the case of structures, however, far greater advantages are generated from changing the propulsion principle. We shall first consider the advantages deriving from redesign and then go on to look at propulsion systems.

Component redesign using composites

While part-for-part substitution of alloy-based composites is justified for secondary structures (not essential to flight), it is of no interest for primary structures : the cost savings achieved by reducing the weight of vital components are insufficient to offset the unavoidable research investment, new calculation procedures, processing and control and so on. Despite the experience obtained with military aircraft and subsidized from the public purse, simply replacing primary structures by identical though lighter structures is neither technically desirable nor economically viable. But if the structural redesign route is envisaged, the potential advantages are very substantial.

The choice of composites to make the wings of the Falcon 10 and the tail unit assembly for the Airbus A310 and the Boeing 327-300 was based on other criteria than weight reduction. The seemingly decisive factors relate no longer to operating costs but, on the contrary, to production costs. As we shall see, the ease of designing simple parts when using composites and the opportunities for automated production were incontrovertible arguments.

One of the main features of composites is their flexibility of use. In engineering terms, composites offer the optimum solution for a great many applications, since the constraints circumscribing their use are far less tight than with other materials. It was very quickly realized that exploiting this property of composites was more advantageous than simply producing copies of traditional components. The point is that composites adapt to a wide variety of forms while retaining their mechanical characteristics. By orienting fibres in the direction of the stresses to which the part is subject it is possible to reconcile several goals such as aerodynamic behaviour and mechanical strength, which is not always possible with conventional materials. So because they are able to offer tailor-made properties, composites can remove the constraints traditionally imposed on the designer. This possibility has been gradually exploited by manufacturers, so that some aircraft components nowadays are revolutionary in design. Their principal characteristic is that they contain far fewer parts than their metal equivalents. This trend is very clearly illustrated by developments in the design of helicopter propulsion systems. The use of composites for the rotor hub of the MBB BO 105 (Figure II.2.b in Part II and the Aerospatiale SA 365C made it possible to simplify the design by reducing the number of parts. In the latter case, the Starflex system yielded a reduction in part numbers from 293 to 92, and a weight gain of 50 kilograms.

Another example is the Airbus A310 carbon fibre composite rudder unit, in which the number of parts was cut from 2,092 to 96 as compared with the metal equivalent. Many more such examples could be given. In all of them, the principle at work is simplification of elements by integration of parts. At the same time there is an improvement in surfaces - ie aerodynamic performance - which is an important factor in reducing fuel consumption.

In addition to the ease with which composites adapt to complex and aerodynamic shapes, they also have potential processing advantages. The use of composites, for example, afforded a 10% cost saving in the A 310's control surfaces. These cost savings stem mainly from the lower number of production steps. Many structural elements are produced at the same stage, and a reduction in the quantity of parts cuts the number of assembly operations. Table III.1.D shows very clearly the effects of redesigning primary structural elements on the number of fasteners for a variety of applications.

	Metal		GFRP	
	Parts	Fasteners	Parts	Fasteners
Tornado CFC aileron	422	5000	104	2150
Front fuselage	1650	8000	258	1050
Airbus rudder	600	17015	325	4800
Airbus fin	2072	42000	95	3600

Source : MBB, cited by W. HARTMANN and H. KELLERER (20) .

GRFP : Glass Fiber Reinforced Plastic.

Table III.1.D : Integration of parts and reduction in number of fasteners

With composites, the material and the part are generally fabricated *simultaneously*, which considerably reduces the number of production phases and particularly machining operations. It is therefore much easier to automate the fabrication processes and thus to guarantee greater manufacturing reproducibility. The fact that it is possible to use automated equipment is one criterion of materials choice in the aeronautical industry. For example, the choice of a carbon fibre based composite for the Airbus A300 and A310 tail unit came about because of the automation possibilities opened up by that material.

This tendency towards integration does however have one major drawback. While assembly and part costs fall, the material costs are

(20) Op. cit.

constantly rising (as Figure II.5.a in Part II shows). The fact is that materials are becoming increasingly complex and increasingly well suited to the local characteristics sought at different points within the component. It has therefore been necessary to develop highly sophisticated methods for calculating structures in composites, and quality control methods in order to detect faults in these materials. It can therefore be said that a growing proportion of production costs are information costs, generated by the complexity of the materials. Being able to obtain simple shapes with composites costs dear in materials because their complexity necessitates elaborate structural calculations and sophisticated part control throughout the production cycle, and perhaps even throughout the life of the product (21) .

Improving the efficiency of engine systems

Although engine improvement has pursued a separate line of research into energy savings from that of structures, it has often benefitted from the spin-off of structural weight reduction.

One very good example of a chain of induced effects comes from the space industry (22) . For rocket nozzles subject to temperatures between 2000 and 3000°C it is impossible to use metals, and recourse was therefore had to refractory fibre composites. In order to guard against the effects of heat, a sufficiently thick layer of resin (several centimetres) had to be present nevertheless, so as to prevent the nozzles from breaking down too fast during launch. These problems of pyrolysis are considerably eased by the use of carbon-carbon : a thickness of a few millimetres is sufficient, giving a weight reduction of a factor of 10. But, since thrust gain depends on the length of the nozzle it was becoming desirable to lengthen them

---

(21) Growing consideration is being given to "enhancement" of composite parts for aeronautical applications with electronic sensors able to provide a constant supply of information about the behaviour of these materials' principal properties.

(22) Cf. CPE "L'hyperchoix des matériaux", ("Overchoice of materials"), in "Rapport sur l'état de la technique", Sciences et Techniques, N° spécial, 1985 (mars), 48-61. (special issue, March 1985, 48-61).

despite the increase in weight. Further, once the nozzles were lengthened it was advantageous to increase engine pressure despite the extra weight which that entailed. Ultimately, all these effects together yielded a 10% improvement in the efficiency of the Ariane apogee rocket for the same weight.

The same development trend is observable in aircraft engines. Attempts to reduce consumption have generated two types of development in which materials constituted technological stumbling blocks : the improvement of turbojets and the development of new propellors.

Turbojets went through a period of explosive development during the last World War, gradually replacing propellors in military aircraft and subsquently civil aircraft. In this type of engine, output is related to operating temperature which, in the past, has been limited by existing alloys (to 900°C), thus affecting the performance of gas turbines. Improvements in cobalt or nickel based alloys have raised that temperature limit. However, the availability of these materials is attended by risks because of their strategic role. Ceramics and ceramic-ceramic composites afford a real quality leap, with operating temperatures in the region of 1400°C becoming possible. Nevertheless turbojets continue to be relatively low efficiency engines. There are other constraints too on their performance, particularly the compressor, and this explains the interest taken in other modes of propulsion.

In a turbojet compressor, the first stage or "blower" at present represents a technological barrier. Its dilution capacity (the ratio between the mass of cold air sucked in by the engine and the mass of hot gas passing through the turbine) is only five, compared to 20 or 30 in the case of a propellor. This blower characteristic is a factor in increasing the output of the jet engine. So the idea is to replace the blower vanes by a propellor and do away with the fairing on the first stage of the compressor. In the past, propellor driven aircraft have also had fairly low efficiency because of the limited speed at which propellors could operate. Thanks to carbon fibres it is now possible to make blades which can achieve transsonic speeds at their extremities. A great deal is at stake, since this mode of propulsion is expected to produce average savings of more than 25% in the fuel consumption of a turbojet. Boeing has already announced that the Prop-fan 7J7 will make its first flight in 1991. This twin-propellor engine, favoured by NASA, is built by Pratt and Whitney-Allison.

Other current projects include those of General Electric (which has a 35% sub-contracting agreement with SNECMA), Rolls-Royce, and a Soviet project. Boeing has to compete against the Airbus A320 and appears to be banking on the effect of its announcement to make the airlines think twice before deciding to purchase the Airbus 320. However that may be, there are substantial technical problems still to be solved, particularly the ability of the fuselage to withstand the impact of a blade snapping off at speeds up to MACH 1.1.

III.1.2.2. The industrial use of new materials in motor vehicles : present situation and prospects

Engine developments

Underlying the use of engineering ceramics in automotive engines is the hope of considerably improving the energy efficiency of this essential part. In a present-day diesel engine, scarcely 38% of the calories contained in the fuel are converted into useful work. The remainder is dissipated, being divided among the heat emitted in exhaust gases (33%), direct heat radiation or evacuation by the cooling system (29%) and, marginally, friction in the pistons. Making appropriate use of the antifriction properties of ceramics would substantially reduce the amount of energy dissipated through mechanical friction. But it is mainly because of their thermal properties that ceramics are of interest for applications in car engines. Thermomechanical ceramics are compatible with operating temperatures above 1000°C. In some formulations (zirconia, sialon, silicon nitride), they also have low thermal conductivity. These properties make possible the partial or total "adiabatisation" of the combustion chamber and its immediate surroundings. This means that when substituted for steels and alloys in all the hot parts of the engine (piston crowns, cylinder liners, cylinder block, etc) ceramics would make it possible to limit or even entirely prevent thermal exchanges with the cold parts. The result would be to increase the useful energy by recovering the energy formerly dissipated in the form of heat. It is estimated that recovering all the energy dissipated would double the engine's energy output!

The engine makers' present objective is merely to achieve an improvement of between 30 and 50% in engine efficiency by the year

1990. Through the use of thermomechanical ceramics alone, a car which at present consumes 8 litres per 100 km would then consume only 6. But lower vehicle fuel consumption, which is of benefit mainly to the user, is not the only advantage afforded by these new materials. "Adiabatisation" of the engine permits substantial simplification in design and in the process of engine manufacture by doing away with the cooling system. Apart from the saving in capital investment which that entails, there would no longer be any need to provide for the circulation of water in the engine block and cylinder head or to provide a radiator or water pump. Such simplifications are reflected in important fabrication and material economies. In addition, the low density of ceramic materials and the elimination of the cooling system generate weight reductions and improve vehicle aerodynamics as engine bulk is also reduced.

The many advantages of thermomechanical ceramics explain why all the motor manufacturers, in conjunction with firms making engines and engineering ceramics, are actively engaged in programmes for the replacement of metals and alloys in the engine. In this respect Europe seems to be somewhat behind its American and Japanese competitors, despite significant achievements in scientific research. Apart from the Federal Republic of Germany, where cooperation between ceramics engineers (Feldmuhle, Anna Werke, Rosenthal), motor manufacturers (Mercedes, VW) and parts suppliers (KKK for pistons, Mahle for turbochargers) is already well advanced and where the Porsche firm has been the first to bring in a ceramic engine in a sports model, the absence of national master-plans is keenly felt in France and Italy. After 17 years of substained effort, the British government has stopped funding research into thermomechanical ceramics. The results obtained have encouraged the United States and Japan to launch major industrial research programmes as a result of which these two powers now dominate the ceramics market (23) . The British government recently reacted by encouraging the setting up of two ceramics clubs, one around

---

(23) In 1980, the Japanese market in engineering ceramics accounted for almost half the world market. In Japan, it is estimated that car engines will contain between 7 and 8 kg of ceramics by 1995, whereas French forecasts for 1992 are for only 1 or 2 kg. Concerning the situation in the United States, see CPE : "Les céramiques avancées aux Etats-Unis" ("Advanced ceramics in the United States"), Bulletin du Centre de Prospective et d'Evaluation (CPE), 1985, 22, 73-84.

Rolls-Royce for gas turbines, and the other around Ford and British Leyland for diesel engines (24) .

The potential market for thermomechanical ceramics is gigantic. Even if every engine were to incorporate only 1 kg of ceramics, which is not much if Japanese forecasts are anything to go by, it should be appreciated that <u>every</u> major engine manufacturer produces about two million engines each year. However, the incorporation of ceramics into engines will take place gradually. After current experiments on vehicles produced in small quantities, for low-risk replacement solutions (exhaust pipe linings, diesel engine precombustion chambers), we may expect to see the use of ceramics taken further in diesel engines for heavy goods vehicles. The impact of ceramic engines is likely to be more marked in this type of vehicle than in diesel engined cars, so that the movement towards ceramics will probably affect the latter only at a third stage. Petrol engines will probably be the last to come in for this treatment, because the presence of high pressures and temperatures means making modifications and taking certain precautions in order to prevent backfiring due to spontaneous ignition of the petrol mixture away from the ignition point (25) . With a view to this gradual introduction of ceramics, the German firm Porsche has recently brought out a new turbocharged sports model (944T) which represents the first major application of a ceramic engine to a production car (26) . Together with the army, the American engine maker Cummins has successful tested a heavy goods vehicle concept operating with an uncooled 350 hp diesel engine.

Only the Japanese appear to be looking to more widespread industrial application in their efforts to develop a low-power ceramic engine for a large or medium series car (27) . The stricking feature of

---

(24) CPE : "Céramiques thermomécaniques pour moteurs et turbines : le réveil britannique" ("Thermomechanical ceramics for engines and turbines : Britain wakes up"), Bulletin du Centre de Prospective et d'Evaluation (CPE), 1985, 17, 27-8.

(25) L. AUGIER : "Des moteurs en céramique" ("Ceramic engines"), Sciences et Vie, 1985, 809, 88-91.

(26) J.P. GOSSELIN : "Une nouvelle race de Porsche" ("A new breed of Porsche"), L'Equipe, 7 January 1985.

(27) This is probably one of the main objectives of cooperation between the ninth biggest car maker in Japan, Isuzu, and the first world specialist in engineering ceramics, Kyocera, which is expected to .../...

the Japanese scene is the <u>simultaneous</u> nature of the learning processes tied up with technical know-how and industrial know-how. A ceramic engine factory is now in operation in Japan for the sole purpose of establishing and refining the conditions for the industrialization of these engines, even at a time when some of the technical problems are still unresolved. Underlying this policy is the determination to dominate the future ceramic engines market by establishing a lead in industrial organization.

One reason why the introduction of ceramics into car engines is taking place only gradually is that some of the industrial process problems have not been wholly overcome. Thermomechanical ceramics are extremely <u>brittle</u> materials and highly <u>sensitive</u> to <u>microscopic defects</u>. As a result, parts rejected as faulty account for no less than <u>50%</u> of the present cost price of ceramic components. While critical fault thresholds for metals are measured in millimetres, ceramic materials have to meet specifications a thousand times higher. In addition, ceramics present good thermal behaviour in stable conditions <u>but</u> tend to fissure when they are subject to thermal fluctuation or shocks in a working engine. Imperfect control of industrial <u>reproducibility</u> and <u>in-use endurance</u>, because it adversely affects manufacturing and utilization costs, is consequently the prime obstacle to the rapid spread of engineering ceramics in engines. Efforts are being made to enhance reliability by improving the metal/ceramic interface (partly stabilized zirconia), increasing the strength of ceramic materials (ceramic composites) and developing processes (sintering, prestress processes) which eliminate the finishing phase whose major drawback, apart from cost (28) , is to make the parts

---

.../...

result in commercial application by 1990. However, it is important to realize that there is a significant degree of disinformation contained in the announcements regularly made in Japan concerning the ceramic engine (see J.P. GOSSELIN : "Céramiques : le moteur à malices" ("Ceramics : an engine full of tricks"), Sciences et Techniques, 1984, 809, 88-91). It should also be noted that, from the European standpoint, the danger is not so much Japanese as Nippo-American, as evidenced by the fact that General Motors owns 38% of Isuzu's capital, and that Cummins and Kyocera are collaborating on the development of adiabatic ceramic diesel engines for heavy goods vehicles.

(28) Ceramics being extremely hard materials, expensive tools have to be used in finishing them. For example, the hardness of silicon carbide is just below that of diamond, and the material can be worked only with diamond tools.

brittle by increasing the risk of microcracks. The application of injection moulding processes in the near future should also make it possible to obtain complex-shape silicon nitride components at only a third of a cost of the equivalent metal parts.

Inside the engine, thermomechanical ceramics complement existing technologies, and this is tending initially to assist their penetration. For the time being, there is no move to replace steel by ceramics in cold engine parts, so that no fundamental change in the basic structure of the engine is called for. In the long run, however, it is probable that ceramics will be fully utilized for their specific properties, and this should make engine redesign possible (29) . The increase in engine running temperatures afforded by the use of thermomechanical ceramics underlines the potential benefit of replacing the traditional piston engine by a turbine engine. In addition, the temperature and pressure drop throughout the exhaust circuit could be recouped by a turbocharger and also by a second turbine directly incorporated into the transmission drive. Such redesign in favour of the turbine would remove the problem of friction, so that the lubrication system could be dispensed with. Again, the advantages thus secured in terms of investment, machining, weight and aerodynamics, not to mention the advantages to the user, justify the interest which car manufacturers are taking in thermomechanical ceramics. They are already having to address the question of engine redesign, because present-day lubricants cannot be used at the temperatures at which a ceramic engine operates. The research being done to develop a mineral lubricant represents only a compromise solution pending complete engine redesign.

---

(29) The pattern of innovation taking over technical products, which proceeds through complementary followed by substitutions, culminating in redesign, is outlined by E. ZUSCOVITCH and J. ARROUS ("La diffusion intersectorielle des matériaux synthétiques, évolution et bilan économique" ("Inter-sectorial dissemination of synthetic materials - development and economics") in : "La chimie en Europe", 141-328, P. COHENDET (ed), Paris, Economica, 1984), and is discussed in terms of ceramic materials by D. KIENY and E. ZUSCOVITCH "Les céramiques techniques : entre l'apprentissage et la variété" ("Engineering ceramics : between learning and variety") in : "Les nouveaux matériaux" ("New materials"), Strasbourg, Seminar of 13 December 1985, doc. FAST n° 69, Brussels, Commission of the European Communities, 1986.

Developments in bodywork

A look at the way in which cars have evolved from the early days reveals that manufacturers only recently began to give any particular importance to bodywork. Awareness and exploitation of the status-symbol character of private cars, and the need to vary models for reasons of competition, are factors whose full effects did not begin to be felt until the sixties. In fact, it was not until the energy crisis that interest was once again taken in aerodynamic performance and consequently in body shapes. We have already had occasion to describe the principal technical characteristics which give plastics their superiority as constituent materials for body parts. We shall not return to this aspect, but instead concentrate more particularly on the problems of <u>industrial application</u> which arise when replacing metals in bodywork.

First of all, let us look at the scale of the changes which plastics have brought about in vehicle shapes. At the present time only plastics <u>reinforced</u> with long fibres of the kevlar or carbon type present characteristics meeting the two principal requirements for a <u>complete</u> change in vehicle structure : they must offer sufficient stiffness in the <u>load-bearing structure</u>, so that the stresses transmitted by the mechanical organs can be accomodated, and they must deform sufficiently in the <u>bodywork</u> to absorb kinetic energy in the event of a collision. Apart from their very high intrinsic cost such materials involve long fabrication cycles which are difficult to automate because of the need for careful orientation of the reinforcing fibres. For a long time to come, therefore, the load-bearing structures of motor vehicles will rely on metal technology (30) . The improvements afforded by the optimization of structural calculations and by recourse to special steels give added force to that conclusion. Yet potential future progress in improving reinforcing fibre fabrication processes and the processing techniques for long-fibre composites (31) should

---

(30) Note, however, that there are already some commercial applications of long-fibre composites in certain body parts, eg in the door of the Peugeot 205 GTI.

(31) The du Pont company has a virtual monopoly of kevlar fibre. So we should not look to that quarter for any marked price reduction.

.../...

not be overlooked. Nor should the possibility of unreinforced plastics becoming established as constituent materials for load-bearing structures be underestimated, in view of the appearance on the scene of new high-strength RIM plastics. Nonetheless, as a general rule the advances in question still have to be confirmed industrially, and until they are the overall structure of cars will rely on a combination of a steel load-bearing structure and a bodywork made of "short" glass-fibre reinforced plastics.

Cars with all-plastic bodies were designed and marketed as long ago as 1953, in the United States (General Motors Corvette) (32) and later in France (the Matra Murena), Great Britain (Lotus Esprit) and the Federal Republic of Germany (Porsche 924). A more recent example is the Renault Alpine V6 which has an all-plastic body comprising only 40 parts. However, all these examples are luxury or sports models in small series. Only the Americans, namely General Motors (Pontiac Fiero 1983, Pontiac P. 1984) and Ford (Flain 1984) and, to a lesser extent, Renault-Matra with the monocoque Espace model, have

---

.../...

Carbon fibres are currently produced from carbonization of an organic precursor : polyacrylonitrile. This polymer costs 25 F/kg on average and loses half its weight during carbonization. So the minimum cost of carbon fibre is equal to twice the cost of organic fibre plus the cost of the energy needed for carbonization. It is therefore unlikely that the price of carbon fibre could be brought down below 100 F/kg in the framework of existing technology. Research into new precursors and alternative carbon fibre fabrication methods is therefore clearly of interest. General Motors' research technologists have announced the experimental development of a manufacturing process based on natural gas and hydrogen - both of which materials are cheaper and more abundant than polyacrylonitrile. As regards the industrial processing of long-fibre composites, the same manufacturer is testing a method for production of a chassis structure by computer controlled biaxial filament winding. Other processes whereby the position and orientation of fibres can be better controlled are also being industrially developed. One example comes from a British firm marketing an inexpensive process which uses ordinary moulds and expansion agents (see CPE : "Composites bien armées de fibres" ("Composites well reinforced with fibres"), Bulletin du Centre de Prospective et d'Evaluation (CPE), 1985, 19, 89.

(32) It is often said in the specialist press that we do not know how reinforced plastics will perform over a period of time. The Corvette provides an interesting example in this connection, though not one from which any general conclusions can be drawn : 75% of those which took to the road in 1953 were still going in 1985!

so far managed to show that an all-plastic body can be economic in a medium series vehicle.

Although it is easier to make continuous surfaces and to modify shapes with plastic moulding than with steel sheet pressing, plastic materials do suffer from the handicap that their production cycles are long. That disadvantage is offset by the substantially lower cost of presses, but only when production rates are sufficiently low to make a large numbers of presses unnecessary. Above 400,000 units per annum for the BMC processes, and 150,000 units per annum for the SMC process, investment in presses works to the detriment of plastics. This very important problem of production rates is compounded by the heavy cost of detooling which a full-scale transition to plastics technology would require. Very often, metal stamping presses have been in operation for a long time and are fully amortized (33) . Lastly, the need for metal and plastic body parts to present a uniform appearance works to the disadvantage of plastics. On-line painting of body parts is generally done by an electrostatic process requiring a temperature of some 150°C. With the exception of polyesters, the plastic materials used in car manufacture do not stand up well to such temperatures and are all poor conductors of electricity. The problem of electrical conductivity is solved by applying a special pre-paint primer, and the problem of heat often means that body parts made of plastics have to be painted off-line. In both cases the parts in question are less economic to produce than their metal equivalent.

Overall, in the present state of the art, the balance between steel and plastic substitutes is still slightly tipped against plastics from the standpoint of industrial viability. The trend towards substitution is nonetheless likely to continue because the mechanism is closely dependent on the <u>progressive character</u> of the technologies involved. Steel and aluminium producers are well aware of this, and are

---

(33) In the course of an interview with one of the leading European motor manufacturers, we learnt that some body press shops dating from the pre-war period are still in operation. Likewise, R. THALLER ("Pour une économie de la diffusion des innovations technologiques" ("Towards an economy based on dissemination of technological innovations"), Thèse de l'Université de Lyon II, Lyons, ECT, 1986) reports a case of a bumper chromium-plating workshop surviving despite the undeniable advantages offered by plastic moulding techniques. So the amount of immobilized fixed capital and its amortization undoubtedly work to the detriment of plastic technology and hamper its spread in the automotive industry.

endeavouring to counter the advance of plastics, either by joining them (sandwich sheets) or by improving the characteristics of their metal body pressings. For example, the British company BL Technology, in collaboration with a subsidiary of Alcan, has developed a bonding technique in order to produce vehicle structures using aluminium sheets economically. Furthermore, the aluminium industry is emphasizing the recovery and recycling possibilities of this material, and pointing out that the only feasible way of recovering plastic materials at the present time is anti-economic because it means burning them (34). However, the reactions of the metal and alloy producers to competition from plastics belong to that category of minor improvements which temporarily block the establishment of a promising but as yet insufficiently industrialized technology. As such, they do not cast doubt fundamentally on the introduction of plastics into vehicle bodies, for they do not offer the _decisive_ advantages which plastics do secure - _the redesign of the motor car and its production process_.

The introduction of plastics into vehicle bodies was a gradual process, as their continued advance will be. To begin with, the plastic part replaced the metal part, performing an equivalent function and benefitting from some of its characteristics. Certain investment savings can be achieved at this stage, as in the case of the first plastic bumpers, which enabled costly chromium plating workshops to be discarded (35). At a second stage, plastic replaced metal by _integrating_ a larger number of parts and/or functions. The benefits obtained at this point are perceptible both in terms of investment cost and in terms of component assembly, handling and management costs. The development of bumpers towards a unit incorporating a growing number of parts has already been referred to. But the most significant example of parts integration remains the rear hatch of the Citroën BX, where the use of reinforced polyester reduced the number of parts to 3 against 27

---

(34) A big step will have been taken towards solving the problems of production rates and recycling when it is possible to use thermosets instead of thermoplastics. As the mechanical properties of thermoplastics are inferior to those of thermosets, this advance is likely to be achieved by means of a high-performance but inexpensive reinforcing fibre.

(35) Since chromium is also regarded as a strategic raw material, the example of bumpers also reflects the idea that the new materials place a question mark over the practical significance, and even the economic relevance, of the concept of strategic raw materials.

in the equivalent metal component (36) . Finally, at a more advanced stage currently being researched, entire bodywork elements as opposed to just single parts will be made of plastics. The degree of integration envisaged is such that plans are already having to be laid for the development of bodywork towards an "egg" conception (cf. Figure III.1.c) or towards a modular conception (cf. Figures III.1.d and III.1.e).

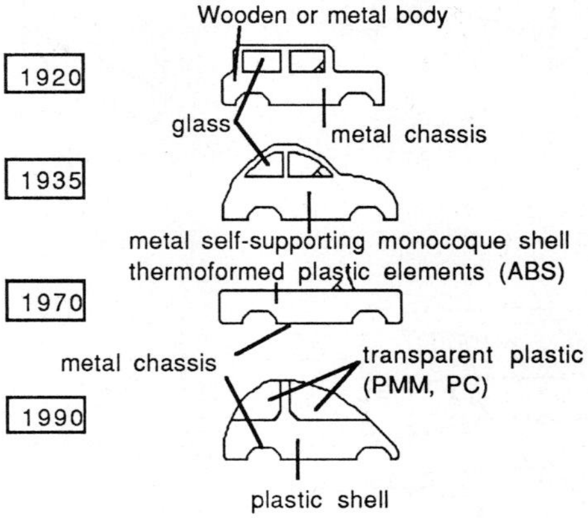

Source : M. REYNE, cited by R. THALLER (37) .

Figure III.1.c : Evolution of bodywork towards an "egg" conception.

In the modular conception, the metal cage performs the load-bearing function, with plastic body parts being bonded to it. The dashboard can be assembled before installation and the roof can receive its interior lining before being bonded to its uprights. Finally, the rear assembly, which is fixed to metal uprights, is modular. In fact, the changes envisaged go beyond bodywork alone and make it possible to envisage redesigning the actual car assembly process.

Since their origins, cars have been assembled from outside (load-bearing structure and chassis) to inside (mechanical parts and

(36) For reasons of production capacity, this innovation has been marketed only on the "sport" version of the BX.

(37) Op. cit.

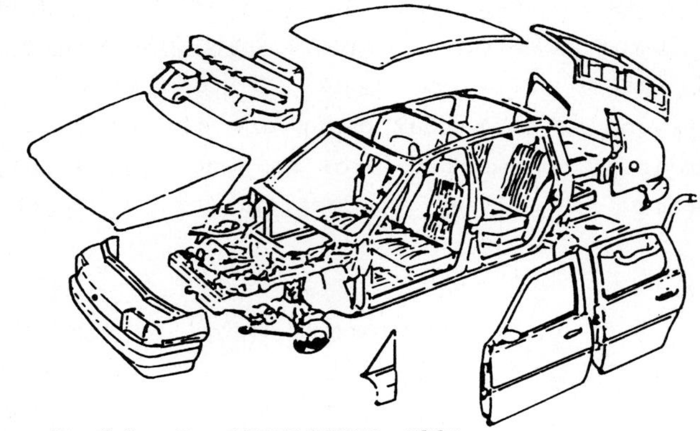

Source : Fiat cited by F. CORNACCHIA (38) .

<u>Figure III.1.d</u> : Modular conception : prototype VSS (Fiat).

Source : Fiat cited by F. CORNACCHIA (38).

<u>Figure III.1.e</u> : Modular conception : CRF-CNR car, 1984.

interior fittings). With modular bodywork or the "egg" conception, the opposite approach to assembly - <u>from inside to outside</u> - makes sense : at the <u>end</u> of the assembly process, the body would be added to the load-bearing cage structure <u>after the internal equipment had been</u>

(38) "L'utilizzo delle materie plastiche nella struttura della vetture" ("Use of plastic materials in motor car structure") in : "Convegno internazionale...", op. cit.

fitted. This about-turn in the way a car is built is expected to yield substantial benefits, as automatic processes are improved and the production system made more flexible. This redesigning of motor vehicles would make it possible to alternate tasks which are hard or prohibitively expensive to automate such as, for example, fitting the steering wheel, which is now done by a manual operative. Since it is generally estimated that 50% of the manufacturing cost of a car is taken up by assembly costs, every operation that can be automated produces substantial savings. Furthermore, the ability of reinforced plastics to provide tailor-made properties and be shaped into complex parts can be systematically exploited in such a way that every plastic material is adapted to the function which it can best perform. The corresponding design and computation techniques, like finished part analysis, already exist. Industrial processes for plastics are also much more flexible than those for metals. Moulding is a much simpler and more manageable process than forging or sheet steel pressing. Presses are less expensive and moulds can be changed relatively quickly. Another advantage is that plastic parts are bonded or glued, not riveted or welded, which means fewer work stations and a much simpler production line. But flexibility also means product adaptability to fluctuations in demand. Plastics technology offers a marked improvement in this respect, because it makes it possible to

Source : R. THALLER (39) .

Figure III.1.f : Modular conception : possible body shapes.

---

(39) Op. cit.

assemble different bodies from a small number of standardized load-bearing structures which can be produced in large quantities (Figure III.1.f).

While it is true that at the present time plastics emerge less favourably from comparisons of industrial profitability in long run part production, that is mainly because such comparisons do not take the evolutive nature of the technology into account. In view of the lead time involved in designing and manufacturing a motor car, every manufacturer owes it to himself to see how technology is developing and to anticipate the results of that development so as to be better prepared for it. In this respect, it seems that the prospects opened up by the new materials point to a veritable transformation of motor vehicle technology geared to relatively recent developments in demand.

III.1.3 THE TRANSFORMATION OF AUTOMOTIVE TECHNOLOGY AND ADAPTATION TO DEMAND

The introduction of engineering ceramics into engines and of engineering plastics into bodywork are not proceeding at the same pace. The scientific, and especially technological, development of thermomechanical ceramics is less advanced than in the case of engineering plastics, which have already been used for a great many commercial bodywork applications. To this must be added a historical factor peculiar to the motor industry. From the outset, the industry has given precedence to some technical requirements over others. Its innovative choices focussed first of all on the propulsion system (internal combustion engine), followed by the chassis and major components, and lastly by bodywork. This explains why engine production is now largely automated (40) , whereas the assembly of cars still occupies the bulk of the labour force employed in the industry. Having reached a high degree of technical and indutrial maturity, the engine

---

(40) The Fire 1000 engine, a joint development by the Italian manufacturer Fiat and the French manufacturer PSA, is the most recent example of automation in practice in the manufacture of car engines.

is less able to incorporate major innovations rapidly, unlike bodywork which has not changed radically in design since the industry was in its infancy. In addition, bodywork redesign goes hand-in-hand with manufacturing processes which are already fairly well perfected and which permit a new approach to the industrial process of car assembly. This difference of maturity as compared with the industrial application of new materials explains why the transformation of automotive technology is following the trend towards redesign which plastics technology makes possible. This transformation, which is still too recent to be amenable to analysis in every aspect, and because it is taking place gradually, underscores the importance of anticipation. That such anticipation is not mere speculation is borne out by the fact that it can be analysed in terms of changeover. It can be demonstrated with this type of analysis that practical decisions taken at the present time are shaped by anticipating a changeover in automotive technology and indeed are taken in order to prepare for it. There is one instance of such anticipation, neither technical nor industrial, in the trend of car demand towards customized demand. In fact, the adjustment of automotive technology to demand goes beyond the framework of transformations prompted by plastics technology, since it also has to do with regulations relating to the environmental norms laid down by the authorities.

III.1.3.1. The role of anticipation or the logic of changeover

Both in engines and in bodies, penetration by the new materials reflects a logic of incremental development eventually culminating in overall redesign. These materials are introduced first of all by reason of their complementarity with existing technical artefacts and technologies (because they improve them). At this level, problems of industrial reproducibility are of crucial importance because, unless they are resolved, the cost of introducing new materials cannot be brought down to a level at which the intrinsic features of these materials are sufficient incentives in themselves. But even at this introductory stage tensions arise. They derive from the properties which make the new materials incompatible with the design of technical products and their traditional technologies. The problems of lubricating ceramic engines and painting plastic body parts generate tensions of this kind. It is in fact these very same specific properties which enable the engine, the bodywork and the entire car assembly process to be redesigned. They will in due course enable the new technologies to break away from the ties of complementarity which act as a constraint on the speed at which the new materials move into engineering applications. As matters stand, we are still at the stage of technical possibilities in this regard. But the car makers are taking these possibilities into account and including them in their projections. As we see it, technical redesign anticipation is an essential factor because it gives that anticipation a role to play in structuring the economic decisions that will be taken subsequently, even if that role is not always fully realized by the people involved. What needs to be emphasized is not the simple process of updating technical forecasts in accordance with the latest advance in the onward march of science and technology. The acts of anticipation which are involved in the process described above bring a measure of discontinuity in the conception of technical artefacts and their technologies and for that reason can be analysed in terms of changeover. What immediately underlies the behaviour of those who are concerned with the introduction of new materials is the acceptance of a compromise between the conditions of industrial reproducibility which dominate the short-term viability calculation and the prospect of a changeover which dominates the (hazardous) calculation of long-term viability. The very existence of this type of compromise, the result of which affects the manner in which resources are allocated, has numerous implications.

In the first place, the existence of this compromise means that industrial decision-makers are prepared to sacrifice part of their short-term profitability in view of the important advantages they expect to gain from changeover. This sacrifice is only partial, for several reasons : first, because there are thresholds - notably financial thresholds - which cannot be crossed without major risk secondly, because changeover is a gradual process the very duration of which requires an effort to keep the production costs of innovation down : lastly, because changeover itself entails significant extra cost in training staff and acquiring new know-how, making new fixed capital investment and scrapping existing fixed capital investment, redefining industrial relations and preparing the market - to which must be added the specific strategic cost which has to be borne by the firm which first takes the risk of making the changeover.

In the second place, the existence of the compromise in question has a marked influence on the speed at which a development results in an innovation. Obviously, if it were not for the redesign potential thermomechanical ceramics and plastics would not be receiving the same amount of research attention from car manufacturers as they are.

Finally, there is a third implication concerning the way in which this compromise evolves in time. Initially, anticipation of changeover remains confined to a passive structuring role, from the standpoint of strictly economic decision-making. Anticipation within the technical horizon of industrial decision-makers does not directly influence production options, but affects the direction of research and development options as long as the prevailing objective is to control technological and industrial reproducibility. As one comes closer to that goal, and in fact long before it is fully attained, the tensions arising from the problems of equating the intrinsic properties of the innovation and the existing design of technical artefacts and their technologies oblige those involved to take microdecisions which prepare for the changeover and cumulatively bring it about. At this point in the analysis, a look at the facts provides a satisfactory picture of the logic of changeover and one which illustrates the foregoing general considerations. For this purpose we shall focus on the introduction of plastics into car bodies. Thermomechanical ceramics have not yet reached a sufficient degree of technological and industrial maturity for redesign anticipation to play an active part by causing the kind of changes in microdecisions to which we have been referring.

The introduction of plastics into car bodies is an already fairly advanced phenomenon. The relationships of complementarity maintained between the new technology and the old are now being left behind, as is evidenced by the examples of all-plastic bodies for small and medium series production cars and the first experiments with modular design prototypes. So anticipation in respect of the redesign of both product and processes has taken relatively clear shape and is already influencing economic choices. Thus if only short-term viability were taken into account the rationale for certain decisions would not be understood. This applies first and foremost to decisions about training and acquisition of new know-how.

One of the biggest European motor manufacturers decided in this context to move the injection moulding machines installed by his usual bodywork parts supplier into his own assembly plant. The reason given for this transfer was that the assembly line operators were rejecting too many of these parts because thay had no experience of working with these components and broke them in the course of fitting, because they were handling them in the same way as the equivalent metal part. The other reason for this microdecision was bound up with the increasingly important role which assembly is coming to assume by virtue of plastics technology (41) . Every element in the system must therefore be fully under control if advantage is to be taken of the synergies afforded, particularly from the standpoint of automation (42) . The decision that was taken was not a "major" decision from the company's point of view, although it has important implications. But it is quite significant, because it illustrates the intention to develop know-how in preparation for the changeover which plastics technology is going to bring about in car manufacturing, a sector mainly structured since its origins in terms of metal-working. In that sence the decision in question constitutes a micro-changeover. There was, moreover, some resistance to preparation for working with plastics although it was perceived as necessary. It was found that operators working with plastic parts systematically overestimated defects in them, and that there was no objective basis to confirm their attitude. So changeover also produces harmful psychological effects which ought not to be underestimated, making the workforce feel that the transition to plastics detracts from their identity and their skills (43) . Citroën, the big pioneer in plastics technology in Europe, had first to train its managers and workforce in order to carry through the many innovations introduced into its BX model. Another aera in which training is needed is that of design. With metal, a simple strength calculation decides the choice of material and shape, environment being only secondary. The situation is quite different with plastics, because they require three-dimensional methods for the calculation of stresses and strengths. The design must be just right, bearing in mind the importance of the environment (evolution of mechanical characteristics) and in order to avoid reworking finished parts. On the other hand, the

---

(41) Moreover, few automobile industry suppliers can afford the heavy investment which the high work rates in this industry entail.

(42) The same manufacturer has also taken the decision to produce plastic compounds.

(43) Operators trained to weld, rivet, drill and machine metal sheet attach a value to those tasks which they do not attach to the task of glueing plastic parts together.

assimilation of new know-how without recourse to in-house training seems necessary for fabrication techniques which mark a complete break with traditional methods. Thus the jobs of injection moulding machine operator and mould maker are new occupations in the car industry. In view of this gradual changeover, a fresh approach to industrial relations is also called for.

Car manufacturers had been in the habit for some years of bringing all their weight to bear upon equipment suppliers and sub-contractors and making the most of competition in order to obtain low prices for components which did not produce themselves. Such a policy is less and less suited to the new needs, because sub-contractors and especially equipment suppliers are having to produce increasingly complex components and consequently need to have sufficient secure markets to finance the substantial R & D costs. Furthermore, collaboration between equipment suppliers and manufacturers is becoming necessary and indeed vital. This is what underlies the new forms of industrial relationship characterized by industrial partnership, which could be defined as a close, single-purpose association between all those able to contribute to defining and making a given component. The new famous case of the rear hatch of the Citroën BX is an excellent example. This was the result of collaboration between the manufacturer and a glass producer (Saint-Gobain), a glass-fibre weaving company (Vetrotex), a chemical firm (Rhône-Poulenc), a maker of injection moulding machines (SMTP-Billion) and a mould maker (SEVA).

Finally, changeover also has strategic implications which affect the pace at which it proceeds. While a changeover is economically justified by the expected advantages of redesigning products and processes, it still places a heavy burden on the firm which initiates it. We must distinguish here between the anticipation of changeover, which leads to the taking of cumulative microdecisions, and the actual phenomenon of changeover itself. Insofar as it results from a macrodecision at company level, the latter is directly bound up with the market risk specific to any important innovation. This explains why some manufacturers are cautious and why those who have progressed furthest in introducing plastics are willing to spread their expertise so as to encourage their competitors to move on to the same technological trajectory. For instance, the PSA group, which has carried out the most significant bodywork innovations in Europe through the intermediary of Citroën, is adopting a strategy of collaboration, notably with the Fiat group and seemingly with the German VW group. The implicit aim is to form a sufficiently large nucleus in terms of market shares to spread the risks and influence the market in the desired direction. Thus on the one hand there is a step-by-step approach, and on the other a coming together of different parts into a coherent unit dictated by technology. However, there are manufacturers like British Leyland which are continuing to develop traditional technologies. In view of the considerable upsurge in plastics technology and the relatively stagnant overall demand for cars, the future of such a strategy is by no means certain. As for the Japanese manufacturers, the reason why they were rather late in introducing plastics into car bodies have already been mentioned. Nevertheless, their capacity for industrial organization can be relied upon to apply optimally the processes developed elsewhere. Moreover, it is not easy for one manufacturer to pursue two major lines of innovation, and the Japanese car manufacturers appear to have opted for ceramic engines rather than plastic bodywork.

III.1.3.2. Adjusting to new patterns of demand

Engineering ceramics and pollution abatement

The pressure being brought to curb atmospheric pollution, which has already proved effective in the United States and Japan since the mid-1970s, has resulted in comparable decisions in Europe. In view of the increasing pace of forest die-back due to acid rain and growing pollution towns, these decisions are forcing the European car industry to make the use of catalytic exhausts general. This is the purpose of the measures taken by the European Economic Community on abatement pollution norms which were recently backed up by two draft directives aimed at emissions from diesel engined vehicles.

The great advantage which underlies the principle of the catalytic exhaust is that it enables toxic gas emissions to be reduced by a two-fold reaction involving the reduction of nitrogen oxide and the oxidation of carbon monoxide and hydrocarbons. In practical terms, this catalytic process requires the use of active phases (catalysts) mounted on a suitable substrate. The most commonly employed technique is to deposit platinum, rhodium and cerium oxide on a ceramic monolith (cordierite). Consequently, as the material providing the substrate for the active phases in the catalytic exhaust, ceramics are making an important contribution to adapting cars to new pollution norms. But their function in this connection is not limited to that aspect alone.

The catalytic exhaust system is attended by numerous problems. It is subject to rapid wear as a result of vibration, and above all to poisoning of the active phases. In the United States, where it has spread fastest, experts are beginning to build up evidence to show that the lifetime of catalytic exhausts is very much shorter than the average life of the vehicle to which they are fitted. This finding gives ammunition to the present opposition on the part of French, British and Italian manufacturers aimed at curbing the rate at which catalytic exhausts are introduced, whereas the German manufacturers are tending on the contrary to step up the pace under pressure from their government. In addition, the catalytic exhaust adds 10 to 15% to the average price of a private car and also places Europe to some extent in a position of economic dependency : the ceramic monolith substrate is covered by an exclusive patent held by the American firm Corning Glass, and the precious raw materials needed for catalytic exhaust systems to operate are mainly imported from South Africa and the Soviet Union (44) . Furthermore, the lead which is present in petrol and engine oil additives settles on the active phases and stops them functioning properly. In order to prevent poisoning of the catalytic reactions, it is therefore absolutely necessary to use a lead-free fuel, particularly in view of the known danger which lead presents to the human organism and the environment. But the presence of tetraethyl lead additives is associated with higher engine compression ratios and enhances efficiency and therefore performance at constant cylinder capacity. It is true that the lower engine compression ratio implied by lead-free petrol could be compensated by increasing the cubic capacity. But that solution would run counter to the goal of reducing vehicle consumption (45) . It is here that thermomechanical ceramics can play a part, preserving intact present performance in

---

(44) See Part IV.

(45) It would also be technically possible to adopt high octane rating components (ramified and/or aromatic compounds) to replace tetraethyl lead. This would necessitate new oil refining plant, thus pushing up the price of fuels.

terms of speed, acceleration and engine response without any increase in cubic capacity being needed.

When thermomechanical ceramics are used as lining materials in cylinder head exhaust ducts, the exhaust gases are not cooled. This means that, even at low speed, both catalysis and turbocharger response time are improved, while at the same time the loss of engine output resulting from the introduction of catalytic exhaust is compensated. This is exactly the solution which the German manufacturer Porsche chose three years ahead of its competitors. By using thermomechanical ceramics, Porsche has succeeded in demonstrating that performance and catalytic pollution control are not totally incompatible (46) . Thermomechanical ceramics also offset a great many other ways of improving engine efficiency : "adiabatisation" of the combustion chamber (47) , lighter turbocharger turbines, use of a second turbine in the transmission drive shaft, etc. So even disregarding the problem of removing lead from fuels, thermomechanical ceramics contribute to pollution control simply by improving engine efficiency and thus reducing fuel consumption.

Engineering plastics and customized demand

We have already had occasion to emphasize the importance of anticipating redesign in the process of introducing plastic materials into car bodies. The influence of such anticipation would probably be much less if manufacturers could expect demand to go on growing over the years ahead. As that is not the case, the only thing which could change the dismal market prospects would be for the economies of the

---

(46) The route followed by this German manufacturer, which is still a costly process, brought Porsche a victory in a 24-hour endurance race in the United States in 1983. After the race the catalyst was as good as new. See J.P. GOSSELIN : "Une nouvelle race de Porsche", op. cit.

(47) However, adiabatisation of engines, insofar as it means replacing the pistons by a turbine, is a departure in the opposite direction to that intended. Raising the engine operating temperature would actually result in nitrogen oxide emissions in excess of the present pollution norms.

developing countries to take off, which is unlikely in the relatively near future. Two phenomena of major importance should be borne in mind in this connection. Starting in the early 1970s, ever fiercer and increasingly international competition has bolstered the tendency towards standardization of the shapes and technical characteristics of cars produced in the west. Further, the saturation of demand in the developed countries has virtually reduced the whole market to the renewal of existing vehicles. In parallel to these developments the pattern of demand has shifted markedly. Customers have become more critical and behave less impulsively, looking for a combination of durability, safety and reliability, while rejecting over-standardized products. The growing importance of the quality-reliability-durability criterion in the choices made by consumers is borne out by all the surveys conducted in Europe by manufacturers and the various associations concerned. It is also illustrated by the success of those manufacturers (BMW, Daimler-Benz, Porsche, Audi, Volvo, Saab) which have based their brand image on quality and robustness, and by the public's disinclination to buy cars from manufacturers which, like Renault, have had rising levels of rework operations in their plants and of warranty claims. One interpretation of this trend in demand is to say that the principal characteristics demanded of cars nowadays are variety and versatility. Manufacturers have various possible means of adapting to that demand : in particular, they can step up quality control or offer special active safety equipment (ABS anti-lock braking systems, proximity radars). However, it seems that the necessary adaptation will have to focus mainly on vehicle shape, since the priority factor determining the choice made by the purchaser continues to be the appearance, design and style of the vehicle. It is here that plastics technology helps to match the vehicles on offer to the new patterns of demand.

The renewed interest taken by car manufacturers in bodywork is partly explained by the fact that plastics are the only materials that enable a theoretical shape to be mass produced. This characteristic, which is particularly important when it comes to improving aerodynamic performance and may perhaps one day free manufacturers from the need for expensive wind tunnel testing, also offers many opportunities for producing attractively flowing, curved shapes. Plastic car bodies are more rounded, more pleasing and free from sharp angles or joins which break up both the airflow over the surface and the aesthetic continuity of shape. Nor is it inconceivable that manufacturers may be able to create new colours by exploiting the "built-in" colours of plastics, once they are no longer obliged to emulate the surface appearance of painted sheet metal.

In a highly competitive market, plastics technology offers the undeniable advantage that "face-lift" and restyling operations on vehicles can be achieved faster and less expensively than metals technology permits. Modular design will make it possible to use vehicles in probably unprecedented ways - the motorist may adapt the shape of his vehicle, depending on the season or the purpose for which it is to be used, by interchanging certain body parts. It is probable that damaged body parts will be replaced rather than repaired. This should result in savings to the consumer, since plastic parts will be simple to fit, and in some cases the motorist will even be able to do the job himself.

Technological flexibility in the assembly of plastic body parts enables manufacturers to adapt rapidly to changes in demand. The car industry, whose earliest products were tailor-made, will rediscover a similar capacity. Since those early days the industry has been structured around metal working and has become a mass production industry. With the plastics technology now available it can envisage combining these two aspects, mass production and customized demand through technological flexibility and the production of small series. In 1986 the United States, where the concept of "a new model every year" first arose, produced only 11 models in longer runs than 200,000 units a year. This development, which heralds a change in the car industry towards the production of small series, boosts the spread of plastics technology because low production rates are favourable to it. Furthermore, the transition towards an ability to adapt to customized demand is underpinned by the complementary features shared by the new materials and the new information technologies. For example, Mitsubishi recently developed a computerized system which provides the potential customer with information in an interactive mode about colour, body trim and certain options such as, for instance, power-assisted steering. With this system, the future purchaser knows instantaneously how much his particular package of features will cost and can vary it as he wishes. This is undoubtedly an excellent means whereby the manufacturer can satisfy his customers while safeguarding his markets, tying demand, or at the very least information about demand, to himself. There can be no doubt that such methods, which bolster the car industry's capacity to satisfy customized demand, are going to grow in many directions.

CHAPTER III.2

New Materials and Packaging

The packaging sector occupies a pivotal position in the economy. Lying at the crossroads of production and consumption, packaging materials have to incorporate not only technical developments in products and processes, development constraints in the distribution and transport sectors and changes in consumer tastes, but also environmental constraints and the need to provide a vehicle for information about the nature and quality of products. Consequently, packaging is one of the most significant symbols of changes in production methods and lifestyles, and it is for this reason that the profound transformations now affecting production processes and consumer habits are bound to up-end structures and strategies in this sector. The principal feature of this upheaval is once again the phenomenon of <u>growing variety</u> in the production and utilization of packaging materials.

For a very long time the packaging sector was divided up into separate markets each of which was dominated by a particular material (glass for bottles, cardboard for cartons, tinplate for cans, aluminium for aerosols, etc). This is less and less the case : for any given market, several materials are often in fierce competition with each other. It is uncertain how that competition will turn out, and many key questions concerning the future of this key sector are still largely unanswered. (Will there be a development towards multimaterials or towards plastics ? Will specific local characteristics due to different lifestyles remain, or will the various types of packaging be internationally standardized ? Will the present production structures - contents manufacturer - packaging machinery manufacturer - container manufacturer - continue, or will there be a move towards more integrated industrial structures ? Will the authorities pursue a laisser-faire policy or will they lay down strict health and environment norms ?

In endeavouring to identify the overall prospects for the development of packaging materials, we shall begin by examining the ways in which this sector responds to each of the principal particular constraints to which it is subject (the requirements of the ultimate consumer, the requirements of the authorities with regard to health, environmental or information norms the requirements of distribution and the requirements of production (cf. Figure III.2.a), and then go on to present an overall summary of the manner in which the sector is evolving, highlighting those technical innovations in the design of packaging materials which may enable the verious constraints affecting this sector to be eased.

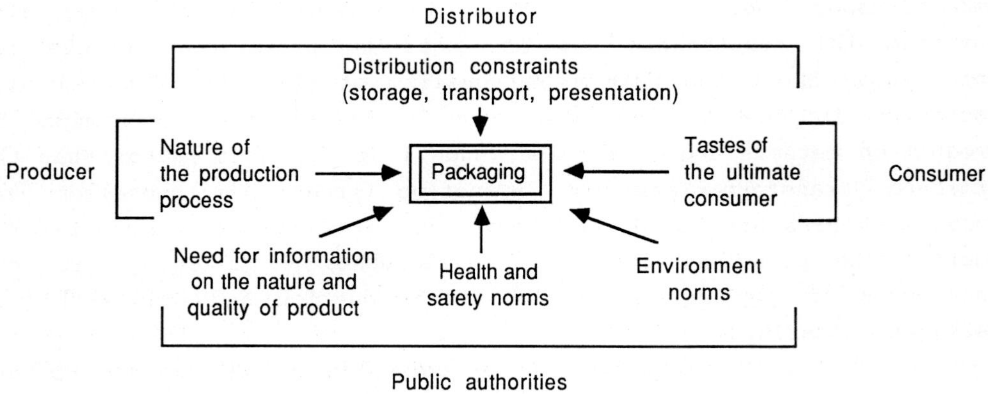

Figure III.2.a : The various constraints on packaging materials.

III.2.1. THE VARIOUS CONSTRAINTS ON THE PACKAGING SECTOR

III.2.1.1. The consumer

The packaging sector caters for two distinct types of customer whose requirements are fundamentally different :

- Firms, which are customers for industrial packaging either for intermediate products or for the distribution of bulk products (cement, flour, etc).

The packaging in question is relatively simple and consequently fairly standardized in relatively few different forms (wooden or metal containers, large-capacity sacks, transport crates, corrugated carboard boxes, etc). What firms mainly require of this type of packaging is that it shall be simple to use and suited to their own type of production process.

- The final consumer, who is a customer for commercial packaging designed for presentation and retail sale. Here there is a wide range of different packages using a great many materials and generally having a fairly high unit value in relation to the value of their contents. Within this category, packaging for foodstuffs represents between 60 and 65% of the total, packaging for non-food products with a relatively short shelf life (cigarettes, cosmetics, etc) between 15 and 20%, and packaging of consumer durables about 20%. In many cases, the commercial packaging is the main medium for differentiating a product from its competitors. Thus it is a function of marketing to define the end-user's requirements. The shape of the package, ease of opening, aesthetic aspects and handling comfort are some of the principal parameters involved. The importance of marketing in the packaging sector is illustrated by the part it has played in the success of the product concerned through the development of the Coca-Cola or Perrier bottle shape, that of the aluminium drinks can with a ring-pull for easy opening, or the shape of certain perfume bottles.

The distinction between industrial packaging and commercial packaging is economically important : on the one hand, sales of industrial packaging (whose main purpose is to protect the contents during transport and storage) faithfully reflect the economic pulse and react immediately to fluctuations in industrial activity; on the other hand, sales of commercial packaging are linked to final consumer demand and appear to be much less dependent on the general economic situation (some two-thirds of commercial packaging is taken up by foodstuffs which are fairly immune to the vagaries of the economy). That is why the commercial packaging sector has been less affected by the recession than some other industries.

Broadly speaking, the strong substained growth of the packaging industries over recent years is largely explained by the fact that packaging foods has become a general practice, now covering virtually all intermediate and end-products. The spectacular development of new forms of distribution (eg supermarkets) has been among the crucial factors in that growth. However, since the early 1980s the volume increase in packaging sector sales has reached a certain saturation level. There are a variety of reasons for this : practically no product is now sold unpackaged; there are even a number of attempts in the food sector to return to the retailing of loose items; lastly, there is a general trend towards lighter packaging materials - which we shall consider below - so that the total volume of products to be packaged is unlikely to increase (at least in Europe, through there are still strong growth prospects in the developing countries). Despite this saturation of the overall market, it should be borne in mind that changing lifestyles will boost the market substantially in certain particular applications (one example being the development of microwave ovens, as a result of which a quite specific type of material has to be chosen for the packaging of deep-frozen food).

III.2.1.2. Constraints imposed by the authorities

The public authorities have a considerable role in the development of the packaging market. It is precisely because packaging occupies a pivotal position in economic activity and reflects much of what is happening in society that industrial concerns in this sector have to take account of a great many norms and regulations of every kind. There have always been health and safety standards; since the mid-1970s, like other sectors, packaging has had to bow to the requirements of energy saving; and since then too, it has had to take the problems of environmental pollution increasingly into account; lastly, there has been increasing interest over a number of years in packaging serving as an effective medium providing information about the nature and quality of the product.

Health requirements : mean that packaging must be as safe as possible from penetration by micro-organisms (bacteria, moulds in the case of foodstuffs), outside attack (rodents, insects, etc), gases, radiation, chemical reactions (corrosion), interaction between package and content (any packaging can potentially shed substances into its contents : glass sheds chromates and silicates, crystal sheds lead salts, tinplate sheds tin, aluminium alumina salts, PVC vinyl chloride monomer, and so on).

The attitude of the authorities decisively affects the choice of packaging materials in this sphere. The United States Food and Drug Administration published a communiqué in 1975 which was interpreted as a pure and simple ban on PVC for food packaging. In fact, it was a proposal to forbid rigid PVC packaging for foodstuffs. American PVC does actually contain far more vinyl chloride monomer than bottles made in Europe, and as a result authorization has nowadays to be obtained in the United States in order to market semi-rigid bottles and films. The Food and Drug Administration was previously in the habit of authorizing everything that was not prohibited. It is now moving towards the European tendency, ie a positive list : only products which are explicitly authorized can be used in contact with food.

Safety requirements seem to be more lax in Europe than those obtaining in the United States. American legislation has shown particular concern for children by insisting, first in the case of pharmaceuticals and then in the case of household cleaning products, that containers must be difficult to open. An order was issued for medicines to be packaged in containers with instructions on how to break them open. Some recent malpractices which had a considerable impact on public opinion in America (aspirin capsules containing cyanide which caused several deaths) may perhaps account for the difference in stringency on this point as between American and European norms.

Energy saving requirements : the problems of "waste" regularly stir up debate as to whether permission to produce non-returnable containers should be withheld, and on a full-scale return to the deposit system (which applies almost exclusively to glass). Such a measure would have the advantage of reducing Europe's energy bill. In reality, the potential savings would not be immediate. It would take eight years to reduce energy consumption by 20%, and it is even probable that the energy bill would increase for the first few years. Among the other positive aspects would be the creation of unskilled jobs in bottling plants (potential job creation in France alone is estimated at 3,000). On the other side of the balance, such a measure would have the following negative effects : the liquids packaging industry would be hard hit (European production of glass containers would drop by about 50%; bottlers' investment capital would be tied up for 10 years in returnable bottles, crates, lorries and storage areas; the production of food-grade PVC would decline substantially); and there would be job losses in glassworks and in plastics processing (the number of jobs lost in the whole of France is estimated at 11,000, of which 5,500 would be those of skilled operatives). There would be other effects, although these would be hard to gauge : industrial companies with agriculture-related products would probably stand to gain, while those bottling close to the production site or whose packaging activities are concentrated at one works only would probably lose out. Lastly, though the effects are also hard to measure, it is not sure that a return to the deposit system could help to abate the pollution problems which we shall be discussing. Consequently, the position of the authorities in this area is extremely hard to define in the absence of in-depth studies. It does seem, however, that with the present relative easing of the energy constraints a return to the deposit system is not essential. Only if there were to be another acute and prolonged energy crisis would this measure again merit consideration.

Environment protection requirements : unlike the energy constraint, the <u>environment pollution</u> constraint is quite immeasurable.

In a sense, packaging can even be regarded as the most visible sign (though probably not the most serious one) of pollution. A measure of the scale of the problem posed by throw-away packaging is given by their relative weight in household refuse - 45% of the total weight, of which 5% is made up of plastic packaging, 20% paper and cardboard, 6% wood, 6% glass, 5% metal and 3% various other materials. There are two possible approaches to the recycling of packaging materials : either physical recycling or energy recycling.

<u>Physical recycling</u> is primarily a matter for consensus among consumers, public bodies, local authorities and industrial companies. It has been possible to make economic arrangements in some countries for the recycling of glass, paper and certain metals on the basis of such a consensus. But there are still huge differences between countries : why, one may ask, should it be possible to recycle aluminium economically in the United States or Sweden, while the same thing is proving extremely difficult in Europe where only an insignificant proportion of the 3.5 milliard aluminium cans produced is recycled ? Yet the United States example shows how beneficial it can be to plan metal recycling by getting everyone involved in the economic process to cooperate : everyone stands to gain (including, it appears,

the New York tramp who helps to keep his city clean by collecting up aluminium cans at 20 cents a time!). It should be noted that, unlike the deposit system for containers, the recycling of packaging materials does not cause the packaging companies to produce less. It may be reflected further up the line by a decrease in the production of primary metal. This is an aspect which might even take on strategic importance in Europe, if the production of primary metal (particularly aluminium) were also to take place outside Europe. In all the above cases, recycling has been organized right at the end of the consumer process with bins and other containers specific to this or that material. There is, however, another method of physical recycling based on the recovery of materials from among other waste. This approach is particularly attractive in the case of tinplate. Articles made of tinplate are incinerated together with household refuse and then separated as they leave the incinerator by the inexpensive method of magnetic sorting. This property of tinplate has been exploited in order to defend it against the inroads made by aluminium into drinks cans. Clearly, however, apart from this very special case the physical collection of other materials at this stage raises difficult problems because the materials are more and more mixed up together.

<u>Energy</u> recycling is actually one way of overcoming the limitations on physical recovery. All packaging materials in the form of waste constitute a significant source of energy which can be exploited either by incineration or by turning it into solid fuel or by chemical breakdown. Incineration, the commonest method, is likely to go on growing, mainly because of the increase in the proportion of plastics in waste. The energy produced by one tonne of plastic waste in household refuse is estimated at about 0.8 tonnes of heating oil. The net calorific value (NCV) of household refuse is thus tending to rise with the growth of plastic packaging waste (PVC has a calorific value of 5,000 Kcal/kg as against 4,000 Kcal/kg for wood). So it is becoming less and less necessary to add coal or oil in order to burn waste (the calorific value of the annual household waste produced by ten people living in Europe today can be estimated as sufficient to provide hot water and heating for one person for one year). However, the large quantities of hydrochloric acid given off when household refuse is burnt (mainly because of the PVC present) raise a great many objections from scientific and ecological circles.

Another way of tackling environmental problems is through biodegradability. Pressure from ecological organizations has even prompted such countries as Italy to make biodegradability a requirement for a certain number of packaging papers and plastics, with effect from 1991.

The EEC is already pursuing a very determined policy on all the problems relating to energy savings and pollution. In 1985 the EEC Environment Council gave its agreement to the substance of a Community directive on liquid food packaging. The measures proposed (to be reviewed every four years) are designed, among other things, to enhance consumer education, to foster the recycling or reuse of packaging, particularly throwaway packaging (by keeping the deposit system where economically justified), to develop selective collection and effective methods for the extraction of packaging from household waste, to increase wherever possible outlets for the materials extracted from packaging, and to promote the development and marketing of new types of packaging.

Packaging as information medium : packaging is being required increasingly to serve as a vehicle for information about product composition, "buy before" and "eat before" dates, product price (the use of bar codes is constantly spreading), instructions for use and hazards (medicines and cleaning products). The public authorities keep a particularly close watch on the standard of information given on packaging, though again with varying degrees of strictness. American regulations on the composition and hazards to the consumer of certain products appear to be stricter than in Europe at the present time. Moreover, labelling is one of the crucial ways in which one product is distinguished from another. All these requirements have repercussions on the choice of materials and their processing method (materials are selected for their ease of direct printing, integral colour, use of inks, etc. 30% of Europe's ink consumption now goes into packaging - and the inks must meet strict criteria in respect of resistance to light, aging, non-migration in the packaging, non-dilution in the product and so on).

III.2.1.3. Production and distribution constraints

Ever-greater process automation is placing severe constraints on packaging materials, which need to be simplified and adapted to the nature of the production process. In particular, it is becoming harder and harder on an automated production line to separate the product production stage from the packaging stage. This production stage integration is likely to have a significant impact on the strategy of indutrial firms and to shift some part of the decision-making to the contents manufacturer. This hypothesis is all the more likely as there is a marked tendency towards the simultaneous design of the product and its packaging. For example, in the agri-foodstuffs sector the use of perfectly hermetic packaging may render the use of preservatives unnecessary.

Transport and distribution requirements often coincide with those of production in prefering simple packaging : the development of big retail outlets has given a powerful boost to packet shapes which are simple to handle and to stack on shelves. The need to minimize wasted space in containers requires strong but lightweight packaging materials (able to shape themselves to the most complicated articles).

III.2.2. IMPLICATIONS FOR THE DEVELOPMENT OF PACKAGING MATERIALS

The constraints to which packaging materials are subject have been described above. The growing number of those constraints has ultimately brought about a redistribution of the patterns of materials used in packaging. Until the end of the 1970s, each major type of material clearly occupied a relatively well protected "niche" in the packaging market, as Table III.2.A shows.

For the last ten years or so, ground has been constantly won and then threatened again. Among the first major confrontations between materials were those which took place in drinks distribution (competition between glass, metal and plastic). They have tended to become widespread. Although the traditional materials have made great efforts to resist the trend, the most remarkable phenomenon observed over the last few years is the breakthrough by plastics and packaging "complexes" (See Figure III.2.b and Table III.2.B).

Material	Type of product	Area of use
Paper	Kraft - sacks	Cement - livestock feed
	Kraft - bleached	Bags
	Calendered paper Non-embossable paper	Food
	Technical and special papers	Food, deep-frozen products
	Ordinary thin papers	Food
Card	Folding boxes Cartons	Food
	Crates	Fruit, vegetables
Wood	Pallets, crates	Intermediate products
Tinplate	Preserve cans Drinks cans	Food products
	Aerosols	Household chemicals cosmetics
Aluminium	Trays Preserve cans Drinks cans	Food products
	Aerosols	Pharmaceuticals
Glass	Bottles Lightweight glass	Food products and pharmaceuticals
Plastics :		
PVC	Bottles	Food products
Polyethylene	Bags, shopping bags	Retail distribution
Polypropylene	Films, packets	Food, cigarettes
Polystyrene	Boxes	Packaging eggs, etc.
TPE	Trays, bottles	Food products

Table III.2.A : Materials and their uses for packaging.

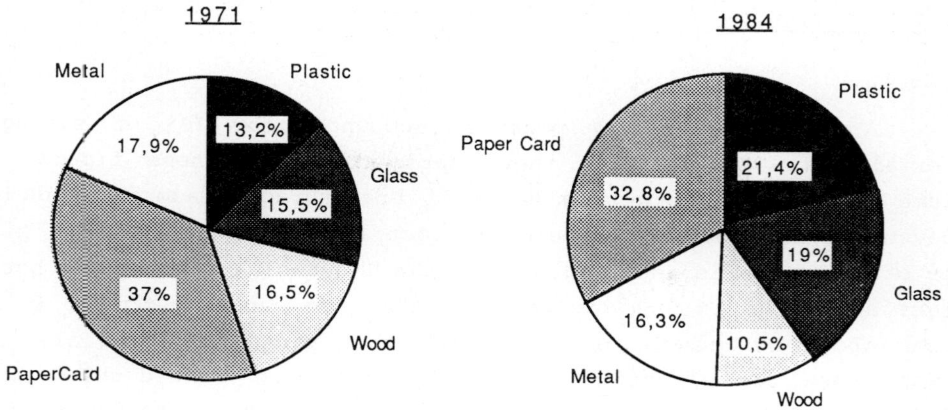

Source : DRAPT Study on weight reduction in industrial products, 1985.

Figure III.2.b : Tonnage distribution of materials used in packaging

Country	Tinplate	Aluminium	Glass	Paper as such, bags and sacks	Corrugated card	Flat card	Plastics	Cellulose Film
Germany	630	104	2900	600	2100	500	1200	16
France	600	52	2700	530	1600	390	800	14
U.K.	800	43	2000	440	1450	500	750	22
Italy	500	70	1800	300	1300	400	600	12
Spain	290	24	1000	200	1000	230	330	8
Belgium	270	10	400	150	300		215	4
NL	300	7	600	200	350	160	210	6
Sweden	100	7	100	120	200	100	200	2
Norway	60	5	60	60	200	40	100	0.5
Denmark	100	10	120	125		80	90	2
Finland		7	50	50	150	75	100	1
CH		11	200	100	300	80	120	4
Austria		4	250	100	200	70	100	1
Greece			70	50	150	60	60	1.5
Ireland			90			50	40	2
Portugal			250		150	60	70	2
Total (rounded off)	3700	370	13000	3000	10000	2800	5000	18
U.S.A.	6000	1300	15000	4000	15000	3500	4500	45

Table III.2.B : Estimated consumption of packaging in Western Europe and the United States in 1984 (thousands of tonnes).

III.2.2.1. Plastics

In volume terms, plastics accounted for 13% of packaging materials in 1971. In 1985 they represented 21%. These figures for volume actually fail to show the real breakthrough made by plastics, because tonnage does not exactly reflect the extent to which a given type of packaging is used. For example, a lightweight 50cl glass bottle weighs about 200 to 300 grams, while the same capacity bottle in PVC weighs about 20 grams. Compared with other packaging materials, the growth index for plastics in packaging has more than doubled since 1970, while during the same period the index for glass has risen by a factor of 1.8, the index for corrugated cardboard by a factor of 1.4 and that of aluminium by a factor of 1.3, while that for packaging paper has declined slightly and the index for cellulose film has fallen back sharply. The breakthrough achieved by plastics is accounted for by the following factors :

- They meet the majority of the requirements described in the first section

* Their resistance to deterioration (resistance to outside attack), impermeability (to gases and water), corrosion resistance and natural asepsis (when processed at temperatures above 150°C, plastics become aseptic packaging materials) offer guarantees of health and hygiene.

* Plastics have often been accused of costing too much energy and creating serious pollution problems. But a look at the overall energy budget casts doubt on those arguments. If, for a given material, account is taken of the cost of winning or importing the necessary raw material for its manufacture, production and processing as well as the ultimate recycling of the material where possible, then the total cost in the early 1980s for one litre of packaged liquid was about 25 centimes for PVC, 35 centimes for returnable glass and 50 centimes for non-returnable glass or metal can. Concerning the problems of environmental pollution, we have seen that plastics help to raise the calorific value of household refuse but also its chlorine content. Insofar as the multiplicity of materials in household waste makes it less and less easy at the present time to recover various materials (with the possible exception of magnetic materials), incineration will

increasingly establish itself as the best solution, and as a result society will doubtless show greater tolerance towards plastics.

The only eventuality which might curb the development of plastics here would be the emergence of a new oil crisis, as a backlash, the issuing of more stringent regulations on returnable containers, so as to save energy, and on the collection and recycling of used packaging materials.

* Because they are light (making transport quicker and cheaper and storage easier), easy to handle and impact-resistant, plastics are well suited to the requirements of retail distribution.

* Plastics lend themselves well to bulk colouring (obviating a number of production steps), printing and marking. Here, however, a contradiction is apparent between consumer and producer requirements. It seems that consumers prefer products to be presented with a label rather than with information printed direct on to the material (which is advantageous in terms of production costs).

- <u>They have the advantage of important innovations in processing and presentation</u>

There are several recent packaging innovations which give plastics a market advantage.

* <u>Controlled atmosphere packaging</u> consists in incorporating mixtures of gases (usually CO_2) which retard bacterial development. This innovation has the advantage of extending the shelf-life of products (and therefore benefits both the retailer and the producer since, at a time when the emphasis is on production management, lengthening shelf-life enables the producer to respond better to demand). In the case of fresh meat, for example, the usual three-day freshness limit has been extended to seven or even 10 in some cases. The appearance on the market of so-called high gas barrier plastics has proved to be a decisive factor in the development of controlled atmosphere packaging.

* Aseptic packaging (a typically European process) involves sterilizing the packaging materials by means of hydrogen peroxide (oxygenated water) applied in the form of a wash or spray. As in the preceding case, processing techniques depend on the use of high barrier

plastics (especially EVON resins, developed in Japan ten or so years ago).

* <u>For the manufacture of rigid hollow containers</u> (bottles), new blow-moulding methods have been developed (especially in Japan) whereby genuine annealing is possible (as for steel), giving the packaging good high-temperature resistance.

All these examples, which underscore the potential for the development of engineering plastics, also show the extent to which packaging has become a real "function" to be performed, requiring simultaneous thought to be given to product, processes, packaging and manufacturing technologies.

The plastic material which has shown strongest growth in recent years (about 10% each year) is <u>thermoformed polyester</u> (TPE). Having excellent physical and chemical properties (high strength, stability, permanence even at high temperatures) and taking print readily, it has taken over for the packaging of a great many food products (meat, cheese, sweets, coffee and others besides) where it has ousted other rival plastics (polyethylene and polypropylene) and in hot-stamp films, where it has replaced cellulose film. It is now on the point of achieving a major breakthrough in hollow containers (especially trays for microwave ovens where aluminium has the drawback of reflecting radiation, and large-capacity TPE bottles for fizzy drinks where the market growth rate is estimated at 20% in the United States and 20 to 30% in Europe).

The European packaging materials situation is fairly satisfactory : of all packaging materials, plastics are the only ones in which European consumption (about five million tonnes) is slightly higher than in the United States. In Japan, the consumption of plastics for packaging is only about 1.6 million tonnes. Because of the importance of the European chemical industry, Europe is generally represented in the most advanced plastics technologies. For instance, the Belgian firm SOLVAY is one of the four leaders in high barrier plastics (with DU PONT DE NEMOURS in America, and KUMARAY and NIPPON GOSHEI in Japan).

III.2.2.2. Packaging "complexes"

Packaging "complexes" (combining several different materials) are another way of responding to the different specific requirements of the packaging sector. The normal principle, for a given shape, is either to combine several parts made of different materials or to superimpose several layers of different materials to obtain a final material which incorporates the different functions of the starting materials. The "clip can", for example, is a complex made of card, metal and plastic. Having a plastic ring bonded to a (metal) rip top, it offers security against opening while in the distribution circuit but is easy for the consumer to open and close. Its main feature is modularity : as it comes in a wide range of cross-sections (20 in all) between 40 and 165 mm, simply varying the height yields a maximum of possible combinations. It is an interesting solution for those having to pack standard weights of products which vary in density, such as powders, pastes and so on.

One of the most remarkable examples of a multilayer complex is diazet, a paper coated with polyvinylidene chloride which, in addition to paper's traditional properties, possesses those of plastic and is totally impermeable to gases and fats. It is used in agri-foodstuffs and can withstand a temperature range from -20°C to 90°C, so that it is suitable both for deep freezing and for packaging products intended for hot countries. In the United States it is even used to pack fruit juices because it resists acids, alcohol and micro-organisms.

There are a great many examples of complexes that could be quoted. One feature of them is that in the majority of cases they involve a combination with plastics (though of course this is not invariably the case : there are for example paper-metal complexes, and the development of the famous tetrapack container for milk is indeed one of the very rare cases where a new material has ousted plastics). Moving to a complex often affords an opportunity for a manufacturer of paper, card or glass packaging to discover plastics. For this reason, one of the main determinants in the future of packaging will be the possibility of making complexes which guarantee a technically superior and economically more satisfactory answer than any other plastic solution. The goal here is to develop processes which combine several materials in a competitive way (while keeping bonding and assembly costs to the minimum).

III.2.2.3. Other packaging materials

Other packaging materials are not of course passively submitting to the inroads made by plastics and complexes. Remarkable process has been achieved (thanks to CAD) in the manufacture of glass, where weight reduction has been possible by optimizing thicknesses. Wooden packaging has also managed to benefit from technical advances in the areas of throughput, packaging design and finished products. But the most spectacular reaction in recent years has come from tinplate which, in Europe, has managed to keep and even win back shares of the drinks can market coveted by aluminium. This perfect illustration of cutthroat competition between materials in packaging and the reaction of the traditional materials is examined in greater detail below.

III.2.2.4. An example of competition between packaging materials : the case of the drinks can

The fiercely competitive war being waged in the drinks can market (33 cl) between tinplate and aluminium, with plastics now insinuating themselves into the picture, exemplifies the whole complex mechanism governing developments in the choice of materials. Every determinant factor in that mechanism is present here : consumer taste, differences of lifestyle between one region and another, energy and environmental constraints, progress in packaging technologies, pressure from lobbies, and the importance of industrial structures. (It should be emphasized that the metal drinks can market is only one section of the packaged drinks business. In the remainder of the market (bottles) the competition is played out between glass and plastics).

Tinplate dominated the whole of the drinks can market in the early 1970s. It was at that time that the American aluminium producers led by REYNOLDS, foreseeing the growth of an important market, launched an unprecedented offensive. That offensive was a multi-pronged attack :

- marketing policy was employed to demonstrate the properties of aluminium - hygienic, hermetic, stainless, pleasant appearance (attractive to a young clientele), light, etc.
- development in 1970 of an easy opening system.

- development between 1972 and 1973 of a process for the production of a two-part aluminium can.This was a crucial advance in terms of technical integration, because at that time the corresponding process for the tin can was a three-part process (so that aluminium afforded important production management savings).
- establishment of an aluminium recycling circuit in the United States. Supermarkets have can disposal machines and the disciplined consumer receives a bonus (10 to 20 cents) per can returned. In 1985, a firm like ALCOA Recycling had 2,500 recycling points in the United States and recycled 250,000 tonnes of aluminium.

The campaign was a great success. The aluminium offensive secured a foothold on the basis of a range of criteria as against tinplate. The growth during that period of large retail outlets, for which the metal can was a practical drinks container, boosted the market strongly. In the United States, aluminium rose from 10% of the can market in 1970 to almost 95% in 1985 (in a market where average annual growth has been 10% since 1970 and which represented about 60 milliards cans in the United States in 1985) (see Figure III.2.c).

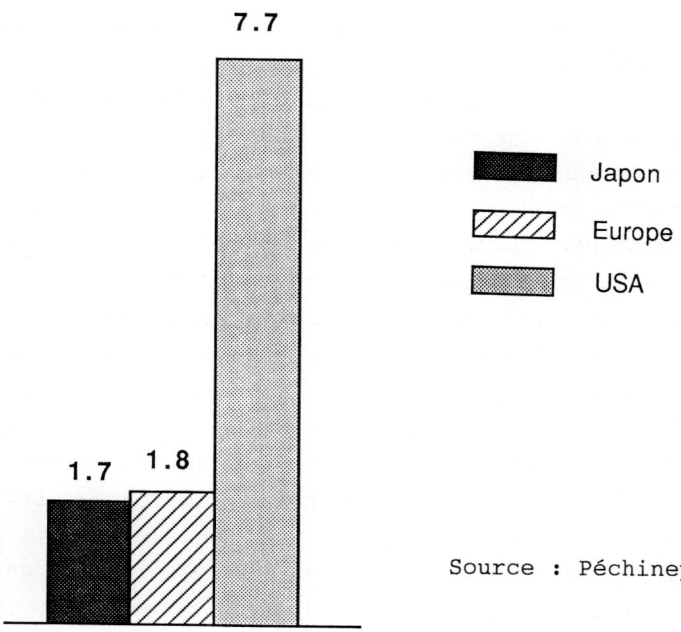

Source : Péchiney Euroconsult.

Figure III.2c : Total aluminium consumption (kg/capita) in packaging in 1984.

Tinplate seemed doomed. But it was destined to benefit from a very special situation in Europe which enabled it to launch a counter-offensive in the late 1970s, based on the following arguments:

- In the early 1970s, when aluminium was penetrating the can market in the United States, drinks in Europe were being sold mainly in glass bottles. The drinks can was extremely slow to take off.
- The crisis in the European iron and steel industry which began in the mid-1970s led to the mobilization of the steel lobby, which perceived the threat posed by the aluminium can. The producers negotiated in particular with the big contents manufacturers (fizzy drinks, beer and so on). Their strategy was assisted by the fact that Europe's tinplate producers are virtually organized on a cartel basis, unlike their rivals in aluminium. In some cases they did not hesitate to suggest that an aluminium can production line should be changed back (at their own expense) to a tinplate can production line (this type of changeover takes about three months to complete). At the present time in Europe there are 36 modern production lines making cans from tinplate as against 19 in the case of aluminium.
- The steel producers developed a two-part tinplate can copying the aluminium process as well as the easy opening top.
- Tinplate was very substantially lightened to bring it close to aluminium in performance. In the space of ten years the weight of a can has been brought down by 48%.
- The fact that tinplate can be recycled by magnetic recovery after waste incineration was emphasized.

This property was even made the theme of an advertising slogan ("you can recycle tinplate") on certain European drinks cans. European tinplate manufacturers claim that they are able to recover about 20% of their production through the incinerator circuit.

This counter-offensive has yielded spectacular results : tinplate cans now hold 70% of sales in a European market estimated at 12 milliard cans (and growing by 6 - 8% a year). Aluminium holds the largest part of the remaining 30% (plastics already having a small market share). There are very large disparities between European countries. Sweden, for example, has been prompted by the official regulations in force to deliberately choose aluminium, which now accounts for practically 100% of the drinks can market (80% of that aluminium is recycled). Conversely, in countries like Germany or France the market share of aluminium cans is no more than 10% (and there is practically no organized system for recovery of cans at the present time, despite the efforts made by certain aluminium producers). It is true to say that in Europe today, aluminium's advance in drinks cans is being strongly curbed by the growth of tinplate cans. This "old" material (dating from 1811) has managed to regain its competitiveness relative to a newer material by imitating it, by playing an additional trump card to set itself apart (magnetic sorting for recycling) and by taking advantage of a more favourable industrial environment.

However, the outcome of this rivalry between the two metals, aluminium and tinplate, is complicated by the advent of plastics. Plastics have virtually taken over the market niche for fizzy liquids in containers larger than 33 cl (especially 1.5 litre TPE bottles). But they are now invading the actual drinks can market where metals have hitherto had a monopoly. The Thermoforming Company has recently developed a process for the manufacture of thermoformed, bi-oriented TPE drinks containers, presenting the same exterior appearance as metal cans but with the following additional advantages :

- the investment necessary is far less than that for stamping metal sheet, is easily adaptable to the production of large-capacity containers of any shape, and permits totally integrated production line set-ups.
- the containers are lighter and stronger, immune to corrosion and inert in the presence of most chemical agents.
- there is great scope for decoration (an argument used particularly by the Coca-Cola company which is considering changing over the plastics for its drinks containers).

The bright development prospects for plastics in this sector doubtless explain the recent association between the British can maker METAL BOX and the American producer ALCOA, in the form of a 50/50 partnership to produce plastic containers. Paradoxically, it seems that just when the battle between aluminium and tinplate is reaching its height, it would be in the most immediate interests of these two metals to form a common front against the penetration of plastics or (as in the case of tinplate) to use them as lining materials.

III.2.3. CONCLUSION

The European packaging industry is in the throes of profound change. Traditionally, packaging was part of the industry deriving from each of the main basic materials (steel, glass, wood, aluminium, etc) and few firms producing packaging employed more than one material. Today, two main trends are emerging :

- packaging companies are diversifying into the processing of several materials. This diversification (for essentially economic reasons) most often ends with the incorporation into the company of a plastics processing line. But the move towards multinational groups is being made cautiously : producing and processing companies are still, at the present time, applying their special know-how to one basic material.
- for technical reasons, "two or even three basic materials come together in packaging more and more often. This is particularly common in flexible or complex packaging, where a growing number of big processors are abandoning their speciality and, instead of working principally on this or that material, engaging in a general way in processing, printing, soft package complexing, combining paper, card, plastic film, aluminium foil, cellulose film, coating and coextrusion" (48) .

Thus the packaging industry is becoming less and less a branch of the major basic materials industry and moving towards an industrial activity performing the <u>function</u> of product packaging by integrating all the constraints described above. In the course of this development, it is likely that increasing attention will be paid to the <u>simultaneous</u> design of content and container. It is a development necessitated in particular by the constraint imposed by new forms of automation (flexible and integrated processes, industrial quality requirements and so forth). That is why much closer cooperation with the contents manufacturers is probable and even necessary. In this respect, Europe reveals marked differences from the United States, where it is the contents manufacturer who is the packaging innovation leader : in Europe, the power in this area lies more with the packaging manufacturers. It may be that recent trends will bring Europe closer to the American situation. At all events, those trends ought to encourage the constitution of big European packaging groups. Concentration is likely in this sector. One of its advantages might be to enable a larger number of materials research centres to be set up and perhaps to make this sector (which is and will remain one of the

---

(48) "Les industries d'emballages en Europe de l'Ouest", ("Packaging industries in Western Europe"), Pardos, 1984.

biggest users of materials) an essential link in the spread of materials innovations.

Another important implication for this sector - the simplification of packaging - derives from the constraints of automated production lines for products, including their packaging. There are other imperatives pushing in the same direction, namely the need for rapid distribution, limited storage and ease of transport. However, while simplification is to be expected in packaging shapes, the functions performed by packaging (impermeability, leak-proofness, safety, etc) will become more and more complex so that the largest possible number of constraints inherent in this field can be incorporated. It is at this level that the most important technological challenge for this sector arises (See Table III.2.C).

There are two possible alternatives :

- either packaging complexes will succeed in meeting economically the two-fold constraint of simpler shapes and more complex functions, in which case every material still has a potential future in packaging, with groups originally relying on one dominant material (glass, tinplate, paper, etc) even maintaining their specificity by offering a number of complexes based on their original specific product;
- or plastics will triumph and leave only a few isolated niches for other materials.

The answer to all these questions is assuredly not purely technological. It will depend in large measure on the public authorities which, by setting norms in respect of safety, environment, health, package standardization, energy savings and so on, can cause the big markets to shift very fast from one material to another. Whatever innovations the future may hold, this sector has significant implications for employment in Europe. Packaging is a sector which employs a large number of people in Europe and is relatively immune to job shifts as compared with other regions in the world. While technology and patents can be exported, while it is always possible for foreign firms to set up business in Europe (and the major world manufacturers have already done so), packaging is difficult to export and consequently the bulk of production takes place locally. Assuming that the packaging sector will continue overall to grow at the same pace as the rest of the European economy, it should continue to be an important area of job creation, though skills will have to change considerably in keeping with the main trends analysed above.

It is of course not possible to describe in detail all the individual developments which are at present discernible, but a number of examples are given below.

Films and flexible packaging
- further developments of shrink and stretch films, and take-off of linear low-density polyethylene film.
- appearance and development of recent processes such as coextrusion and of barrier materials, ionomers, ethylene-acrylic acid copolymers to make packaging offering a lasting barrier function.
- continued development of directional PP film, still to the detriment of cellulose film.
- substantial future research and development in the packaging of fresh meat, both wholesale and retail, with vacuum or controlled atmosphere packaging using barrier films.
- development of metallized polyester and probably PP films, though not necessarily to the exclusive detriment of complexes based on aluminium foil.

Pots and trays
- consolidation and development of integrated packaging systems for fresh dairy products, yoghurts and desserts, and possible extension to other products such as fruit juices and ready cooked dishes. In the medium term, strong possibilities of partial replacement of tinplate preserve cans by barrier materials using new types of film, designed for forming-filling and sealing of aseptic packages in a single operation.
- development of thin-wall injection-moulded PP pots, mainly to the detriment of PS pots.
- development of freezer-to-cooker cartons treated in various ways, in parallel with the medium-term boom in microwave cooking.

Bottles and hollow containers
- development of coated card "bricks" for packaging food liquids other than milk (fruit juices, wine).
- development of directional PVC containers.
- development of thermoplastic polyester (TPE) containers (250,000 tonnes in the United States already; 35,000 tonnes in Europe, most of it in the United Kingdom, and likely to increase to 150,000 tonnes by 1988).
- improvement of barrier properties through the development of new coating, complexing and coextrusion techniques for bottles and hollow containers.
- virtual ceiling on all other hollow containers, including glass.

Transport and industrial packaging
- development of HDPE pails of every kind, particularly as containers for paints and glues.
- development of large flexible containers made of plastic flims, particularly for heavy products and powders to be transported by sea.
- general spread of containerization.

Source : "Les industries d'emballage en Europe de l'Ouest", Pardos Marketing, April 1984.

Table III.2.C : Examples of foreseeable short and medium term developments in packaging materials

CHAPTER III.3

New Materials in Building and Public Works

There are two ways of apprehending the building and public works sector :

- The first is to consider that it undeniably remains a traditional sector. There are many grounds for this attitude, including dispension of firms, occupational compartmentation and diversity of products and application contexts. This does not mean that the sector is non-innovative, but it does appear to borrow most of its innovations from other sectors since the stimulus to innovate stems from demand rather than from any internal dynamism. Consequently, the changes induced by innovations often seem slow and uneventful, particularly as regards innovations concerning new materials. The introduction of new materials is apparently not governed by any specific mechanism and therefore cannot be analyzed as an established set of problems, in contrast to the situation in the transport sector. The upsurge of new materials in building and public works thus reflects the acceleration or intensification of developments specific to the sector and unrelated to these materials.

- The second is paradoxically to regard it as a modern sector, in view of the materials which it applies and its production structures. True enough, the concept of building in itself typifies the idea of a composite object incorporating extremely varied materials from time immemorial; the sector is a decisive proving ground for materials or methods incorporating advanced materials; the structure of the many different firms involved illustrates more vividly than any other the concept of a flexible network of highly decentralized firms; the sector constantly and characteristically endeavours to achieve adaptation of skills and compatibility between the various trades;

lastly, it is a field in which quality control and codification of operative capacity have always been valued.

These two possible approaches to building and public works highlight the need for caution in forecasts concerning the sector. The very idea that it is a homogeneous sector encompassing "building" activities (where the need to solve "personalized" problems often predominates) and "public works" activities (where concentration on scale-oriented efficiency predominates in many applications) may be subject to reservations. With due regard to this need for caution, the evolution of the sector nevertheless reveals certain dominant underlying trends.

- Economic recession and above all population downturn in Europe have caused an appreciable fall in building demand. To a certain extent the combined effect of these two factors in the mid 70s significantly helped to alter the sector's technological trajectories. The innovative solutions devised in the late 60s in this sector (eg the design of "sealed" prefabricated panels incorporating internal "technical ducting" for electrical circuits, telephone wiring, etc) required an expanding market to become viable. The subsequent collapse drew attention away from these innovations towards another approach geared to fulfilling two main requirements, energy saving and renovation of old buildings (cf. section III.3.1. "New materials and adaptation to demand"). The prospect of considerable "extensive" development thus yielded without warning to a need for redesigning "from the inside".
- Cyclical economic factors greatly accentuated this trend. Rising interest rates to begin with dampened incentive to build new housing, while tax incentives for energy saving were conducive to investment in renovation. In this context, the materials applied have generally speeded up what were already long-established trends in technologies; enhancing the light weight and variety of materials (cf. section III.3.2. "New materials and building sector technologies"). Managing the potential of new materials naturally causes firms to alter their strategies (cf. section III.3.3. "New materials and strategies").
- Building, and public works in particular, are moreover activities which are closely dependent on official intervention. Government intervention, which is obvious in the case of public works, is also decisive in building as such, whether it involves decisions to build major housing estates, the introduction of legislation and issue of building permits, laying down norms of materials, the determination of guarantee periods, taxation policy or the fixing of preferential

rates. This marked official element in the development of the sector has come to support the idea that building and public works activity is one of the key features of government economic policy, which is perhaps why the authorities pay very special attention to unemployment problems in this sector. In a difficult transitional period, questions of employment and skills take on a crucial significance, especially as the building and public works sector traditionally relies on a large and comparatively unskilled workforce (cf. section III.3.4. "New materials, skills and employment).

III.3.1. NEW MATERIALS AND ADAPTATION TO DEMAND

In general, notwithstanding major efforts to modernize, the sector continues to be widely regarded as a traditional one. Not that it is unprogressive; the innovations which it achieves, rather than being the outcome of the sector's own dynamics, are responses elicited by demand. In this respect, the underlying tendency for some years has been towards a definite rise in the quality required. For building sector industries whose profitability and competitiveness have been severely affected by recession, cost considerations nevertheless remain decisive. The crucial issue is therefore ability to provide higher quality products and services at equal or lower cost. That is the reason for using new materials having the advantage of offering gains either in terms of lower utilization costs or in terms of higher utilization value of products.

III.3.1.1. The fall in utilization costs

The fall in utilization costs varies as between building (particularly housing) and the construction of roads, engineering works and public amenities. In the former, the importance of norms and energy saving programmes is conspicuous. In the latter, the fall in utilization cost applies to the maintenance and life of roads and engineering works.

Energy savings in building

In the countries of Northern Europe, concern for energy saving in buildings is by no means recent. In Sweden, for instance, 96.5% of houses are made of wood, a material possessing good heat insulation properties (49) . In the other European countries, the increasing burden of heating costs, even before the energy crisis, alerted the authorities, the building trade and tenants' associations to the need for technical energy-savings measures. The 1974 energy crisis finally caused a general mobilization in Europe in the interests of cutting the energy bill. The authorities were aiming at a reduction in the trade deficit while the building sector, confronted with the drop in demand, needed to restore competitiveness by proposing solutions which would generate significant savings in respect of charges. As certain energy costs for buildings (lighting, cooking, hot water, etc) are difficult to curtail, heating energy was singled out as a special target (50) .

The solutions proposed initially concentrated on improving available interior insulation materials and techniques (51) . In the case of walls, for instance, the insulant is applied to the inside surface. Pitched roof insulation is laid in the roof space and flat roof insulation must be thicker. Such methods nevertheless encounter technical and economic problems. Cold bridges in floors cause major heat losses which cannot be eliminated. Furthermore, very thick insulations causes distortion of the material or crazing of walls in case of temperature fluctuations or thermal shocks. Very thick walls also waste living space and thus raise the price of buildings per

---

(49) P. CHEVILLEY "Le pari des 100 000 maisons en bois" ("Gambling on the 100,000 wooden houses"), Les Echos, N°14557, 24 January 1986.

(50) J.P. CAUSSE and P.F. GOBIN : "Les matériaux", report of the materials studies group commissioned by the Ministry of Research and Industry, Paris, Science and Technology Section, Atomic Energy Commission, 1982.

(51) The whole of this passage on insulation techniques and materials was directly inspired by P. CHEMILLIER "L'évolution des techniques de construction" ("Development of contruction techniques") in "Les évolutions techniques dans le bâtiment : bilan et perspectives" ("Technical developments in the building sector : achievements and prospects"), 7-49, Paris, Centre Scientifique et Technique du Bâtiment (CSTB), 1982.

square metre. Lastly, once a certain thickness is exceeded, added insulant has no further effect and generates no further heat savings since the proportion of leaks accounted for by cold bridges becomes decisive (Figure III.3.a). Concurrently with these problems, a number of new products are appearing. Their speciality is to serve several combined functions in addition to insulation. This is true of concrete blocks with built-in insulation, autoclaved cellular concrete, light aggregate concretes and wall panels made of materials with good insulation properties and incorporating load-bearing structural uprights. The role of the new materials making up these products is to eliminate cold bridges through <u>external</u> insulation, while at the same holding up the structure and protecting it against thermal shocks and water seepage. This allows structural performance specifications to be

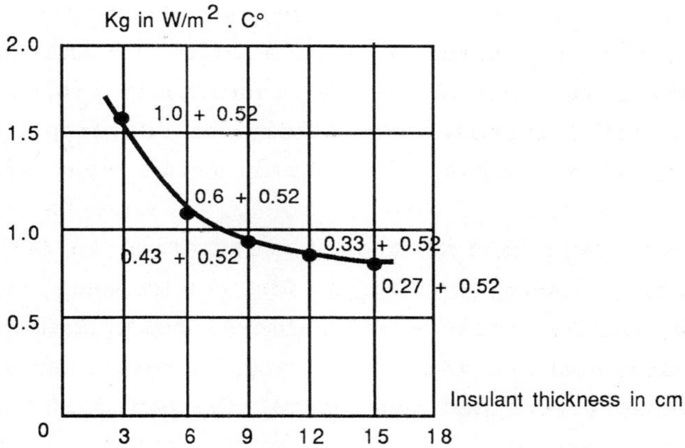

This CSTB sketch shows the decisive influence of cold bridges once a certain thickness of insulant is exceeded. At five points on the graph the weight ratios of the continuous section of the partition (first figure) and the cold bridges (second figure) give the global weight coefficient for a standard aggregate concrete wall 4 m wide by 2.50 m high and including a window space of 2.45m². The inside lining is a glued-on insulating complex consisting of expanded polystyrene. Note the asymptotic shape of the curve.

<u>Source</u> : CSTB, quoted by P. CHEMILLIER, op. cit.

<u>Figure III.3.a</u> : Effectiveness of insulation.

adjusted downwards (lesser thicknesses, less elaborate joints), which results in many types of savings. These advantages, to be incorporated as from the design stage, have made it possible to develop a technique which involves reducing the load-bearing structure to a bare minimum and covering or encasing it with a very good insulant. In this way, a walled structure is replaced by a framework of uprights or beams. Wooden frame houses (whose framework itself has insulating properties) and houses with light concrete, steel or aluminium frames come under this technique.

As well as by external insulation, heat loss can be reduced at the transparent surfaces which are required to serve the contradictory dual function of letting in sunlight and keeping back the infrared rays emitted by the ambient space, ie the larger the transparent surface the greater the heat loss. The use of polymer-derived materials such as polycarbonate (PC) and polymethyl methacrylate (PMMA) offers a partial solution to the dilemma thanks to improved insulation of glazed surfaces or production of transparent insulating panels. As they are usually translucent rather than transparent, they are intended mainly for industrial and agricultural applications. In the housing sector, types of glazing with still better insulating properties than the conventional double glazing are likely to develop thanks to the selective use of glass with a low-emission metal layer (52) .

Finally, the reduction of energy consumption in buildings depends on better control of heat generation. The use of heat recovery devices and heat pumps, dual or multi-fuel boilers, solar energy and improved boiler efficiency are all ways of saving energy. Insulation and heat generation have long remained two separately developed aspects; housing is not yet apprehended as a total system of production and consumption of energy flows. The more and more obvious need to deal with these two aspects of the energy problem concurrently is reflected in the emergence of a new specialist trade, heat engineering. Designing a building as a regulator of interior volumes (especially thermal ones) in a given physical environment accentuates the role of new materials (thermomechanical ceramics in particular), often the only ones adequate to increasingly demanding requirements.

---

(52) P. CHEMILLIER, p. 23, op. cit.

The role of norms and legislation

Norms have a considerable role in construction, understandably so in a sector where no mistake are permissible. However, an unduly stringent set of norms can of course strongly discourage innovation, particularly in the use of new materials. For instance, the 30 years guarantee which used to be applicable in France was undeniably a great obstacle to the testing of new ways with new materials (even more so because knowledge of the ageing of certain new materials is generally unavailable). French legislation has since evolved towards the recognition of an ordinary guarantee for building materials. While helping to lift a barrier to the spread of innovation, such a decision tends to place the onus of litigation more on the materials manufacturer, who is therefore likely to be more involved in the search for approaches which are economically viable and meet quality specifications.

The establishment of Europe-wide service norms and compatibility rules (particularly for sizes) would be a desirable undertaking in ensuring market permeability, and at the same time a most difficult one as norms now often go out of date faster than they can be defined. It would therefore appear that for the next decade the "turnkey" market, by nature a "closed" one, has the best development prospects.

Maintenance costs of civil engineering works

The reasoning behind the use of new materials in civil engineering is clarified by the resultant decrease in maintenance costs. In many cases there is also a resultant reduction in the costs to society of the use of public structures.

Considering the lifetime of structures like bridges, roads, motorways and mains, a good knowledge and command of the microstructure of the materials employed are needed to obviate excess maintenance costs chargeable to the community. Public works specialists must therefore pay very close attention to the evolutive characteristics of materials in the light of the applications. This factor is especially important in that the building and public works sector does not have a well-established tradition of using new materials. Lower manufacturing cost therefore does not necessarily mean that a material costs less over its whole lifespan than a more expensive one. New materials must be tested in situ even after the indispensable industrial mastery of their microscopic complexity has been attained. The new materials, belonging to several families, have made their appearance in public works and are applied equally in construction of new works and in maintenance proper. There are general signs of a vigourous advance by plastic, elastomeric and composite materials alongside the development of new steels and concretes (dispersoid steel for bridges and frameworks, fibre-reinforced cement for pipes).

The production of expanded polystyrene filling material saves weight on landfill operations by a factor of 100 while averting subsidence of bridge approaches in marshy areas. The use of a steel fibre reinforced elastomer to make engineering rollers in the Federal Republic of Germany has considerably improved their strength, heat resistance and chemical corrosion resistance as compared with steel ones. Whereas these required regular maintenance and cleaning, use of the elastomer completely eliminates maintenance costs. This method is eminently suited to harsh environments and may also spread in building in earthquake-prone areas.

At a time of credit squeeze, local authorities endeavour to delay road resurfacing for as long as possible. When climatic conditions are unsually difficult, the accumulation of water under poorly drained roadways precludes any postponement. An intermediate elastomer layer between the ground and a bed of granular material significantly improves drainage. It is also feasible to carry out maintenance on roads and motorways by simple, quick and economical methods whereby cracks are covered by a continuous elastic membrane made up of a bitumen-rubber binder and a gravel layer.

However, the future of roadbuilding seems to belong to "road carpet". It is a skidproof, waterlight, very tough and easily applied prefabricated material which is soon ready for use after application. Is is made of a PVC film, coal oil, gravel and a plastifier. It would lead to a marked improvement in road transport safety (better grip, better braking, elimination of aquaplaning). This method already has numerous applications in factory floors, cattle trucks and lorries, swimming pool surrounds, airport runways and boat decks. The material is light (7 kg per m^2 as against 250 kg for conventional materials) and 20% cheaper. Conventional materials of the bitumen, tar or macadam type are furthermore highly pollutant and dangerous to man. Factory production of "road carpet", thanks to pollution control and the enclosure of any pollutants in the plastic bonding agent, points to a reduction in the ill-effects of road building and maintenance.

III.3.1.2. Increase in utilization value

The new materials are capable of providing varied responses to such demands as comfort, safety, convenience and more generally quality of life. Here are a few examples of these capabilities.

- In detached and group dwellings, the new materials make for greater diversity of shape and appearance in buildings. Architectural concrete is an example. It yields a variety of façades which can make any building a work of art. Polymer concrete is another instance of a material whose properties are more suited to decorative than to structural functions (53) . Made of a thermosetting resin and mineral additives of different textures and origins (silicate, carbonate, sand, etc), it can assume various shapes and imitate materials such as marble, stone or cement concrete. Polymer concrete also competes favourably with cement concrete in non-structural applications. Although it costs more, it generates savings at the various production and application stages (Table III.3.A).
- Another advantage of using new materials from the utilization value angle is the increase in usability of interior space in buildings. A light framework in conjunction with movable panels allows users to modify the indoor space virtually at will, not only in dwellings but also in industrial and administrative buildings. Thus the Renault centre at Swindon, England has a modular design catering for irregular ground level and future extensions (Figure III.3b). This design relies

---

(53) Polymer concrete has poor creep properties at temperatures over 30°C and its price is 5 to 25 times higher than for standard (cement) concrete owing to the high and variable price of the synthetic resins employed. On the other hand, it has good tensile and flexural strength and weathers well (waterproofness, resistance to freezing and thawing, etc). Cf J.F. FERRIOT, "Les bétons de résines" ("Polymer concretes"), Document series n° 3, Centre d'Etudes et de Recherches de l'Industrie du Béton Manufacturé (CERIB), 1983 and CRIQ "Polymer Concrete", Quebec Industrial Research Centre (CRIQ), Ministry of Industry, Commerce and Tourism, Directorate General of Industry, 1984.

in particular on spheroidal graphite cast iron, foam fillers and glass fibre, and gives the building functional flexibility.

Factor	Result in favour of polymer concrete	Reason
Saving of material	50-65% of volume	Thinner partitions due to far higher strength; smaller dimensions (internal diameters achieved through better surface quality)
Weight	50-65%	Achieved through saving of material (comparable density to cement concrete owing to the density created by a high proportion of filler).
Less moulds and storage space for finished components	30-40% depending on mechanization of production	Can leave the mould after 4-5 minutes; fin ished components trans portable after 1 or 2 hours.
Saving on handling costs	20-30%	Due to lower weight and smaller dimensions.
Reduction in cost of transport to site	20-30%	Due to lower unit weight smaller loading area, simplified loading.
Reduction of construction time and costs	up to 80%	Due in many cases to elimination or simplification of foundations, elimination of coats of paint and insulation, high dimensional accuracy, simpler handling (usually without auxiliary equipment).

Source : Respecta Company (F.R.G.), mentioned in CRIQ, op. cit. p.29

Table III.3.A : Advantages of polymer concrete over cement concrete for prefabricated building products.

Extensions do not require any structural disassembly, simply the addition of further modules (54) .

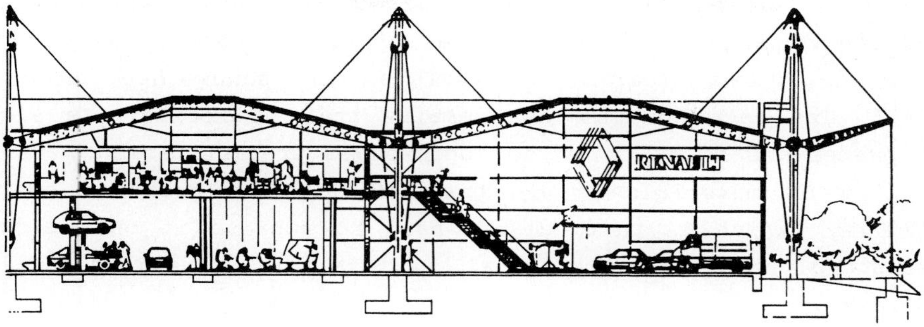

Source : C. COMANI and M. SALMON (55) .

Figure III.3.b : Cross-sectional detail of the Swindon Renault centre designed by Norman FOSTER.

- New materials also offer the possibility of designing integrating multifunctional fixtures. A multifunctional staircase combines lighting, storage and refrigeration functions. The hand-rail incorporates a sprinkler system for buitl-in plant containers. Another example is the movable refrigerator cupboard-cum-table. Such attempts at integration, which are constants in the history of prefabrication, are revitalized by the use of new materials.

These trends towards functional integration are nevertheless to be interpreted with the utmost caution. Perhaps more than any other, the building and public works sector has been the proving ground of very

---

(54) "Norman Foster à Swindon : finesse et tension pour le centre Renault" ("Norman Foster in Swindon : finesse and tension for Renault centre"), Le Moniteur, 20 April 1984, 90-1.

(55) "La robotica e suoi riflessi sui caratteri degli edifici industriali" ("Effects of robotics on the character of industrial buildings"), in : "Politica edilizia e gestione del territorio : scienza, tecnica e dimensione urbana" ("Building policy and regional planning : science, technology and urban dimension"), Bologna, Edizioni Ente Fiere di Bologna, 1985.

advanced functional integration approaches such as "sealed prefabrication" elements in the form of sandwich sheets incorporating multifunctional technical ducting (electrical wiring, telephone cables, etc). The comparative failure of such methods to date may be due to the conservatism of existing trades, levelling-off of demand which has not produced the required effects of scale, or to the absence of any real technical expertise (certain materials in sandwiches used to sublimate). The fact nevertheless remains that the building's sector highest preference seems to go to "open-ended prefabrication", ie the availability of materials or systems offering the greatest overall compatibility (which is particularly hard to achieve if only because the dimensional tolerances are not all uniform for the various families of materials).

- In other instances, reliance on new materials finally allows <u>safety</u> to be enhanced. "Road carpet" laid on key stretches of the road network should lead to a fall in accident rates and in the corresponding health expenditure while at the same time affording more motoring conditions.

Given that the new materials generally increase utilization value, this factor clearly needs to be allowed for in an overall manner; housing should be approached as a system in which the various aspects of utilization are mutually coherent. This is where seals of quality are meaningful. In France the "Qualitel" seal of approval guides consumer choice and promotes quality housing. It takes account of 11 basic criteria (noise, floor covering, charges, electrical systems, etc). It was introduced in 1974 at the instigation of the building trade and consumers' and home owners' representatives. French government subsidies for the construction of accomodation comprising over 50 units are subject to the award of the seal. While the authorities were inclined to withdraw from the financing of construction operations over the period 1970-1980 (56) , their action has become more selective and is aimed at saving energy and promoting higher quality. From the angle of materials, this trend accentuates the need for operational norms to define the utilization and compatibility features of components. (New materials are, however, still open to

---

(56) EURO-CONSTRUCT : "Les innovations potentielles dans le bâtiment en Europe et leurs conséquences sur l'emploi" ("Potential innovations in building in Europe and effects on employment"), FAST series n° 5, Brussels, Commission of the European Communities, 1982.

doubt regarding their durability and compatibility not only with each other but also with well-tried materials).

III.3.2 NEW MATERIALS AND BUILDING SECTOR TECHNOLOGY

Two technological trends are establishing themselves in building and public works. Firstly there is the use of lighter materials, allowing numerous savings on the structure of buildings and engineering works and also on the cost of various site operations. Secondly, there is the appearance of varied materials associated with more flexible industrialization up the line and, further down, with the definition of an open-ended construction system based on "compatible" components.

III.3.2.1. Lighter materials

The trend to lighter materials is appreciable in building and civil engineering. Lightweight material often incorporate several functions at once, so that multiple gains are derived from their use.

Where concretes are concerned, two families are conspicuous owing to the advantages of lightness and insulation which they hold over conventional concretes; these are lightweight concretes and fibre concretes.

- Even though some lightweight concretes like shale and expanded clay concretes have long been known (USA, 1918), only after the energy crisis did the range of lightweight concretes really expand. They are chiefly foam concretes made of an aerated cement mix which may incorporate fine fillers, adjuvants and fine sand, and concretes containing light aggregates (pumice, pozzolona, wood) or very light aggregates (perlite, glass, polystyrene) (57) . This development results from the quest for materials which are light, heat insulating

(57) The rise in the price of quarried ingredients due to environmental protection measures has made it worthwhile to resort to concrete ingredients of volcanic or artificial origin.

and easy to apply and transport. Lightweight concretes lighten the structures of buildings and constructions and afford appreciable savings, particularly on steel reinforcement, the size of load-bearing components and speed of application. The supporting structures are thus of a different design from those made of standard concretes. Lightweight concretes, which in certain applications compete with standard concretes or dense concretes containing standard aggregates, set out to integrate contradictory functions : lightness and insulation on the one hand and structural strength on the other. Indeed, lighter concretes feature lower mechanical strength but also lower heat conductivity. Owing to their still inadequate mechanical properties, lightweight concretes are chiefly used in work on refurbishment and horizontal or low-rise construction. These properties could be brought up to a satisfactory load-bearing standard through the production of composite concretes combining lightweight and traditional ones (58) . Another possibility is fibre reinforcement.

- Fibre concretes have the same characteristics as lightweight ones while improving some of their mechanical properties. The fibres used, whether natural (silk, sisal, flax, cellulose, rock asbestos, etc) or artificial (blast furnace slag, steel, etc), are intended mainly to alter the fracture behaviour of concrete while making thinner and therefore lighter sections feasible (59) . The fibres withstand tensile load and give the concrete a high distortion capability associated with the control of microcracks which the fibres prevent from spreading and so delay fractures. This makes it possible to combine the incompatible properties of tensile strength and lightness.
- In other lightweight concretes which are appearing and making their mark, instead of improving the bonding agent (usually cement) or experimenting with various types of fibre reinforcement, the bonding agent is changed. The polymer concrete previously mentioned illustrates this method. Another instance of replacement of the bonding agent is the use of sulphur. Although difficult to produce, sulphur concrete has the advantage of almost zero porosity and high resistance to acid;

---

(58) A. VENUAT, "Du béton mousse... au béton de polystyrène" ("From foam concrete to polystyrene concrete"), Cahiers Techniques du Bâtiment, 52, 55-63, 1983.

(59) A. VENUAT "Des fibres pour alléger et renforcer le béton" ("Fibres for lightening and reinforcing concrete"), Cahiers Techniques du Bâtiment, 58, 59-70, 1983.

however, it tends to become brittle with time. It may be of interest to countries without natural limestone and clay resources.

The lightening of materials for load-bearing structures also applies to steels. The rise of new steel families such as dispersoid steels makes for lighter supporting structures thanks to the influence of two factors. During the liquid phase of steel manufacture the enhancement of mechanical properties makes it possible to use less steel in load-bearing structures. In the solid phase, the use of more adaptable rolling mills ensures a more varied and accurate response to dimensional requirements. Such requirements could not be met without these rolling mills, and some of the potential savings to be made from structural weight reduction would be lost. From the weight reduction angle, it is therefore clear that variety can be generated either at the stage where the properties of the material are obtained or at the shaping stage.

The problems of weight saving in construction and civil engineering chiefly concern structures under load or dynamic stresses. In addition, however, there is the development of light materials which are not load-bearing, plastics in particular, which when used in partitions, for instance, bring about indirect weight savings in the supporting structures. We are paying less attention to this aspect of weight reduction simply because load-bearing structures represent the bulk of structural masses.

III.3.2.2 Varied materials

New materials are expanding the variety of materials available for building and public works. The above mentioned innovations in the development of lightweight concretes are significant in this respect. Variety means that it is possible to increase the number and combinations of components for which rules of compatibility and utilization norms must be defined. It also means that the demand for greater morphological, functional and aesthetic variety can be met. Here the new materials are an incentive to industrializing the production of construction components in the direction of greater industrial flexibility so as to meet customized demand.

The idea of modelling the industrialization of the building and public works sector on that of the automotive sector is not new.

Automation in building components plants was intended to replace on-site work through a production line type of organization. Government-backed efforts in this direction in the early 60s concentrated on the rationalization of prefabricated parts plants. This trend was accompanied by very considerable rigidities in production processes and standards of products. The underlying model is that of the closed system, the only one compatible with the mass production demanded by postwar construction requirements and population growth. The failure of the closed system became apparent in the early 70s owing to the resultant living quality and social problems, not to mention the unforeseen rise in maintenance costs which the energy crisis was to accentuate. Concurrently with the inversion of population curves and the drop in solvent demand for housing due to economic recession, this failure considerably slowed down industrialization of the building and public works sector. As a result, building remains a 70% non-industrialized activity. Having regard to the large-scale shift of demand towards higher quality, the goal of present and future industrialization is flexible automation of prefabricated parts manufacture. The "integrated" plant must operate to produce standard components which allow automation, but must operate flexibly and so with smaller production runs. This implies a new approach to design, manufacture and computerization, ie using CAM/CAD techniques and devising methodologies for the specification of components and their assembly (60) . In that respect, new materials and composites in particular have a major part to play in the light of their attractive potential for integration. The aim is to convert from an output of identically shaped objects where size is the sole variable to runs of products of varied shapes and sizes but nevertheless based on one model fabricated and assembled with the same apparatus (61) . An example of these analogical runs is given in Figure III.3.c. These series of compatible components banish the rigidities of the closed model and point to an open system of construction capable of meeting varied

---

(60) P. ANDREOTTI and F. GUELFI : "I processi di automazione e la "fabbrica edilizia" : alcune esperienze di imprese cooperative" ("Automation processes and "building factory" experiments in cooperatives"), in : "Politica edilizia...", op. cit., pp. 103-9.

(61) L. MORRA "Produzioni automatizzate e flessibili di componenti per l'edilizia" ("Automated and flexible production of components for building"), in : "Politica edilizia...", op. cit, pp. 111-24.

customer requirements. Components with varied performance characteristics can be assembled on site to provide a range of "ad hoc" combinations of finished components. This follows the growing trend of building site activity towards pre-arranged and sequentially organized assembly operations.

MORPHOLOGICAL VARIANTS OF DETAIL AND COMBINATION OF MATERIALS AND ELEMENTS				DIMENSIONAL VARIANTS		
DETAIL FORM	Blind	Symmetrical window	Asymmetrical window	12 m	18 m	24 m
SOLID STRUCTURE	Blind	Symmetrical aperture				
INTERIOR CLADDING	Tiles	Plasterboard	Rendering			
INSULATION	Rock wool	Polyurethane	Expanded polystyrene			
EXTERIOR CLADDING	Striated concrete	Formed base concrete	Washed sandstone			
DOOR AND WINDOW FRAMES light alloy		Single casement window opening SX				
WOOD		Single casement window opening DX				
ELECTRIC FITTINGS	Switch	Ducting	Power point			

Figure III.3.c : Open-ended design for an analogical series of external covering components.

Components are thus to be designed according to the type of use and the potential for weight reduction, insulation and variety offered by the new materials. This implies defining compatibility rules and determining the "package" of services to be offered. A European normalization policy would significantly aid the progress of the open system of construction in providing an adequate market for producers of compatible components. However, as indicated above, implementation of such a policy appears very difficult.

III.3.2.3 More materials produced by the chemical industry

The pursuit of lightness and energy savings and growth in the variety of materials are the factors which have precipitated the intrusion of chemicals manufacture into the building sector. The use of polymer foams for insulation, polycarbonates for glazed surfaces and plastics for light frameworks associated with moveable partitions are examples underlining the very significant increase of the share acquired in the building sector by materials and products originating from the chemical industry. This tendency naturally affects the way in which certain building trades are developing, eg plumbing which increasingly uses plastic pipes and applies methods offered by chemicals (glueing, for instance) more than traditional metalworking methods (soldering). Zinc, the noble material for rain gutters, is consequently about to be supplanted by PVC.

The chief applications of industrial chemistry materials in building appears to be in the following areas :

- "Bituminous products represent the main sector of innovation, especially elastomeric bitumens. The development of resin-based materials and techniques employing sprayed polyurethane save time and money. For roof insulation, use is made of heavy-duty rigid polyurethane foams obtained by spraying on a blend whose two components react chemically. This compound can be created anywhere under any conditions. The results in terms of coverage and waterproofing are remarkable, and the compound suits industrial roofing, terraces, hanging gardens and disjointed or cracked surfaces. Furthermore, the speed of application (600 m^2 per day) and drying (three hours) allow re-waterproofing in a single application.
- Foams have other uses, eg draughtproofing, insulation of encased piping and securing window frames.
- Polyurethane also finds many applications in rigid "sandwich" panels made of an insulating sheet with a synthetic "skin". It is possible, for instance, to rapidly convert traditional cladding into a sandwich type by using polystyrene panels coated with a sheet of reinforced polyester. The material obtained in this way is very light (2 kg/m^2)

and completely resistant to atmospheric corrosion. To produce high-strength panels, a three-dimensional metal lattice is incorporated into the foam.

- Where the renovation of buildings calls for wall insulation work from the inside, rigid panels are supplemented by an additional textile layer which gives this type of covering a decorative advantage. The outside layer is made of polyurethane and the intermediate one consists of a plastic film which holds all the components together. (This type of product compares to a good quality wallpaper in cost and is of course washable).
- Where waterproofing is concerned, new vinyl adhesives are appearing. They are used in conjunction with sub-layers reinforced by synthetic fibres such as dralon. The technology of now-woven materials allows the use of very thin outside coverings. The application of a thin insulating film to patches which have lost their waterproofness suffices to obtain rotproof surfaces unaffected by contamination from the defective coating. The insulating film consists of a polyester fibre web surfaced with an elastomeric "bitumen sealant" (62) (see Figure III.3.d).

III.3.2.4. Increase in the proportion of "finishings" to "fabric"

The boom in finishing work compared to structural fabric work is probably one of the decisive lines of development in building and public works. Just a few years ago, structural work accounted for almost 80% of final value. The increase in finishing operations naturally coincides with an increase in the "services" content of structures. Among these services, informational and electrical networks in general play a vital part. The idea of "informed building" or "smart housing" corresponds to a resolute research strategy on the part of certain manufacturers who have decided to invest in "domotique", or domestic information processing. Seven European mass electronics producers (Thorn EMI, Mullard and GEC in England; Thomson in France; Philips in Netherlands; Electrolux in Sweden; Siemens in Germany) have

---

(62) Excerpt from "Les Echos", supplement 9, October 1985.

accordingly proposed a scheme under the EUREKA project to rationalize the connections between all their appliances in order to create a real European standard and make all appliances mutually compatible. In so doing the Europeans are attempting to counter American and Japanese ambitions while there is still time (the latter are known to be working on the development of the "home interactive system" as a standard in domestic processing). The home data processing field looks towards the development of a real integrated network of services in a building (allowing, for instance, remote switching of appliances or alerting the security services in case of break-in).

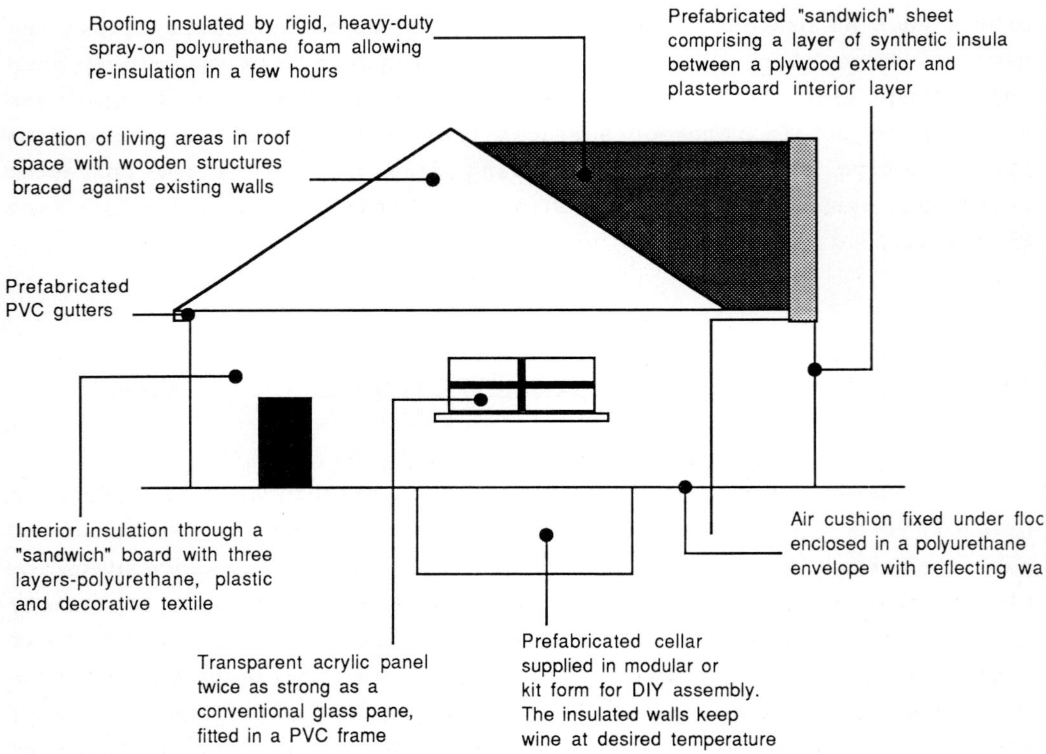

Figure III.3.d : New materials in housing.

The increase in the proportion of finishings generally facilitates penetration by the new materials. As pointed out in the recent scientific and technological guidelines of the French Ministry of Town Planning, Housing and Transport, "finishings, where components fulfil vital but not directly visible functions, are more easily penetrated by new materials (insulants, waterproofing materials, wall and floor coverings, plumbing, taps, etc) to the extent that new materials permit progress in the application of finishings without unduly upsetting sacrosanct values or professional methods. Thus we are witnessing the appearance of plastic doorframes, doors and fastenings, flexible rapid-connection polymer piping, higher-performance cellular plastic insulation and so on".

III.3.3. NEW MATERIALS AND STRATEGIES

The strategies under discussion are of various kinds and do not involve the same agents, but they have a common feature. The trend of demand is such as to compel both builders and materials producers to review their position in the sequence. Chemical and metal manufacturers have consequently tended towards <u>downward</u> liaison to improve their knowledge of demand, whereas building firms are moving more and more in the direction of a division of labour marked by the <u>predominance of services</u>.

<u>III.3.3.1 The strategy of downward integration</u>

The building sector sequence is exceptionally long and complex, firstly because of the many and varied agents involved and secondly because of its low degree of integration. It has hitherto been dominated by the builders, although there are signs of this preponderance being eroded by contractors (architects and consultants), those who set consumer standards, and a few major materials producers. The last-mentioned have been spurred by the energy crisis to liaise <u>downwards</u> in order to deal more satisfactorily with the constraint imposed by the general fall in overall demand, and also to respond more adequately to the qualitative changes in demand for which other firms were less prepared for reasons relating to finance and expertise. Accordingly, in 1983 the French Péchiney group set up Péchiney-Bâtiment by merging two subsidiaries. As yet, this aluminium producer has no intention of "building", but rather of supplying construction kits, ie a group of products (frameworks, cladding, roofing, lightweight facades, windows, etc) and assembly instructions which can be adapted to various construction types. Suitable forms of cooperation are being set up with building firms and clients to supply the requisite information on changing needs. Similarly, the Italian chemicals group Montedison has launched a major scheme ("Progetto Cultura") involving designers from every sector. The aim is to work out new designs for building materials and products relying on plastics. Without going so far as to absorb building firms, this strategy presupposes the acquisition of knowledge and skills beyond the traditional ambit of the materials producer. Being in control of a variety of materials and products, the major chemical and metallurgical groups can only exploit it by placing themselves in a position to gather more information about demand in order to fulfil customer needs, but also to anticipate or even stimulate them.

III.3.3.2 Evolution of building firms

The evolution of building firms varies considerably depending on national situations. European building firms generally have nevertheless been hard hit by the recession in their own sector (Table III.3.b).

The major groups have not always been able to make up for the decline in domestic activity by conducting large-scale works abroad since this field of activity has also declined with the appearance of keen competition from countries in the process of industrialization. The major groups are turning more and more to activities with high added value, ie mainly service activities in construction work organization and planning. Having previously tended to produce everything, they now tend to become suppliers of organizational services ("constructor managers") organizing the work of others in a context where urban expansion is giving way to restoration, rehabilitation and reuse. What these groups have to offer is all-round proficiency in construction work as such, roadworks, sanitation, infrastructure and amenities. Only these large firms are used to combining civil engineering expertise with construction skills. This proficiency becomes decisive in the context of work on the installation of the new telecommunications networks and the evolution of urban settlements towards a network-type interlinkage (63) . The trend of large European building and civil engineering firms towards the supply of construction-related services is accentuated by a number of factors.

Percent variation of production in real terms on previous year	European Community (c)					
	1981	1982	1983	1984	1985 (a)	1986 (b)
1. Building	-4,6	-3,0	-0,4	-0,5	-1,3	-0,6
1.1.Housebuilding	-7,7	-2,6	0,9	-1,9	-4,6	-3,0
1.2.Non residential	-4,1	-2,5	-3,1	0,4	1,7	2,3
2. Civil engineering	-3,7	1,2	-2,5	-2,0	-2,9	0,4
Construction	-5,3	-0,9	-1,0	-1,0	-1,6	-0,2

(a) Estimate - (b) Forecast - (c) Without Greece.

Source : EEC (64) .

Table III.3.B : Trend of production volume in the European building and public works sector.

In many cases these firms need not call in private architects because they can rely on their own architectural or technical departments. Few architects at present are able to exploit the potential of new materials. Owing to the privileged relationship maintained by building firms with materials producers, the former can manage in the design context the variety generated by the latter. Furthermore, the division of labour in construction has developed so as to diverge from the

---

(63) OTEB : "Prospectives des techniques de construction pour les années 2000-2020" ("Outlook for construction techniques for 2000-2020"), excerpt from the research report submitted to the CSTB gathering of 18 March 1986, Paris, Observatoire Technico-Economique du Bâtiment (OTEB), 1986.

(64) "L'activité de construction dans la Communauté Européenne", ('Construction activity in the European Community"), Brussels, International European Construction Federation (FIEC), 1986.

traditional division between structural fabric and finishings. The latter type of work is being supplanted, as demonstrated above, by the concept of development which comprises all operations to adapt buildings to the physical surroundings (land development and site preparation) and urban environment (cables, mains and roads) (65) . Concurrently, building activity proper is tending to become more and more an assembly activity lending itself to sequential organization. This rationalization causing the disruption of traditional trade sectors, is the main source of productivity gains which the building industry means to tap in the near future. It requires the adoption of new techniques (laser cutting, site robots, bonding and assembly operations for lightweight structures) but above all scrupulous on-the-spot job preparation. The French group Campenon Bernard has accordingly created a subsidiary whose amazing dynamism at a time of recession stems from a decentralized network of associated firms and from its overriding concern for pre-arrangement of on-site tasks, even if it takes three months (66) . Consequently, construction is becoming an increasingly intellectual activity and requires firms unaccustomed to such methods to adopt techniques of analytical accountancy and time units. This again means that once rationalized, the construction process will be able to achieve greater profits by using innovations in materials and so to exploit more systematically the gains generated by the new materials. Rationalization of the construction process thus permits more effective diffusion of innovation in a world where the segregation and plurality of traditional trades has been a strong deterrent to innovation. Conversely, owing to their variety and microscopic complexity, new materials require greater attention to the design of structures and this accentuates the proportion of services by comparison with manual work. The big construction firms will probably take the largest share of land development activities in the broad sense, design activities and preparation of on-site work, leaving the assembly work to specialized SMEs. At the same time, these SMEs can be expected to increase their specialization in other fields. Once marked by organization into trade sectors and hence by a bi-specific

---

(65) OTEB, op. cit.

(66) P. CHEVILLEY : "Campenon Bernard Construction : une recette anti-crise" ("Campenon Bernard Construction : how to beat the crisis"), Les Echos n° 14615, p. 10, 17 April 1986.

relationship between <u>one</u> skill and <u>one</u> material, they will evolve towards <u>functional organization</u>. The best example of this is provided by the firms specializing in <u>heating engineering</u>. Mastery of all aspects of the thermal systems represented by buildings requires a combination of different trades (heating technicians, insulation specialists, glaziers, etc). Here the trade is defined by a relationship between <u>several</u> skills and <u>several</u> different materials.

III.3.4. NEW MATERIALS, SKILLS AND EMPLOYMENT

The enormous variety of operations covered by building and public works activity reflects extremely variable demands as to the type and skill standard of employees. This sector has always been labour-intensive, particularly as regards the construction of the buildings which are our main concern here. In relation to the problems raised by new materials, there are two themes concerning skills and employment. To the extent that new materials require command of <u>new know-hows</u>, they bring about <u>development of trades</u> and <u>learning processes</u>. Furthermore, <u>rationalization of on-site work</u> emerges as a major phenomenon which affects skills and employment in the building sector and is the counterpart of the trend towards the use of new materials.

In the buildings construction sector industrialization takes only a comparatively small share, mainly in the production of prefabricated components. Flexible automation of plants producing components has adverse effects on employment. However, the use of new materials entails a technological as well as an industrial form of microscopic mastery and hence presupposes advanced know-hows which often subsume older forms. As the plants evolve towards flexible automation it becomes necessary to recruit or train specialists with a combination of skills in building materials and new information technologies. A similar conclusion is reached concerning building wholesalers. The sector has always been familiar with a wide range of materials. Use of the new materials accentuates this situation and thus increases the need to adopt computerized systems for managing the variety of materials. In construction, a marked development of specialist know-how concerning the control of energy is to be expected. The stakes are particularly high where insulation is concerned. At present, insulating work is often assigned to traditional tradesmen such as glaziers or roofing specialists who have thus been able to adapt to new techniques while identifying primarily with their traditional materials. Yet it is not certain that they will retain possession of the insulation sector in future. Insulation is tending more and more to be incorporated and conceived at the prefabricated components manufacture stage. Furthermore, new occupations such as heating or air conditioning engineering have emerged, incorporating insulation work in the overall function which they perform. Thus different fields of know-how are competing to take over a new task. In so far as heating engineering firms or the heat engineering departments of general construction firms (67) look after the "energy" function as a whole, they can probably dominate in the long run by relying on the support of prefabricated components manufacturers. The firms and departments concerned will in turn need to achieve a blend of traditional forms of know-how and compete with them directly. Lastly, the new information technologies enable traditional trade sectors such as heating installation to increase their level of specialization as they evolve. New materials are also frequently unfamiliar to designers; they require different methods of calculating specifications. The advance of information technologies in architect's offices should make it possible to release the creative energy of architects and to encourage the use of new materials through the application of suitable design methods.

The construction sequence involves two different groups of agents. Before the building-site stage, a group of highly skilled workers has the task of designing the structure and organizing the tasks. On the actual site, a mixed group of more or less highly-skilled workers (tradesmen) and unskilled workers (labourers) is assigned to the execution of the pre-arranged tasks. The rationalization of tasks is expected to be the main source of productivity gains to be made by the construction sector as a way out of its present slump. This presupposes re-allocating construction time in favour of preliminary organization of tasks. The objective is an arrangement that will permit the sequential execution of tasks on a pattern which remains of the Taylorian type : jobs are broken down into units and interlinked. Behind this intention lies the idea of depersonalizing tasks, hence the

---

(67) For instance, Philipps Holzmann, the leading construction firm in the Federal Republic of Germany, recently acquired a 25% share in Scheu & Wirth, a firm specializing in energy-saving technologies.

preparation for the use of site robots. In terms of skills, such a development has two major implications. The first is the growing need for engineers and technicians capable of organizing site operations in advance. This should favour the introduction of new materials in so far as the emphasis is on ease of installation. Obstructions to the dissemination of innovations created by trades whose skills revolve around traditional materials could thus be removed. The second implication is that the conduct of site operations will rely on an increasingly specialized workforce of semi-skilled workers. In practice this means a downgrading in relation to the traditional trades. The tasks performed by this new workforce will consist chiefly of assembly operations using simple bonding and fastening procedures, flexible and mechanical joints, etc. The future of traditional trades can only be assured through development in restoration work, as against the construction of new buildings.

In general, the dual trend towards industrialization of components and rationalization of site operations is definitely moving towards fewer jobs in building, all the more inevitably as this sector needs to achieve large productivity gains if it is to emerge from the current slump. Supply of the assembly kit construction system and growing simplicity of component application have enhanced the homeowner's ability to do minor construction jobs himself, and even to carry out large-scale upkeep and repair jobs which form the largest market in the sector. This factor accentuates the observed trend towards job losses. The use of new materials of course creates needs for new jobs, but in most cases these new needs are connected with higher skills or new expertise.

PART IV

Functional Materials

As the vehicle for a particular physical property, (conveying electricity, light, etc), the functional material plays a key role in the technical system. Every major technological breaktrhough involves devising and using a specific functional material : silicon for informatics, optical fibres for telecommunications, catalysts in chemistry, etc. Thus at any given time, functional materials define the limits within which the technical system can evolve. It is, for example, impossible today to envisage going below a size of $10^{-8}cm^2$ for a transistor using silicon technology, and this places a limit on the possibility of miniaturization in informatics. By its very nature, the functional material has a number of specific features which distinguish it from other materials.

The majority of functional materials are high value-added materials and the market for them is inherently a world market. Insofar as these materials are normally used as components in various industrial systems and articles, they do not directly reflect the tastes and requirements of the ultimate consumer, unlike structural materials. This aspect is enhanced by the fact that a large proportion of these materials are easy to transport because of their small size and that the cost of transport is in any case negligible when compared with the price of the component. Moreover, although they are strategic in their applications, these materials normally account for only a small part of the value of the articles or systems in which they are incorporated ("chips" account for only a few cents in the total value of a computer).

Generally speaking, there is a very limited range of materials able to meet a given physical characteristic. In electronics, for example, only semiconductors (and even then, by no means all of them) are able to provide the "transistor" function : in electricity, there are only a few compounds with suitable superconducting properties for use as high-field magnets; in catalysis, it is rare to find more than two elements able to bring about a reaction in acceptable selectivity conditions. It is this very restricted range which explains the frenzied efforts of both fundamental and applied research to find new compounds able to perform these functions. Controlled microscopic combinations of the 92 natural elements are definite, and it would be very rash to imagine that modern science has catalogued them more or less exhaustively.

This "technical" monopoly of the material is moreover reinforced when the material has established itself in industry. The development of increasingly efficient production processes helps to brings costs down and - when both the technical and economic criteria are taken into account - "protects" the dominant material from its competitors. For example, micro-electronics depends today largely on the use of a single material - single crystal silicon - because, in addition to the material's exceptional physical properties, a whole range of production processes (gaseous diffusion, epitaxial techniques, masking, photoetching) have gradually been established which are better and better controlled and have given this material a broad spread of applications.

So any replacement of an industrially established functional material by another material requires a major technological leap. The example of molybdenum-based catalysts replaced by platinum catalysts for petroleum cracking by UOP (United States) just after the second World War is a good illustration of the difficulty of replacing one functional material by another : the obvious technical superiority of platinum (better selectivity, better stability, etc) had been proven since the end of the 1930s. Yet it was not until the early 1950s that the investment barriers (at the time the amortization of one cracking unit was measured in tens of years) and also entrenched habits (the UOP company even went so far as to install complete platinum-based units in some refineries at its own expense in order to demonstrate their advantages) were overcome. The superiority of platinum having been demonstrated (it is true that the war had weakened resistance to it), and as a result of the decisive economic advantage obtained by the first refineries to be equipped with it, within a decade every refinery in the world had gone over to platinum catalysts (entailing huge investment).

The difficulty of substituting one functional material for another is further compounded by another of their properties, namely complementarity; this is explained below.

A functional material normally establishes itself only in close complementarity with other functional materials. Thus the development of optical fibres is closely linked to that of the materials used in connections (lasers in transmitters and diodes in detectors based on gallium arsenide, for example); similarly, the development of electronics (silicon) is closely linked to that of materials incorporated in terminal equipment (eg metallic oxide-based magnetic materials).

For any given type of application, therefore, a coherent set of materials which are compatible with each other gradually comes into being. Once established, that coherent set becomes difficult to "shift". Only in particular cases does a partial substitution, compatible with the rest of the system, prove possible : to some extent the replacement of copper wire by aluminium wire of carrying electricity falls into this category. But the general pattern seems to be a convergence towards complementarity among a given set of materials. This will give a measure of the evident complexity of a move from one system to another, when a major technical change takes over.

More specifically, functional materials are used in accordance with two main classes of application which need to be distinguished :

- components (microprocessors, catalysts, laser diodes, magnetic tapes, optical disks, etc) are the key elements used in industrial processes and articles. They are generally characterized by relatively low price as compared to that of the application which they serve.
- infrastructure materials (electric wire, optical fibres, etc) are the links in the networks of energy and information transport. They are generally produced on a large scale and their relative price is an important parameter.

Over and above each specific application, functional materials are closely involved in defining the basis of the technical system in the sense which Bertrand Gille gives to that expression : understanding far-reaching changes in the technical system involves understanding interactions between functional materials and other elements in that system - information, energy, biotechnologies. Here again there are complementaries to be considered, but this time the links of complementarity are observed between the major elements in the technical system themselves. This "key" to interpreting the evolution of functional materials has been adopted in the following paragraphs.

In particular, it enables us to underline the strategic aspects of functional materials and their crucial role in all technological development. From this point of view it can be clearly observed that certain functional materials are the starting point for veritable technological routes which can help to restructure a whole area of economic activity and which in turn condition the development of other materials "compatible" with the dominant materials in ensuring coherence in the overall system. In some cases, as in semiconductors in electronics, the impact of developing the basic material is considerable and the resultant technological route causes upheaval throughout the technical system.

Lastly, functional materials are characterized by the importance of <u>fundamental research</u> into them. Any country or group of countries which allowed itself to fall behind in research into functional materials would run the risk of becoming dependent upon its competitors in every application involving them. Recognition of the importance of fundamental research into functional materials is particularly obvious in all the major research projects within the EEC (EURAM, ESPRIT, RACE, BRITE, etc) which seek to give Europe once again the structures and essential critical mass in fundamental research to withstand competition from the United States (where research into functional materials is heavily supported by the Materials Research Society vis-à-vis the federal authorities) and Japan (with massive intervention by the MITI). Mastering present developments, but also and above all breaking through new technological barriers, will in fact require ever more substantial resources and concerted, coherent efforts (synchrotron, high-flux neutron reactor, space laboratory, etc).

CHAPTER IV.1

Functional Materials and Information

Exploiting the semiconductor properties of a particular material - single crystal silicon - lies at the heart of developments in the electronics and computer industry. This industry has a turnover of some thousand milliard francs at the end of the 1980s, a figure expected to increase five-fold by the end of the century. Through the scale of its induced technological developments, the size of the markets involved and the considerable social implications, no material better illustrates the strategic role of a functional material than silicon.

Two principal phases can be distinguished in the development of silicon-based micro-electronics.

The first phase runs from the early 1960s to the mid-1980s and corresponds to the establishment of silicon technology (production of the crystal and fabrication of integrated circuits) and of the industries downstream from it (electronics and informatics). The main driving force in that development was the search for greater and greater integration in integrated circuits, accompanied by a spectacular fall in production costs. This move towards integration is certainly not over, and silicon is likely to remain a dominant material for more than ten years to come, but it is now realized that the continuing growth of demand in electronics will eventually come up against the theoretical limits of silicon use.

The second phase started recently, in the mid-1980s, and is likely to be characterized by increasing convergence between informatics and telecommunications. The now inevitable linkage of these two fields is largely an induced effect of the development of micro-electronics : the more micro-electronics decentralized information processing by means of micro-processors and micro-computers, the more urgent it seemed to connect information processing units up to each other. The transmission of information has become inseparable from information processing, and the need to transmit information more and more rapidly may well in the long run, in certain applications, cause other semiconductor materials than silicon, and especially gallium arsenide, to be preferred for information processing.

IV.1.1. THE ESTABLISHMENT OF SILICON TECHNOLOGY

The miniaturization of integrated circuits (assemblies of transistors, resistors, condensers and conductors inserted into a single piece of semiconductor material) is the phenomenon by which the development of micro-electronics and other industries dependent upon it have been conditioned (Table IV.1.A). The speed of a computer (ie its capacity) depends largely on the switching speed of its transistors (the smaller the transistor the faster it is, and consequently the solution is to reduce transistor size to the minimum) and on the length of the connections between the basic electronic components (the greater the integration of aluminium wiring in the circuits, the higher the speed at which information moves). In addition to the physical characteristics of the connections, there are crucial economic factors : the cost of a given system - in terms of the basic chip, but also at higher levels, be it the chip carrier, the printed wiring board, the complete information processing system or even the telecommunications network - owes more to the connections between the components than to the components themselves. And since the cost of connections rises with distance (cf. Table IV.1.A and Figure IV.1.a), integration of connections is also an economic imperative. Lastly, one of the major economic effects of integration is to improve the reliability of information processing systems by reducing the number of interconnections.

Single crystal silicon - as the basic element of integrated circuits - affords a particularly satisfactory answer to all these requirements. Since the early 1960s, it has even been possible to show that there is an empirical law proving that the number of components (transistors) per silicon chip doubles each year : from 1700 transistors in the first 45 bit microprocessor in 1971, we have now advanced to 150,000 transistors in the 32 bit micro-processors or complex circuits.

However, the development of a basic micro-electronics technology around silicon is not due only to the exceptional physical properties of this material. It also derives from the application of high-efficiency processes in the production of silicon crystals and in the manufacture of integrated circuits.

There are constant innovations in the production of silicon crystals for the manufacture of integrated circuits : continuous manufacturing processes are now being employed, manufacture is being computerized (giving the crystal a high degree of reproducibility), and processes which make the manufactured crystals more homogeneous are used. One of the essential aims is to control the quality and quantity of microscopic defects. In fact the latter condition the maximum size of chips, because the larger they are, the greater the likehood of finding defects. With the processes currently used, chips of the order of 1 cm^2 are economic. It is hoped that it will soon be possible to achieve sizes in the region of 10 cm^2, which would ease the technical constraints due to integration.

The manufacture of integrated circuits involves a complex series of operations on the basic semiconductor material. In the case of silicon, MOS (metal/oxide/semiconductor) technology is most commonly used. This type of technology entails alternating masking operations and various physical processes (transcription of circuit design on to the material by photo-engraving, chemical etching and engraving to obtain the patterns, oxidation to isolate circuit layers, controlled distribution or implantation of impurities, evaporation under vacuum or atomization or vapour-phase deposition in order to deposit metals and alloys on the board etc). These highly automated techniques are now fully mastered and contribute greatly to building up experience and consolidating the supremacy of silicon. There is another technical limit to miniaturization that must be pointed out. It is not possible at present, with the light sources used for photo-engraving, to descend to the theoretical limits of silicon because of the diffusion phenomenon due to the excessively long wavelengths of these sources. Major advances in optics (X-ray lasers, for example) are essential if the movement towards maximum miniaturization is to continue.

As a general technology, silicon technology has required the use of a very large number of other materials which are involved at every development level in information technology :

- in integrated circuits, aluminium or copper (conductors) and other elements (chromium, silicon nitride, etc) are present in order to perform various necessary functions.
- in chip carriers, the materials most commonly used as carrier elements are plastics or ceramics, while gold is the best conductor because of its chemical inertness.
- in printed wiring boards, hybrid circuits carriers are usually ceramics. Hybrid circuits are connected up on copper wired epoxy resin supports to form the boards. The latter are often inserted into metal racks and connected to an epoxy carrier by wiring (made from various alloys, eg palladium alloy). The different carriers may be built up into circuit packs connected together by optical fibres or copper wiring.
- in the various systems (computer), a large number of materials (glass, plastic, copper, gold, liquid crystals, ceramics, magnetic materials) perform the various back-up functions to the elementary information processing (recording, transmission or visual display of information, etc).

- lastly, on a broader scale, information processing systems are inter-connected using copper or glass fibre or even plastic materials.

Integration level	Abbrevia-tion	Nb of circuits per chip	Width of lines (micrometers)	Production date
Small scale	SSI	1-100	10	1960
Medium scale	MSI	100-1000	5	1970
Large scale	LSI	1000-10000	3-1	1980
Very large scale	VLSI	> 10,000	< 1	1980
Ultralarge scale	UHSI	?	0.1-0.001	?
Molecular electronics	-	?	0.001	?

Table IV.1.A. : Miniaturization of electronic devices

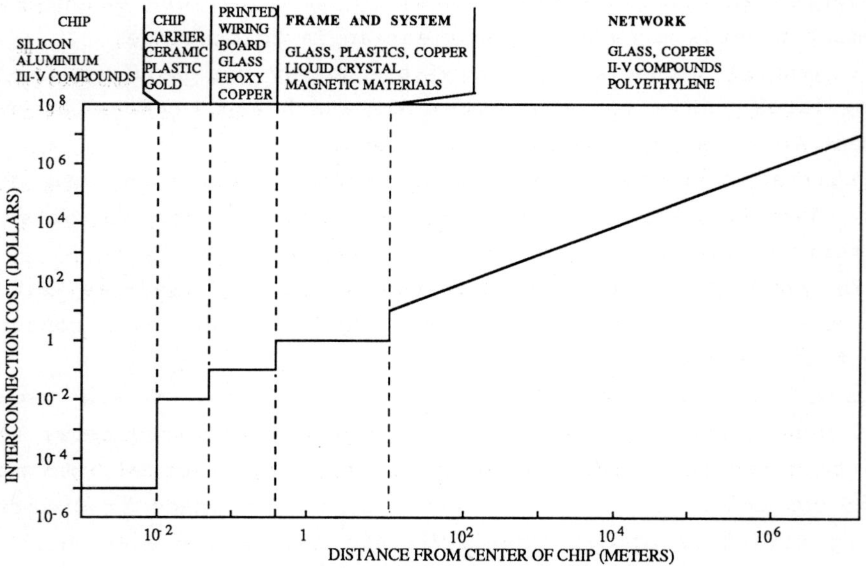

COST OF AN INTERCONNECTION Increases with its length. The chart shows the approximate cost and the main materials used at five levels of interconnection. For example, each of the metal conductors on a state-of-the-art silicon chip costs about 10 millionths of a dollar. In contrast, a single optical Aberguide in the long-distance telephone network may cost millions of dollars.

Source : Scientific American, December 1986.

Figure IV.1.a : Cost of interconnections in information technology as a function of distance.

The extent to which silicon technology has given rise to a whole range of complementary materials designed to provide overall coherence in the processing and transmission of information is therefore quite remarkable. The principal question that arises today is whether this supremacy of silicon will continue for much longer. The question is all the more pertinent as certain physical and economic limits to the use of silicon are now becoming apparent.

The integration of silicon circuits can hardly progress by more than another factor of 50. This limit is due both to a physical factor (it is almost inconceivable that silicon chips could carry transistors smaller than $10^{-8}cm^2$) and to an economic factor (no economy way of manufacturing silicon chips larger than $10cm^2$ is known).

Electromigration (undesirable movement of metal inside connections due to heating, where one chip carries a large number of circuit elements) and short-circuiting into the silicon substrate when the insulating silicon layer becomes thin are two problems which are curbing advances in the integration of silicon-based circuits.

The switching time of elementary silicon transistors cannot be brought below one nanosecond (10^{-9} seconds) and this restricts the use of silicon for ultrarapid electronic circuitry. Consequently, silicon is rather unsuitable for the growing needs of telecommunications which require processing and transport of a growing number of information items in the shortest possible time. Even though the position of silicon still seems extremely strong for the decade ahead, the coupling of informatics and telecommunications may well produce the biggest upheavals in semiconductor materials, preference gradually being given to gallium arsenide or its derivatives over silicon in rapid circuits.

IV.1.2. THE CONVERGENCE BETWEEN INFORMATION TECHNOLOGY AND TELECOMMUNICATIONS AND ITS IMPLICATIONS FOR THE CHOICE OF BASIC MATERIALS

IV.1.2.1. The development of opto-electronics

Until recently, information processing and telecommunications technology developed along separate lines, relatively independent of each other. While the growth of information technology was conditioned by the development of silicon technology, the telecommunications sector was developing light-based technologies (photonics or opto-electronics) which have profoundly changed the possibilities for transmitting information, using light (photons) as a vehicle to carry items of information instead of using electricity (electrons) as in conventional conductor systems based mainly on copper wire.

Photonics covers all the methods, processes or systems whose function it is to study, measure, transform or transmit by means of light (photons). Photonics enables these functions to be performed with four main categories of product : lasers and their accessories; optical fibres and their accessories; systems for the reception, display and exploitation of information; photovoltaic and solar systems.

These major product categories can sometimes be used on their own, but they are often combined with each other to form complex systems designed to serve different elementary functions (like computerized production systems, complex communication systems, etc).

Photonics, through the industrial application of the photon, performs information collection, transmission and visual representation functions. For this reason, like all information technology, it is characterized by universality. This means that it can be used in a considerable number of industrial sectors, where it serves either to improve production processes or in the manufacture of new products.

It is also a technology which has the ability to process materials by use of lasers. In performing this function, photonics generally comes into competition with other techniques (welding, cutting, destruction of objects, etc) but in every case offers the advantage of very high precision which guarantees production quality in the application concerned.

In these two fundamental applications - transmission of information and processing of material - photonics reveals a wealth of uses which gives this technology a vital part to play current technological changes. The later part of this chapter will be concerned solely with the relationship between photonics and information technology, and for this reason we shall prefer the term "opto-electronics" to designate all the photonic technologies discussed.

To date, opto-electronics and electronics have been made compatible by means of essential components in the form of light emitting diodes and photodetectors. The first transform electronic signals into light signals and the second convert light signals into electronic signals. This ability to communicate between the two technologies comes up against a major problem in present structures : opto-electronics offers a potential for information processing and transmission which far exceeds that of micro-electronics in terms of speed and output. In the now unavoidable construction of a coherent information network, extending from the most elementary information systems (personal computer) via the most common local networks (local company networks) to the "meganetworks" represented by the large optical fibreguide lines, the macroelectronics elements form the slowest links in the chain and consequently determine the overall speed at which the information circulates in the network (slow stage). It is precisely at this point that we come up against the limits of silicon technology, and it is on this critical point that a close analysis of the potential offered by gallium arsenide (or other III-V compounds) is unavoidable (1) .

IV.1.2.2. Silicon versus gallium arsenide

In the past, silicon and gallium arsenide were complementary rather than rival materials. Silicon reigned unchallenged in micro-electronics and gallium arsenide in the field of opto-electronics and hyperfrequencies. The relative size of the markets in question gave silicon a considerable advantage (about 95% of the total market for semiconductor components in 1985). The coupling of information technology and telecommunications, and also silicon's own limitations, may cause gallium arsenide to begin replacing silicon. The extent of that replacement is as yet very uncertain, but it requires us to compare the two materials in greater detail.

---

(1) III-V semiconductors are the product of associating one (or more) element(s) from the third column of the periodic table of chemical elements (aluminium, gallium, indium) with another (or several others) from the fifth column (phosphorus, arsenic, antimony).

Technical advantages of gallium arsenide

Electron mobility inside it is about five times higher than in silicon hence a shorter switching time and consequently higher information processing speed.

Because of its non-existent light-emitting properties, silicon is not suitable for making optical components (like light-emitting diodes, semiconductor lasers or light modulators), unlike gallium arsenide. (On the other hand, silicon is the basic element of photodetectors).

Gallium arsenide has better insulating properties than silicon and this delays micro-circuit breakdown.

The frequency limit for the use of gallium arsenide devices is in the region of 60 GHz as against a few GHz in the case of silicon, thus opening up the hyperfrequency market to gallium arsenide (in particular, for reception of satellite signals which are carried by waves in this frequency range).

The energy consumption of gallium arsenide is lower than that of silicon (about one-tenth), leading to far less heating in integrated circuits. Moreover, the range of operating temperatures is far wider than in the case of silicon (350°C as against 200°C).

Various kinds of radiation - electromagnetic, neutron or radioactivity - create fewer defects in gallium arsenide than in silicon (three to four times fewer), which gives gallium arsenide a special advantage in the military sphere (eg for "hardening" nuclear warheads). Two-thirds of the present supplies of this material are reserved for military applications.

Moreover, it is no secret that a nuclear flash at medium altitude would have a disastrous effect on all electrical and electronic networks (completely paralyzing a country like the United States). A light-based network (fibreguide or a light-based information processing system (optical computer) would be completely safe from such interaction.

These two "military" properties of gallium arsenide confirm the strategic value of the material.

Gallium arsenide based semiconductors have the property that there is a gap between the conducting band and the valence band of the electrons which is variable as a function of additives or the thickness of the layers. The generic term used to describe the manufacturing techniques for these components is "gap engineering". There are plans to extend these techniques to silicon, but at the present time that is merely a research hypothesis. Both for semiconductors lasers and for detectors, this variable gap makes it possible to obtain components operating in hitherto unused light frequency ranges, especially infrared. This explains the hopes placed in gallium arsenide and its derivatives infrared optical fibres are developed (glass having better transparency at these frequencies).

DEMAND FOR GALLIUM

Worldwide demand

The world demand for gallium in 1983 was in the region of 25 tonnes per annum. Almost all of this comes from the electronics industry; indeed, it is claimed that gallium has a higher proportion of its production devoted to the electronics sector than any other metal. Its major use at present is in electronic devices which employ gallium combined in semiconductor compounds. In order to achieve the desiral electrical characteristics, the gallium used has, of course, to be of extremely high purity.

The established commercial application include those listed below.

(1) **Light-emitting diodes (LEDs).** Gallium phosphide (GaP) and gallium arsenide phosphide ($GaAs_{1-x}P_x$) serve as the basis for LEDs used as indicators and numeric displays in a wide variety of technical instruments and consumer products. Gallium arsenide diodes are used as infra-red illuminators in conditions of poor visibility. At present, light-emitting displays are believed to constitute the largest single sector of demand for gallium.

(2) **Transistors, diodes and negative resistance devices.** Field-effect transistors (FETs), avalanche diodes, and Impatt and Gunn devices of gallium arsenide are used as amplifiers and oscillators in microwave communications systems.

(3) **Photodetectors.** Photocathodes of gallium arsenide, gallium aluminium arsenide ($Ga_{1-x}Al_xAs$) and gallium arsenide phosphide, and also secondary electron emitters of gallium phosphide, are used in scintillation counters and spectrometers.

(4) **High current switches.** High current liquid metal switches made by GEC use the Ga-In-Sn eutectic alloys as the liquid.

As noted above, the total world consumption of gallium is estimated currently to be about 25 tonnes per annum. Judging from US consumption figures shown in Fig.1 of this table, demand for gallium has been erratic over most of the decade. Since 1978, however, consumption has risen consistently, and the US Bureau of Mines predicts that this trend will continue for the forseeable future : a 6% annual increase demand to 1990 is anticipated.

The existing productive capacity for gallium in the Western world in 1979 was approximately 25 tonnes per annum compared with a demand for only 14 tonnes. This imbalance between supply and demand is reflected in the price of device-quality gallium (6N purity), which fell from \$750 per kg in 1973 (equivalent of £306 per kg at that time) to about \$500 per kg (£295 per kg) in 1982.

New demand

There are several new applications of gallium that are currently entering commercial production which have the potential to boost the demand for gallium quite considerably. Included are the applications mentioned below.

(1) **Integrated circuits.** Gallium arsenide complements silicon in this application, being superior to silicon in certain of its electronic properties; its higher carrier mobility and drift velocity enable gallium arsenide devices to operate faster and to dissipate less power than their silicon equivalents. However, there are high price penalties to be paid for this improved performance; problems with electron states produced at the interface between gallium arsenide and most insulators make integrated circuit manufacture with this material considerably more difficult and expensive than with silicon. High-quality single-crystal wafers of gallium arsenide are also more costly to produce.

(2) **Photovoltaic cells.** Gallium arsenide solar cells have been developed with efficiencies exceeding 20%, which is substantially higher than those obtained with any other material investigated so far. These cells also possess superior high temperature characteristics and their performance is relatively unaffected by exposure to cosmic radiation. The principal factor limiting the widespread use of gallium arsenide in solar cells has been mainly the high cost of manufacture of the single-crystal wafers required for these devices compared with those of silicon.

(3) **Lasers.** Solid state heterojunction lasers of gallium arsenide and gallium aluminium arsenide are used in conjunction with optical fibres in telecommunications. Experimental communications systems based on optical fibre technology are already in operation in a number of countries, including the UK. A large expansion of this field is envisaged throughout the 1980s. However, a move to low transmission frequencies would result in a switch from gallium arsenide to indium phosphide in lasers.

(4) **Superconducting magnets.** The niobium-gallium compound Nb_3Ga is being considered for use in superconducting magnets in electric power generation. The feasibility of producing electricity economically with generators incorporating these high-field superconducting magnets has already been demonstrated in experimental systems at MIT and Westinghouse. Plans have been put forwards for the construction of a 1 000 MW prototype generator at MIT.

(5) **Magnetic bubble memories.** A gallium-gadolinium garnet substrate promotes single-crystal growth of the special magnetic rare earth-iron garnet composites required. Magnetic bubble memories offer high storage densities and they retain their stored information when power to the device is switched off. There is believed to be a fairly strong demand for dense, non-volatile bubble memories, especially for portable equipment, including pocket calculators. This new application of gallium is now

apparently threatened by cheaper substitutes for gallium-gadolinium garnet substrates.

To illustrate the kind of growth in demand that gallium could experience, Rosi has cited a NASA proposal for equipping space satellites with large solar power sources comprising arrays of photovoltaic cells. Just one 5 GW system of this type now under study would require approximately 400 tonnes of high-purity gallium. This alone would represent 16 times the present world production of 25 tonnes per annum. It is obvious that a development of this nature would transform the gallium market.

AVAILABILITY OF GALLIUM

The natural occurence of gallium in the Earth's crust if approximately 10 ppm by weight, which makes it almost as abundant as lead.

The most important commercial sources of gallium, accounting for some 90% of current world production, are the bauxite ores of aluminium, which possess an average gallium content of 0.006 wt%. Extraction from ore with such a low concentration of the element is only economic at today's prices because gallium is recovered as a by-product in the extraction of alumina and aluminium. Fortuitously, gallium concentrates naturally in the sodium aluminate solution used in the Bayer process for recovering alumina from bauxite, from which the metal is recovered by electrolysis. The gallium obtained at this stage has a maximum purity of 99%. Further purification is necessary, of course, to meet semiconductor specifications.

It is estimated that less than 2% of the 60 million tonnes or so of bauxite used for producing aluminium is used presently for gallium, so that fifty-fold reserve is available from this source alone. The fact that gallium is a relatively low-value by-product blunts any incentive for boosting the output. Thus, on average, 5 tonnes of bauxite costing £25 yields 1 tonne of aluminium valued at £600 in 1983 and 0.1 kg of gallium, which for equivalent commercial-grade material (99.9% pure) was worth about £20. In the simple case where these two metals are the only products of bauxite processing, it is seen that the price of gallium would have had to increase by about 30% to effect at 1% rise in the overall value of the products.

The remaining 10% of the world's gallium supplies derive from zinc residues and recycled material. In terms of the primary minerals, the reserves of gallium are vast. In the first place, there are abundant reserves of bauxite in comparison with the present level of exploitation. Furthermore, certain zinc, lead and copper mineral deposits are known to contain gallium concentrations of between 0.001-0.05 wt%, and these constitute additional reserves. Similar proportions of gallium are also present in certain fractions of coal ash. For the two decades before 1970, gallium was recovered commercially from enriched coal ash in the UK as a by-product of germanium extraction. Therefore, the reserves of gallium may be considered to be virtually unlimited in the context of present-day demand.

STRUCTURE OF THE GALLIUM SUPPLY MARKET

As mentioned in the section above, almost all of the gallium currently available is produced from the bauxite ores of aluminium. Consequently, the bulk of the world's supplies are provided by leading aluminium producers. These companies are listed in Table 1 of this table.

Alcoa in the USA is the world's largest supplier of gallium with an estimated output of 5-6 tonnes in 1979. The other major aluminium producers in North America (Kaiser Aluminium, Reynolds Metals and Alcan) do not recover gallium. In Europe, the major producers are Alusuisse (based in Switzerland), VAW Ingal (West Germany) and Pechiney (France). Rhône-Poulenc has also jointed the ranks of the major European producers and already claims to be among the world's largest. Sumitomo is a notable Japanese producer. There is no UK gallium producer of any significance.

Gallium is also produced in aluminium plants in the USSR, Czechoslovakia, Hungary and China. Czechoslovak production is believed to be about 2 tonnes per annum and Hungary is reported to produce close to 200 kg, but no figures for the two major Communist countries are available.

Country	Producer	Estimated output of gallium (tonnes)
USA	Alcoa*	5-6
Switzerland	Alusuisse	
West Germany	VAW Ingal	
France	Pechiney	8-9
Japan	Sumitomo	
USSR	State enterprises	not known
China	State enterprises	not known
Czechoslovakia	State enterprises	2
Hungary	State enterprises	0.2

*Alcoa has announced that it might halt production of gallium.

Table 1 : Major primary producers of gallium in 1979

VULNERABILITY OF SUPPLIES

General considerations

Australia is responsible for approximately one-third of the world's supply of bauxite, which amounted to some 88 million tonnes in 1979. Next in rank to Australia come Guinea, in West Africa, and Jamaica, which together account for a further 25% of the global supplies. Surinam, Guyana and Greece contribute an additional 13% of the available bauxite. As a result, one may reasonably regard the bulk of the West's supplies of gallium ore to be politically secure, both in the short term and in foreseeable future.

The principal dangers facing gallium supplies lie with the extremely small size of the gallium market. This market is small even by comparison with those of other special elements used in the semiconductor industry : the 14 tonnes of gallium produced in the Western world in 1979 compares with an output of about 100 tonnes of germanium and about 2 000 tonnes of device-grade silicon. One consequence of this fact is that small changes in demand for gallium can seriously upset the supply-demand balance. This was demonstrated in the mid-1970s when a surge in demand for gallium arsenide and gallium phosphide for LEDs in calculators and digital watches lifted the price of gallium. In the latter part of the decade this market slackened, as reflected in Fig.1 of this table, with the shift from LED to liquid crystal displays. Moreover, because this metal is a low-level by-product of aluminium, significant delays may occur before production is boosted to meet any sudden increase in demand. The chances of such a situation occuring are fairly high in view of the promising applications of gallium currently being developed.

The position could be further exacerbed by speculative intervention encouraged by the promotion of "strategic metals" as an investment medium. Gallium, like germanium, is particularly vulnerable in this respect : firstly, because it is necessary to make only a very small outlay in order to have a major impact on the market; secondly, because dealing is conducted largely on an ad hoc basis, rather than via long-term contracts which help to stabilise trading of the high-tonnage commodity metals.

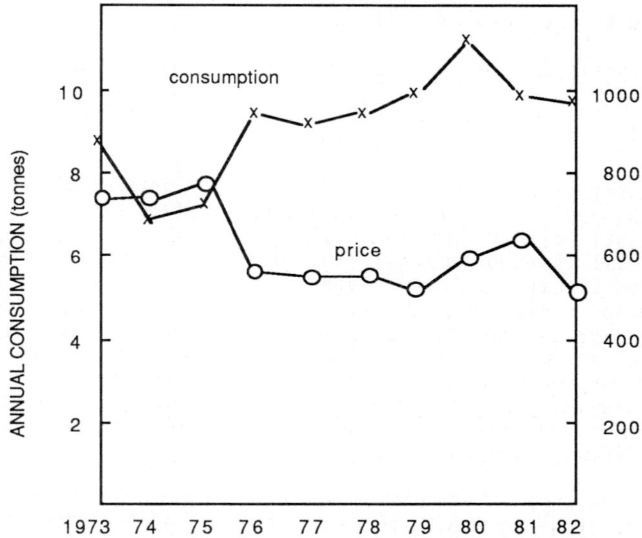

Figure 1 : US gallium consumption with produces prices, 1973-82.

Exerpt from D.M. JACOBSON, D.S. EVANS "Critical Materials in the electrical and electronics industry". The institution of Metallurgists. London 1984.

Table IV.1.B : **Gallium**

Silicon, the second most abundant element in the Earth's crust, accounts for about a quarter of its estimated mass. Only oxygen exceeds it in abundance. However, elemental silicon is never found free in nature : it always occurs as an oxide (eg quartz, amethyst, agate, flint and opal) or in silicates (eg granite, hornblende, asbestos, feldspar, clay and mica). The mineral sources for silicon production are generally early palaeozoic deposits of quartz and quartzite. These are distinguished for both their high purity (95-99% silica) and for their natural abundance.

Silicon has gained substantial worldwide importance as the basis for semiconductor devices. The production of semiconductor-grade silicon requires several stages of processing, in the sequence presented below.

(1) Direct reduction of silica ore by a carbonaceous fuel such as coal, coke or charcoal in an electric arc furnace. The purity of the silicon obtained in this primary smelting stage depends largely on the grade of silica ore and the purity of the reductant used, but a purity of 98-99% is usually achieved.

(2) The nominal-grade silicon recovered from the smelter is pulverised and reacted with hydrochloric acid, methyl chloride or chlorobenzene to produce a mixture of chlorosilanes. These are then separated by fractional distillation.

(3) The purest fraction of a chlorosilane, usually trichlorosilane ($SiHCl_3$), is decomposed by reacting it with hydrogen to form polycrystalline silicon of high purity.

(4) Polycrystalline silicon is converted into single-crystal silicon by one of two established methods : "float-zone growth" or "Czochralski pulling". The Czochralski technique is the cheaper of the two and is used primarily for growing crystals intended for integrated circuits and small-signal devices. However, crystals of the superior purity required for rectifiers or other power devices must be grown by the float-zone method. Various dopants are added to the molten material at a preliminary stage of the operation to meet the required electrical specifications, defined in terms of carrier concentration, carrier type (p- or n-type, resistivity, etc). Device manufacturers generally require silicon wafers to several different specifications.

The sequence of processing steps is summarised in Table 1 of this table. Most of the output at each stage of the production is used for other purposes. Thus, about 85% of the silicon smelted finds use in steels, aluminium alloys and other metallurgical applications. This is quite apart from the much larger quantity of ferrosilicon produced directly by the co-reduction of silica and ferrous ores. The majority of chlorosilanes produced are consumed in the manufacture of silicones. It is only a relatively small proportion of the primary silicon that is retained for further purification and use in semiconductor devices - in fact, less than 1% of the total. Ultra-pure (>99.999%) polycrystalline silicon may therefore be regarded as the basic starting material for electronic applications.

Power and capital charges are major components of the processing costs at each stage of beneficiation. The approximate price of the material at various stages of the processing chain is shown in Table 2 of this table.

DEMAND FOR DEVICE-QUALITY SILICON

Worldwide demand

Production figures for polycrystalline or single-crystal silicon are not readily available. Some figures have been issued for polycrystalline silicon production in the USA; they shown an annual growth rate of 17% per annum over the decade 1968-78, from 245 tonnes to nearly 1 000 tonnes. The US demand has been increasing at possibly an even faster rate since sizeable quantities of polycrystalline silicon are imported, especially from West Germany.

quartz, quartzite (SiO_2)	
+ carbon	
silicon (metallurgical grade)	about 85% for metallurgical use ferrosilicon, aluminium-silicon
+ HCl	
$SiHCl_3$ + other chlorosilanes (nominal purity)	Majority for use in silicones (corresponding to 14% of the original silicon production)
distillation	
$SiHCl_3$ (high purity)	
+ H_2	
polycrystalline silicon (device quality)	
crystal growth, cutting and polishing	
silicon single-crystal wafers (<1% of the original silicon production)	

Table 1 : The production chain for silicon wafers.

Form of silicon	Price *(£/tonne silicon content)
Quartz, quartzite (>98% pure SiO_2)	20
Metallurgical-grade silicon (>98% pure)	800
Polycrystalline silicon (>99.999% pure)	30 000
Single-crystal silicon wafers (100 mm dia x 500 μm)	400 000

(*1983 prices).

Table 2 : The approximate value of silicon at different stages of processing.

In 1979, however, investment in the US polycrystalline silicon industry was held back, resulting in some shortages. This development was blamed on the government-funded Low Cost Solar Array (LSA) project, one of the main aims of which is to significantly cut the cost of producing high-purity silicon.

The following year (1980) saw a slowing down in the growth of demand, together with a boost in the production of polycrystalline silicon, which eased the supply situation. However, an upsurge in demand coinciding with the US economic recovery of 1983-84 put supplies under pressure again : over that year, the market for device-quality silicon grew by between 30 and 40%. In 1984, the world demand for that material is estimated to be in the region of 5 000 tonnes.

Future demand

Demand for device-grade silicon is expected to continue growing at a much faster rate than industrial growth in general. Forecasts for the years 1980-85 give an annual growth in consumption of silicon and semiconductor products of between 19 and 24%.

The main sectors of future growth in demand for silicon are expected to be in :

(i) integrated circuits, where production will shift increasingly towards silicon wafers of larger diameter

(ii) solar (photovoltaic) cells, which constitute a large area of potential growth, contingent on the development of economically viable silicon solar cells; however, the material specifications might be quite different for this application.

World suppliers of device-quality silicon

Production Country	Trichlorosilane ($SiHCl_3$) producer	Device-grade silicon producer supplied	capacity (tonnes)
USA	Dow Corning Corp*	Hemlock Semiconductors Corp (Dow Corning subsidiary)	1 200
	Union Carbide Corp	Union Carbide Corp	100
		Monsanto Corp*	230
		Great Western Silicon Inc (General Electric subsidiary)	300
		Motorola Inc	100
	Texas Instruments Inc	Texas Instruments Inc	350
Japan	Osaka Titanium Mfg Co	Osaka Titanium Mfg Co	450
	Shin-etsu Chemical Industry Co*	Shin-etsu Handotai (USA and Japan)	100
FRG	Wacker Chemie-tronic GmbH*	Wacker Chemietronic GmbH	2 000
	Dynamit Nobel AG	Dynamit Nobel AG (Italy)	400

*Major manufacturer of silicones.

Table 3 : Leading producers of high-purity silicon and their 1983 capacity with their suppliers of trichlorosilane.

The majority of the high-purity polycrystalline silicon currently available is produced from ultra-pure trichlorosilane ($SiHCl_3$) according to a process devised by Siemens. A modified Siemens process using dichlorosilane (SiH_2Cl_2) is claimed to double the yield of material without detriment to its quality; this route is currently being explored by Hemlock at its new plant. Komatsu in Japan uses silane (SiH_4) instead for its small production and Union Carbide has decided to adopt this process for its photovoltaic-grade material.

The few major producers of trichlorosilane are listed in Table 4. Together with the high-purity silicon manufacturers supplied by them. In a European context, Rhône-Poulenc (France) deserves mention as a producer of trichlorosilane.

The chlorosilanes form the basis of the silicone industry, which is much larger than device-grade silicon manufacture, and so it makes sense that polycrystalline silicon plants are frequently sited close to silicone factories. Examples are Dow Corning at Midland, Michigan and at Carrolton, Kentucky, USA, and Wacker Chemietronic at Burghausen, West Germany.

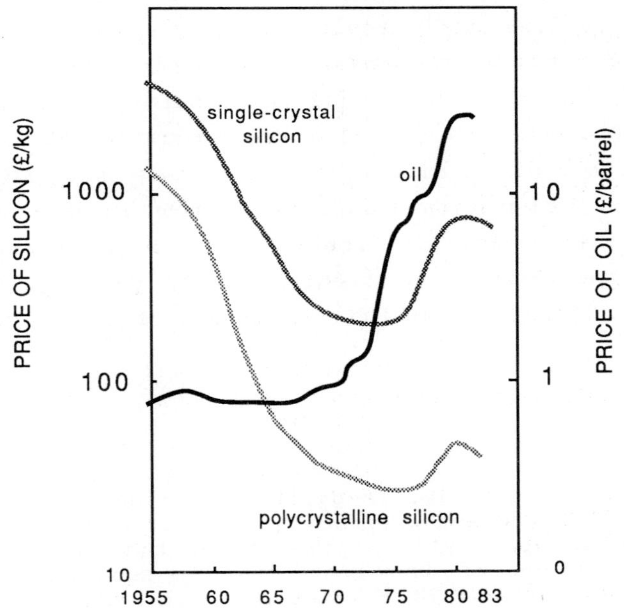

Figure 1

Table IV.1.C: **Silicon**

Technical advances of silicon

Silicon is obtained from the most abundant raw material on earth. Gallium, by contrast, is a relatively rare element (see Tables IV.1.B and IV.1.C).

Silicon is stable and relatively easy to obtain. Combining arsenic and gallium, on the other hand, is a complex and tricky operation. This production difficulty is due in particular to the volatile character of arsenic, which cannot withstand the high temperatures reached in the crystal growing methods used for silicon, and also to the difference in mass between arsenic and gallium, which tend to separate under the effect of weight and convection currents during crystal formation and consequently cause a high incidence of microscopic defects through lack of homogeneity. This technological difficulty could be overcome by fabricating III-V compounds in a microgravity environment in space. This accounts for the interest being shown by semiconductor manufacturers in the crystal growing experiments conducted both by the American shuttles and by Soviet orbiting space stations.

Silicon enjoys the benefits of experience, as a result of which the manufacturing cost of silicon circuits is still very low compared to that obtained with any other semiconductor (2) .

The physical limits of silicon have not yet been reached. In particular, integration density on silicon integrated circuits is still doubling every year. (It should however be observed that with gallium arsenide the density is trebling every year, so that in theory it should equal that for silicon by the end of the century).

---

(2) At the present time, a gallium arsenide wafer costs between $3000 and $5000 (to set against the cost of a 4" wafer of CMOS silicon, which is about $120), the bulk of the cost being represented by depreciation of the manufacturing equipment; for example, the figures for a 1K SRAM (Static Random Access Memory) using gallium arsenide are :

$9,000,000 equipment
ie $1,800,000 amortization
or $1,000 per wafer for a production rate of 15 units per month.

The result is a high cost price - for example, almost $100 for 1K gallium arsenide SRAM at the end of 1985.

These prices are bound to fall as output improves, possibly coming into line with silicon prices in a few years' time, and as the quantities produced increase (according to CPE Bulletin N°. 18, July-August 1985).

From an examination of the respective advantages of both materials, there are several possible development hypotheses that can be offered.

The first is that the status quo will be maintained, in other words silicon will continue to dominate, leaving gallium arsenide and the other III-V compounds with specific markets only (hyperfrequencies, opto-electronics, ultra-fast electronics, etc). This hypothesis is based on the capacity for improving silicon's possibilities still further, particularly by making large chips, developing the "three-dimensional" structure (for the time being, only a "two-dimensional" structure is exploited on silicon wafers), developing "silicon on corindon" technology to counter the semi-insulating properties of gallium arsenide, and so on.

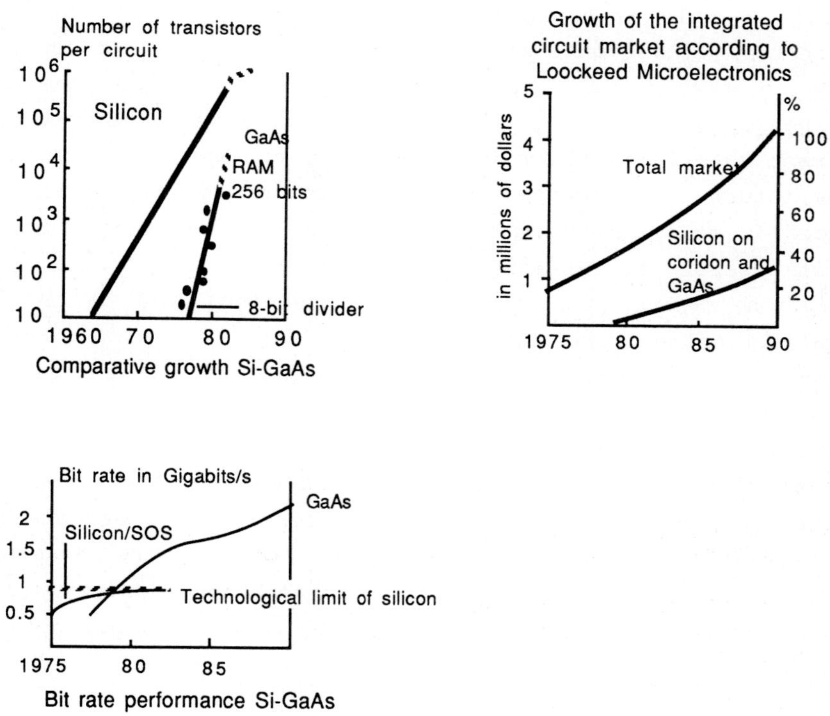

(Source : Sciences et Techniques, May 1985).

Figure IV.1.b : A few key elements in the comparison between silicon and gallium arsenide

The second hypothesis is a slow breakthrough by gallium arsenide, from applications in ultra-fast electronics to more routine applications in micro-electronics, etc., as the benefits of experience gradually assist this technology. This gradual substitution could happen all the more easily as gallium arsenide is compatible with most of the manufacturing processes for integrated circuits involving silicon. The recent processes employed - which, with greater and greater integration, require operation at atomic level, so that it really is possible to talk of "atomic engineering" - also favour applications of III-V compounds. This is true of the new crystal--growing techniques, and especially of <u>molecular beam epitaxy</u> which offers the possibility of producing combinations of crystals adapted to a wide range of uses (introducing a measure of "variety" into the manufacture of semiconductors) and of building compounds up in successive layers only a few atoms thick. This is where hetero-epitaxy comes in : this involves the coherent piling up of thin layers of different materials on a crystalline substrate. This process, which is likely to favour III-V compounds, is very valuable one in making <u>super-networks</u> (periodic networks consisting of alternate layers of different materials) which will be a key concept in micro-electronics in the future. The combined use of silicon and gallium arsenide is not excluded in this development : mastery of the multilayer method might make it possible to fabricate "bivalent" chips made from a combination of silicon layers (for electronic processing) and gallium arsenide layers (for optical processing) (3) .

The third hypothesis is the manufacture of entirely optical circuits containing no electronic devices whatever. Integration of all optical functions (transmitters, switches, amplifiers) on one and the same material will be the key to the development of real "photonic computers". This hypothesis would have considerable effects and would bring about a veritable technological leap in the development of information processing. Information processing systems (computers) would then have far greater compatibility with the major optical fibre

---

(3) It must be emphasized that other, radically different, research approaches in micro-electronics have been envisaged, notably the use of Josephson effect (a superconductor current capable, under the influence of a magnetic field, of traversing an insulating material at the temperature of liquid helium). The advances made by silicon and gallium arsenide appear to have set back the implementation of this research substantially.

networks. Although a great deal of research is being done on this subject, the development of optical components is still in its infancy. The only optical switching devices on the market today are lithium niobate devices. The development of III-V compounds, in the forefront of which are gallium arsenide and indium phosphide, could be strongly boosted by the development of integrated optics (or micro-opto-electronics).

The radically different implications of these three hypotheses explain the widely divergent market projections for gallium arsenide and silicon. However, all the signs are that in fact the three hypotheses will correspond to three successive phases in technological development : initially, consolidation of silicon technology, exploiting effects of scale and the technical "margin" until the limits of the material are reached; gradually increasing use of gallium arsenide beyond its present technological niches (particularly ultrafast electronics, as in the case of the Cray 3 supercomputer), which should enable this material to start winning some market sectors from silicon from the 1990s onwards; finally, it seems certain that the whole logic of the evolution of information processing and transmission will bring the advent of integrated optics sooner or later, though it is extremely hard today to foresee when that period will occur. Constructing the "photonic computer" is primarily a problem of integration. Just as the value of semiconductors in electronics derives from the possibility of using a single crystal to perform functions which previously required a whole assembly of separate components, so the development of the "all-optical system" will depend on the possibility of integrating components which are now produced separately. Even if it cannot yet be stated that gallium arsenide and III-V compounds will be the best possible materials, it is clear that they are playing an eminently strategic role in that development and that no country or group of countries wishing to master future technologies can afford to neglect its options in the slightest degree where this type of material is concerned. Gallium arsenide has already gone beyond the stage of pre-industrial development, and everything points to its being at the heart of future developments in information technology.

However, it must be said that this scenario may be overturned at any time by serious political tensions. Because it has "military" properties greatly superior to those of silicon, gallium arsenide could be in very heavy demand from the armed forces (principally American).

The advantages which silicon enjoys by reason of its abundance and the experience gained with it would then only play a very marginal role.

	1985	1995
All semiconductor components and devices	31.7	205.2
Groups III and V semiconductor components and devices of which	1.3 - 1.9 (4.1 - 5.9%)	12 - 20.3 (5.8 - 9.8%)
** Optoelectronics	0.9 - 1.3	7.5 - 12.8
** Hyper-frequencies	0.2 - 0.3	1.5 - 2.5
* Others	0.2 - 0.3	3.0 - 5.0

(Source : CPE Bulletin, N°. 18, July-August 1985).

Table IV.1.D : World production in milliards of dollars at current prices (assuming 5% inflation per annum).

IV.1.3. CONSTRUCTING AN INTEGRATED INFORMATION PROCESSING, TRANSMISSION AND DISPLAY NETWORK : THE ROLE OF OTHER MATERIALS

Over and above the critical competition between silicon and gallium arsenide within the information technologies, constructing a coherent network for the processing, transmission and display of information depends largely on devising materials. One of the most strategic sectors is the transmission of information, where there is keen competition between traditional coaxial cable, optical fibre and satellite transmission. The development of optical fibres - at least for long-distance communications - is a major factor in this competition.

IV.1.3.1. Optical fibres

The principle of transmitting light by optical fibre has been known since at least 1870. But it was the discovery of lasers which enabled fibre communications to really take off. The particular feature of the laser is that it transmits "coherent" light, that is to say light with a clearly defined wavelength whose signals are emitted in phase, so that the signal attenuation is very much reduced as compared with emission from a conventional source where the "wave packets" have a diffusion speed which varies with wavelength, become superimposed and eventually cancel each other out statistically.

Moreover, the laser emits at very high frequencies and the carrier wave can be modulated over a very wide band. Thus the amount of information transmitted is multiplied, affording important economies of scale. So it is not by chance that fibres are preferred for use in telecommunications networks, for high-throughput links - for example, Paris-Lyon in France where capacity can reach as much as 565 megabits per second. The installation of fibreguides at local level is consequently not justified for transmission of voice telephony (sampled at 64 kilobits per second) but for image transmission - in both directions - which requires a minimum 2 megabits per second with present techniques.

In a fibreguide, the light wave emitted by the laser or by a light-emitting diode is guided inside the fibre as it "bounces" off the sides of the core which has a refraction index higher than that of the cladding itself, which is made of glass. As in all transmission systems, the aim has been to limit the losses caused by these rebounds as a function of distance (the losses can be evaluated in terms of decibels per kilometer (dB/km)). The American firm Corning Glass, which at present holds most of the manufacturing patents, succeeded in 1970 in producing a fibreguide with an attenuation of 20 dB/km. Since 1980 it has been possible to achieve attenuations in the region of 1 to 3 dB/km depending on wavelength, and even 0.2 today.

If the core diameter (30 to 80 microns) is large in relation to the wavelength (of the order of 1 micron), the signal zigzags too much and a problem of phase displacement and interference arises, so that the bandwidth (and consequently signal transmission capacity) is limited. This type of so-called "multimode" fibreguide with a sudden index leap between the core and the cladding was the dominant technology in the 1970s. By slightly varying the refraction index inside the core it has been possible to alleviate the problem : the wave no longer bounces fully off the core/cladding interface but is gradually deviated by refraction inside the core towards the centre. By varying the refraction index inside the core, the signals covering the shortest distance are forced to do so in a "slower" medium; in this way the phasing is respected. This type of fibreguide is referred to as an index gradient one. Multimode fibreguides of this kind make it possible to recover a set of phased signals, which is of fundamental importance in telecommunications : multiplexing mixes several hundred types of signals, and to have some travelling faster than others is unacceptable.

Lastly, if the diameter of the fibreguide is comparable to the wavelength of the signal the laws of conventional optics no longer operate, and it is possible to produce single-phase signals with little attenuation, because the wave no longer bounces on the walls. To achieve this objective fibreguide have consequently been manufactured with finer and finer cores and light sources (laser) with higher and higher wavelengths (1.3 microns). These are the single-mode fibres, which are thus the latest generation of fibreguide, the technology of which has scarcely been mastered yet.

The production <u>costs</u> of fibreguides have now come down to economic levels, as compared to cable, for long distances. As repeaters and amplifiers are more expensive but spaced further apart, fibre is comparable in cost to cable for connection lengths over 1 km. In this case it is the connection components - connectors and transducers - which account for most of the cost price of an optical fibre system.

The problem with fibreguides for local use is that they offer excess capacity over actual needs. Only high-volume data processing transmissions in local networks, or interactive cable television programmes, may need the capacities offered by fibres. Experiments have been done (eg. Biarritz in France, Bigfon in the Federal Republic of Germany, Hi-Ovis in Japan and Milton Keynes in the United Kingdom) to test fibre network services and configurations for local telecommunications use after the year 2000, and have shown a measure of redundancy relative to needs at present prices.

<u>Demand</u> for optical fibres can be broken down into several markets (see section IV.1.4 for a more detailed consideration of strategies) :

* long-distance transmission for postal administration networks, where a distinction is drawn between national and international. At the national level, demand from European postal administrations is shown in the following table :

	85	86	87	88
Federal Republic of Germany	50	86	104	145
Italy	10	12	13	15
United Kingdom	120	?	?	?
France	50	?	?	?

Source : Jeunhomme (1985).

Table IV.1.E : Demand from European postal administrations for long-distance transmissions.

It will be seen that a lead has been taken by the British who as long ago as 1979 launched a £ 5 million programme of fibre cabling offering 450 km of optical cables divided among 15 links. British Telecom expects that 50% of its interurban transmission will be carried by fibre by 1988.

At this level, fibre may be in competition with satellites. It certainly has a much longer life span, but it offers a point-to-point link whereas the satellite is much less dependent on network topology.

The other factor in the development of this type of demand is deregulation. If a carrier has to face up to competition he is obliged to modernize rapidly, even if his equipment has not yet reached the point of complete obsolescence. Deregulation is reflected in a price war which can only be kept up in the long run if the marginal cost of the equipment is low. This is why AT & T, which is simultaneously a dominant carrier and a producer of fibres, has announced the opening of production units with a capacity of 1 million km per annum. According to certain estimates, if the projects announced in the United States are carried through, the transmission capacity of telecommunications networks in that country will be multiplied sevenfold. However, it is planned to equip only one-third of new demand in the United States with optical fibre, the other two-third going to coaxial cable.

The problem is similar where international traffic, particularly across the North Atlantic, is concerned. A TAT-8 underwater cable project (335 million dollars) is currently being implemented by AT & T together with STC, ITT's British subsidiary, and France Câbles et Radio (a DGT subsidiary). This project is in the tradition of undersea cables (the TAT series) managed as a consortium by the carrier. But deregulation in the United States and in Great Britain is bringing new agencies into the picture. Two projects have received licensing approval from the FCC (Federal Communications Commission) in the United States :

- Tel-Optik (600 million dollars) is a joint venture between the American company Tel-Optik and the English company Cable and Wireless, owner of Mercury, the rival network to the British Telecom network. One problem remains unresolved, because Tel-Optik is not a carrier in the United States.
- SLC (Submarine Lightwave Cable Company) in the USA is planning to install a fibre cable with a capacity 12 times as high as that of TAT-8. But SLC has no European partner for the time being, although contact has been established with Bristish Telecom.

* The second type of demand comes from local networks or LANS. The fibreguides are easy to install and protected from interference, and therefore insensitive to an electrical environment. As the connectors are expensive, the LAN configuration is generally star-shaped with a passive conductor network (bus) collecting all the information and transmitting it to all stations, only the addressee decoding the message. Demand for LANs is very uncertain at the present time; there is no market standard, despite the efforts of Xerox and IBM.

* The final type of demand is for cable television. Here again, estimates are being revised downwards after a period euphoria. There is no major country in which projects are not being scaled down. France, which had the most ambitious policy on the use of fibres for these networks, may be be forced to abandon its cable plan in the teeth of competition from commercial television and satellite television. The Federal Republic of Germany has resolutely opted in favour of coaxial, and Great Britain, which tended to adopt mixed solutions, has considerably revised its plans, several concession holders having withdrawn. Only small countries like Belgium, already 60% cabled, and Denmark are pursuing an ambitious strategy, but the use of fibres will in any case be limited in this field.

So the really important market continues to be demand for long-distance transmissions. Some estimates made in the early 1980s in fact forecast that this kind of market would account for 75% of fibre production by the end of the decade and 30% of total transmission demand. It is a field in which the construction of a European-scale infrastructure could have a favourable effect on industry, as has been realised by the initiators of the RACE programme, basing themselves on national market demand. Optical fibres have now reached the take-off phase.

The industrial production of single-mode fibres offers a potential such that demand is now preceding supply. Economies of scale are the dominant economic factor, as mentioned above. Europe may well suffer the same setback in this market as in the case of computers or components, unless it creates a common market and manages to join forces at industrial level.

IV.1.3.2. Other components

Semiconductors for information processing, and optical fibres for information transmission, are basic elements in the development of informatics and telecommunications. But the industrial application of these basic materials requires the development of a very large number of other materials in the form of components, in order to make interconnections between the basic elements, serve as carriers for integrated circuits, carry out manufacturing processes, etc. The markets for these materials are large and steadily growing. The estimates below show that their order of magnitude is higher than that of the basic materials.

In electronics

- About half the total market for materials (which can tentatively be estimated at a total of 10 milliard dollars in 1985) is accounted for by metals, which are mainly used for their conducting properties. In terms of money, the most important markets are those in precious metals (gold and platinum especially) and those in traditional conductors (copper, aluminium, nickel, etc).
- The other half of the market is divided up among the following materials :

* 60% in chip manufacture, of which :
    - 45% for semiconductor substrates (silicon, gallium arsenide, etc);
    - 40% for manufacture of carriers, of which :
        * 45% technical ceramics;
        * 20% various pastes;

* 10% epoxy resins;
* 10% various adhesives;
* the rest being made up of various other materials (rare earths, etc);
* total : 100%

- 15% for products or materials used for preparation but not comprised in the chip (photosensitive polymer, various washing baths, etc);
- total : 100%

* 40% for the manufacture of printed circuit boards of which :
  - 60% for substrates (epoxy resin composites);
  - 15% for photosensitive polymeric conductors;
  - 15% for conductors;
  - 10% for various materials (adhesives, inks, etc)
  - total : 100%
* total : 100%

<u>Thus the price of silicon represents only (1x0.5x0.6x0.45 = 0.135) 13.5% of the total material cost</u>.

In opto-electronics

We can distinguish the following proportions (for a total market in the region of 700 million dollars in 1985) :

. 40% for sources and receivers which convert electricity into light and vice versa :
  * 5% for light-emitting diodes (generally based on III-V compounds);
  * 27% for laser diodes (top-range light emitters, generally based on III-V compounds);
  * 9% for photodetectors (for converting light into electric pulses, generally based on silicon);
  - total : 40%

. 60% for passive components :
  * 18% for multimode fibres;
  * 34% for single-mode fibres;
  * 8% for other passive components, including connectors, couplers, multiplexers/demultiplexers and switches
  - total : 60%

Among all these various components, one of the most remarkable developments is that of engineering ceramics. Their principal applications in electronics and opto-electronics are as follows :

- Dielectric (insulating) ceramics, which represent more than half the market. In the commonest applications, the main condensers are made of alumina or are multilayer (barium titanate). However, active research is proceeding into high dielectric constant ceramics - $PbMgO_3$, iron tungstate, lead tungstate and lead titanate.

- Ceramics for "packaging" computer chips. "Defined as the complete set of processes necessary for the physical, mechanical or electronic connecting of a large number of chips, packaging in fact entails the possibility of building up several layers such as to permit three-dimensional connection. The best example of its application is probably the "multicarrier" TCM (thermal conduction module) which has up to 33 layers of ceramic (alumina) about 25 microns thick and which connects 121 chips together" (4) .

- Ceramics for sensors (eg applications of YAG crystals - yttrium, aluminium and gadolinium oxides - in layers).

IV.1.3.3. "Peripherical" display and recording components

Beyond the circle of functional materials directly linked to the basic materials (semiconductor and optical fibres), the implementation of computer and telecommunications technologies requires a range of materials which perform complementary (peripheral) functions : materials (other than semiconductors) for storing information, materials for displaying information, etc. Consequently, around the "hard core" of the basic materials there is gradually built up a coherent set of materials which have to be optimally compatible with each other. However, the further one moves away from the central "hard core" the more the development of the functional materials encountered

---

(4) "Les céramiques avancées aux USA" ("Advanced ceramics in the USA"), CPE Bulletin, N°. 22, December 1985.

becomes independent of the technical development of the basic materials.

Among the "peripheral" materials relevant to information technologies there are two categories of functional materials which are the subject of particularly intensive research - materials for information display and materials for information recording.

The most promising <u>materials for display</u> are, firtsly, light-emitting materials based on zinc sulphide or rare earth oxysulphides, but above all passive materials (which create a visible dot by reflection or absorption). Among the latter, <u>liquid crystals</u> present the most interesting properties. They are already used in medium-sized display systems with fairly low information density, and could well be done of the essential elements in the development of a highly controversial and highly competitive market - that of <u>flat screens</u>. In the longer term, consideration is also being given to the development of electrochromes (especially with organic matrices) or to building up circuits in thin layers on amorphous materials which hold out prospects for the targetting of large display surfaces.

<u>Materials for recording</u> are in most cases magnetic materials. They are developing in spectacular fashion and is it not inconceivable that thin-layer magnetic recording might equal silicon-based micro-electronic recording in volume terms.

At the present time information is generally stored by transcribing data on to an aluminium disk coated with iron oxide. As the disk rotates (approximately 3,000 rpm), a read-write head (a coil of wire wrapped around a magnetic core, typically a nickel-iron alloy) is brought near the surface. An electric current generates a magnetic field in the coil which records the data by magnetizing a particular area of iron oxide on the disk. Information read-out is achieved by the reverse process. The problem is that it is difficult today to increase the density of data on the disk, because "as the bit density increases, the magnetic field of each bit decreases; in order to detect the smaller magnetic fields, the head must be positioned closer to the disk. Unfortunately, smaller head-to-disk spacings increase the probability of occasional but devastating impacts of the head onto the disk. Moreover, since the surface of the magnetic layer is usually quite rough - of the order of 1,000 angstroms - frictional wear is then quite common" (5) .

One way that is being considered of improving the existing systems is to develop smoother disks or to increase the sensitivity of the magnetic heads. However, the greatest hope lies in the design of magnetic materials which will interact with light (magneto-optical materials), which would have the advantage of entirely eliminating the reading head (information being recorded by simultaneously sending a magnetic field and a laser pulse). Among these materials, amorphous-alloy films (containing rare earths such as gadolinium or terbium) and transition metals (iron or cobalt) are receiving greatest attention.

IV.1.4. TELECOMMUNICATIONS - THE STRATEGIC STAKES

As the telecommunications field is traditionally a key sector of European industry, it was necessary in this chapter on "materials and information" to give a description of the strategic implications, bound up in large measure with mastery of the functional materials we have been considering.

In many respects the telecommunications sector is crucial to European countries :

- First of all in terms of markets, since according to recent estimates telecommunications are likely to represent a proportion of added value in the European Community countries equal to 7% of GNP by the year 1990. By then telecommunications will have overtaken the car industry and will probably become the prime industrial sector (in 1985, annual world sales in the sector amounted to 40 milliard ECUs).
- Secondly in strategic terms,since the establishment of a coherent telecommunications infrastructure is a crucial factor for job creation and competitiveness in Europe, both upstream of this sector (suppliers, particularly of materials, to the telecommunications industry) and downstream, where the development of <u>service activities</u> depends on the quality and the cost-performance ratio of the telecommunications

---

(5) P. CHAUDHARI "Electronic and magnetic materials", ("Les semi-conducteurs et les matériaux magnétiques"), Scientific American, December 1986.

network. Thus, just as economic growth was conditioned since the end of the nineteenth century by the creation of a transport network and of an energy distribution network, so the establishment of an efficient telecommunications network is likely to hold the key to economic growth in the decades ahead, for that infrastructure is the "pivot" around which a growing range of services will be able to develop. The quality and competitiveness of the services sector in Europe will depend on the quality of that infrastructure. Thus, as the introductory RACE report points out (6) , "a large part of this investment can create employment in the Community if Europe's industry can successfully compete with international competition, otherwise a large part of this investment will create employment elsewhere. More importantly, however, it is estimated that for those leading in offering advanced information services the employment benefit may be 10 times as large. For Europe's employment prospects the creation of favourable conditions for new and enhanced services is the most significant employment aspect of advanced telecommunications services. Based on these estimates the overall employment at stake may be as high as 5 million by 1990".

The materials industry is of course particularly concerned in current developments : optical fibres, but also various opto-electronic components and devices, as we have already seen, are the principal materials involved in the creation of the future telecommunications network. However, a great deal of uncertainty still surrounds the size of the markets in question and the position of European materials producers in this field. The principal parameters at work here are of three kinds : technical (the development of digitization being the main aspect, apart from the technological aspects which have already been discussed at length); economic (the phenomenon of integration of services); and strategic (deregulation measures and attempts at standardization, involving the public authorities in substantial measure).

---

(6) RACE Introductory Report, COM(86)547 Final. Brussels, October 1986.

IV.1.4.1. The principal determinants in the telecommunications sector

Digitization

The redesign and development of modern telecommunications are made possible by a veritable break in technological continuity - digitization - which enables the various components parts of communication (voice, data, images) to be represented and transmitted by means of a universal code. Combining digital fluxes from different origins and with different functions in one channel is reflected in better transport economy and better use of existing resources and infrastructures. Digitization guarantees better performance, in terms of reliability (an essential component of a network), maintenance costs and installed capacity.

The problem with digitization is that it is a lengthy and costly operation, requiring major investment. A few figures may situate the orders of magnitude involved : in the early 1980s the Bundespost was not expecting to have completely digitized its network before the year 2000, whereas investment per new subscriber, which of course includes modernization of existing lines, is in the region of 3,000 to 4,000 ECUs. As for computer control of switching complexes, using elaborate software, the investment is all the higher as an effort must be made as of now to integrate the whole range of functions and services which users will expect in the years to come.

There are three ways in which the postal administrations (the carriers) can tackle these financial problems :

- charge heavily for existing telephone services;
- explore new markets by offering more services;
- have recourse to national and international money markets.

It is for these reasons that telecommunications are still largely dependent on the telephone, and it is generally estimated that 85% of postal administrations' income will continue to derive from the telephone during the next decade. The difficulties which new telematic services (videotext, teletext, teleconference, etc) are having in getting off the ground amply bear out these forecasts. The development of this sector in the United States provides another pointer. Deregulation (introducing competition into the private AT & T monopoly, which will be discussed further on) has not produced a frantic race to innovate, as was expected, but instead a fierce price war in the most vulnerable market, that of long-distance communications.

Consequently, carriers are primarily concerned with modernizing the telephone network which continues to be the backbone of their activities. In practical terms, that means digitizing signal transmission - starting with the high-volume channels like, for example, links for the Paris-Lyons kind with 565 megabits per second, it is gradually being extended to medium-volume channels, and then to local transmission as far as the subscriber line - and digitization of the switching function. The latter entails replacing the old electromechanical switches by digital switches, that is to say computers which make the subscriber circuit connections which were otherwise made by mechanical contacts or electrical relays. In any event, it seems that the installation or "broadband" infrastructures should make it possible subsequently to improve the performance of the medium and narrow band services.

Integration of services in one network

As we have just seen, digitization makes it possible to <u>integrate</u> several telecommunications services (telephone, telex, facsimile, packet switching of data, specialized lines rented to big users on a fixed price basis, etc) using just one network. By offering the possibility of using one and the same infrastructure for a range of services (which previously came under different management systems), digitization affords <u>economies of variety</u>. The service integration which digitization allows is without doubt the decisive economic factor in making the considerable expansion of telecommunications services viable (cf. Fig IV.1.c). "It gives the user the advantage that all his terminals and applications are connected via the ordinary system; as the signals transmitted and received by the terminals and those carried by the network are then homogeneous, the number and complexity of the

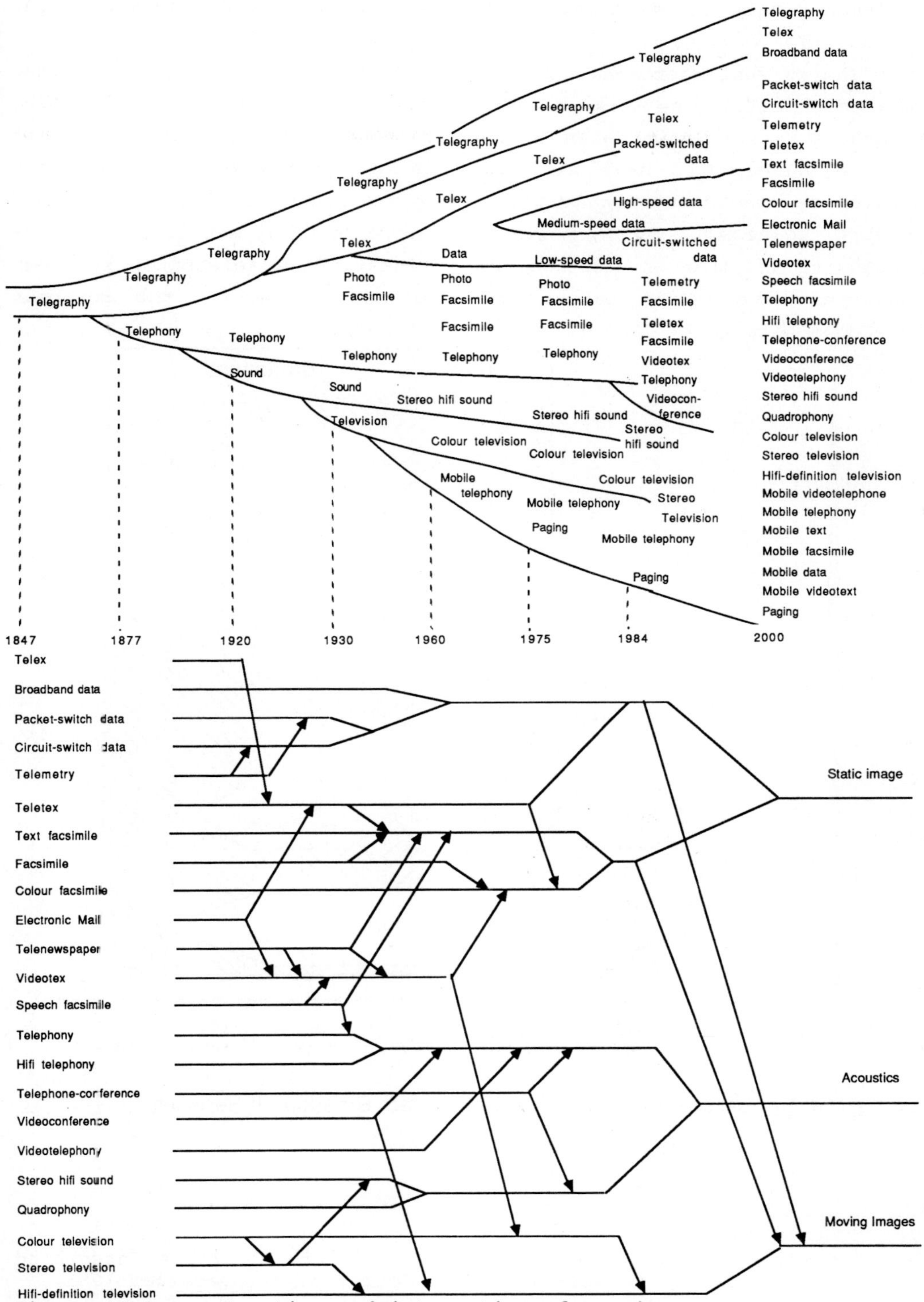

Figure IV.1.c : Creation and integration of services

equipment items in the transmission chain can be limited and costs can be cut" (7) .

The current approach leads to the concept of the ISDN - integrated services digital network - which aims (a) to move the existing telephone network towards an "all-digital" network and (b) to establish other systems which have the advantage of synergy with the switched digital telephone network to which they are connected (packet switched network, Telecom 1 system, value added network (VAN) services - such as processing of data on the network or interactive teletext). European countries have set a target of 5% of ISDN lines in the European telephone network by 1993 (cf. Figure IV.1.d).

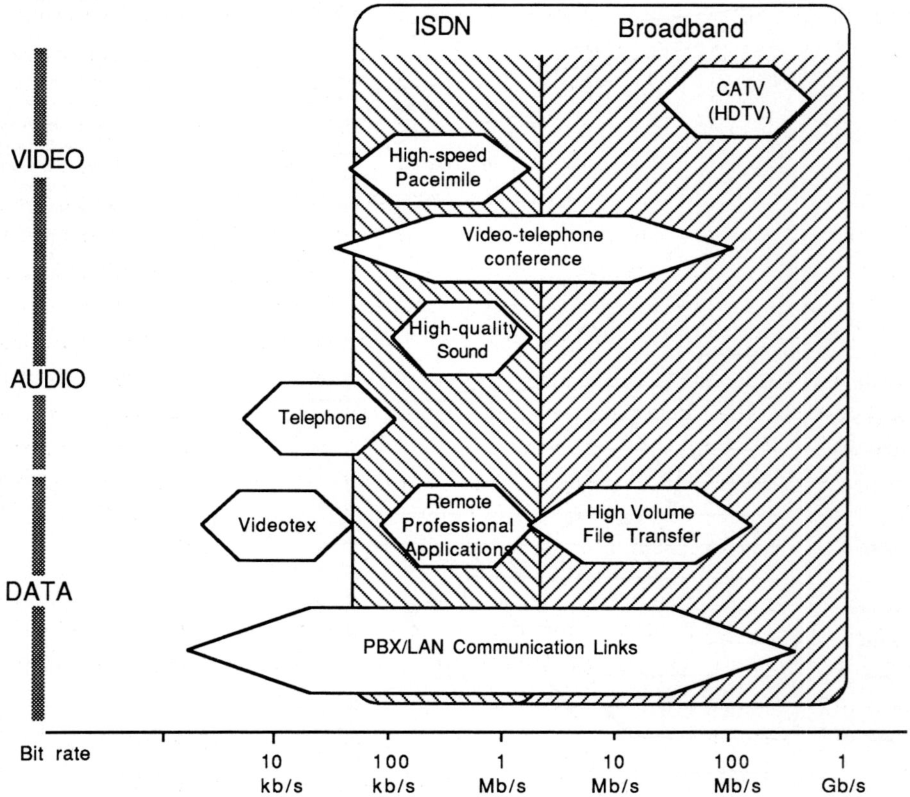

Figure IV.1d: Service-integrated and Integrated Broadband Services

(7) "Le Réseau Numérique à Intégration de Services", ("The Integrated Services Digital Network"), CPE Bulletin, p. 29, December 1985.

One of the most important consequences of the phenomenon of integrated services is, as we have already seen, that it makes the <u>need for the telecommunications and computer industries to get together even more urgent</u> : the computer industry now has the task of developing communication functions between the machines which it sells, while the telecommunications industry is making growing use of electronic components and computers. Finally, office automation is inconceivable without telematics. As a result, firms in one sector devise strategies for penetrating another. The best example is provided by IBM and AT & T. The former is endeavouring to establish itself on the value added services market (abortive IBM/British Telecom agreement), IBM/Nippon Telegraph and Telephone agreement) while AT & T has joined forces with Olivetti to sell its minicomputers and its UNIX operating system. In the United States, IBM is competing with AT & T on its own ground, having bought some of the capital of MCI and Rolm (a producer of private automatic exchanges) while AT & T is developing a new generation of minicomputers with the Convergent Technologies company. So future developments in micro-eletcronics are heavily dependent on the way in which the telematics market evolves (development of complex circuits, processors, digital signals for communication, etc).

These penetration strategies are pursued through a large number of industrial agreements, an analysis of which proves very helpful in understanding the changes now taking place. The outcome is a situation of shared oligopoly between the computer, office automation and telecommunications industries. The features of this emergent oligopoly are as follows :

- very brief product life cycles.
- very high R&D costs.
- internationalized production and marketing (as we shall see later, protected national markets are insufficient to amortize production costs).
- the search for hardware and software compatibility by the end-user, leading to the rapid establishment of legal norms or de facto standards.
- the difficulties encountered by supply in "targetting demand". Spectacular failures are legion, and even the biggest firms are not immune (eg IBM and its home microcomputer).

One of the consequences is certainly that enormous hurdles have to be overcome in order to break into this oligopoly, but another one is intense competition due to innovation and uncertainty about future patterns of demand.

As for the impact of informatics on telecommunications, one of its most significant effects might be the following : the use of semiconductors has indeed increased the modularity of telecommunications equipment, thus reducing the number of intermediates needed between postal administrations and their suppliers. This integration brings down the cost of adapting one network to another, and potential competition among manufacturers has consequently increased. It is becoming difficult, and economically disadvantageous, for the postal administrations to have sole suppliers. The notion of "national champion" is losing out accordingly.

The problem of regulations

In a context of rapid and complex economic change, the question of regulations is one additional uncertainty.

The main purpose of deregulation, presented as a universal panacea, is to speed up the move towards "true pricing" where telephone charges are concerned. However, it is violently resented inside the postal administrations, which have always managed to keep a very large measure of autonomy based on the technical skills of their engineers, the concept of public service raised to the status of a fundamental principle, and powerfully, impregnably organized postal workers trade unions. Although deregulation is not specifically directed against the postal administrations as such, it would force the management to adopt more "commercial" practices. The British paradox clearly illustrates the feeling of the parties involved in the deregulation process.

In Great Britain, a country where the telephone network used to be the most old-fashioned (4,400 Strowger exchanges) whose operating principle dated from 1910, out of a total of 6,700 local exchanges, monopolistic behaviour gave rise to blatant absurdities, particularly where the marketing of terminals was concerned. The policy of deregulation and privatization was fiercely resisted, but the fact is that the British Telecom management itself has now become its most enthusiastic advocate.

It seems quite likely therefore that this movement will spread, bringing about a new pattern of behaviour among networks managers. The Netherlands, Spain and soon perhaps France as well appear tempted, like Japan, to import the British model either wholly or in part. But the process will encounter fierce resistance from among the postal administrations, as it did in the United Kingdom : opening up to competition means above all radically changing ingrained habits. The movement is therefore likely to be very gradual, and thus to displease the American administration, which is pressing strongly for liberalization of the European market.

The point of deregulation is not simply allowing market forces to operate. There is also a commercial battle taking place on two levels :

- to enable the multinationals, most of them American in origin, to lower their communications costs, or even to change over to being providers of information and communication services - this is the purpose of pressures to liberalize transfrontier data traffic;

- to exploit the comparative advantage of American information technology firms - led by IBM - whose efficiency and know-how are at present trammelled by European regulations.

In the long term, deregulation of at least part of the network management activities seems inescapable. Some market sectors, such as the North Atlantic traffic, will be the scene of ruinous competition mechanisms because there is already surplus transmission capacity, bearing in mind the plans announced by the various postal administrations : the European consortium EUTELSAT has announced the launching of two telecommunications satellites, ECS 1 and ECS 2. France already has Telecom 1 in orbit, the Federal Republic of Germany is preparing to send up its Copernikus satellite in 1987, Italy will follow suit with its ITALSAT project planned for 1988, and there are countless satellite projects intended for television transmission (TDF1, TV-Sat, Unisat and Coronet) which can in part replace telecom satellites. Finally, where cables are concerned, the TAT-8 optical fibre project, managed like its predecessors on the basis of a joint agreement between the European and North American postal administrations, is now suffering competition from a similar project on the part of Mercury, British Telecom's English rival, in association with Tel-Optik.

The upheavals are bound to have effects on the long-term planning of European postal administrations. The presence of Mercury alongside British Telecom, for example, has prompted the latter to activate its network modernization plan, with the result that optical fibres are being installed in the long-distance network (between transit centres) on a larger scale than by other European postal administrations. Mercury is also thinking of installing long-distance optical fibre links, because although the fixed installation costs are higher, the marginal operating cost of such networks is very much lower than for coaxials, making it possible to withstand a possible price war and carve out a sufficient portion of the interurban market. On the other hand, British firms are not expecting in the immediate future to install fibre at local level (for cable television for example) where market prospects for the fibre producer are better but the operating economics are very uncertain.

In more general terms, the deregulation of telecommunications will have a decisive influence on the financial situation of the postal administrations. They are already suffering from saturation of telephone demand, and deregulation will skim off their profits on the most economic markets - value added services, interurban telephone traffic, international links, rental of terminals. Moreover, they have embarked on a costly policy of network modernization (digitization). As a result, the resources available for any ambitious industrial policies of support to the electronics branch will be cut to the bone.

On the other hand, the counterpart to deregulation is the search, generally conducted by the multinationals, for new standards which will prevail on the market. In the United States, market deregulation (January 1984) had the effect of allowing the American network to develop to meet the requirements of a very small number of big firms (50% of income comes from 4% of users) which now hold the decision-making power. It is among them that the mechanism (which is likely to hurt many users and competitors) for choosing a standard will operate. In Japan, in spite of the recent privatization of NTT (December 1984), it does seem that governmental policy will seek to ensure that the new infrastructure serves Japanese interests as a whole. This is probably the route which European countries are looking for. A minimum of deregulation seems necessary if the European networks, which are currently very different from each other, are to move towards greater standardization. By contrast, excessive deregulation would have the drawback of encouraging new local solutions and of complicating even further the search for standards. Accordingly, the industry ministers of the EEC have just given their approval to a programme designed to establish common norms for integrated service digital networks.

Deregulation of the telecommunications market may also have important research implications. Prior to deregulation in the United States, ATT's Bell Laboratories were in the position of a public research service, since they were obliged to divulge their research findings. Now the research done by Bell Laboratories has come to be exclusively geared to ATT's needs. Not only is the importance given to fundamental research likely to diminish in Bell Laboratories : the whole of American research and industry in general have lost a quite exceptional source of knowledge. Once again, Japanese strategy is different in this respect, since the NTT research laboratories have remained under public control. Finally, the battle between private and public could well become exacerbated in the actual utilization of equipment, as vertical integration continues to grow between users - especially major users - and the network operator. Private automatic exchanges (or PABX) which are also digitized are now almost indistinguishable from public exchanges. Computers themselves can perform switching functions, with the appropriate software; and this is what happens in local area networks. One major user in the United States, General Motors, is currently defining a market standard for its local network (MAP - Manufacturing Autoprogrammation Protocol). But private logic and public logic have radically different foundations. The former is specific in purpose, the latter aims to be universal. The private user wants primarily to match his communications facilities to his end-purposes, which are specific to him, whereas the network operator is required to offer communications services on the principle of free access and non-discrimination. Defining the boundary between public network and private network therefore lies at the very heart of developments in this sector, particularly as the technology does not enable a clear distinction to be drawn. In fact, the definition depends on a political option.

This idea is perfectly illustrated by the current discussion about value added services. Intermediaries are willing to rent transmission capacity in order to offer improved services (electronic paging, electronic mail, data banks, etc) which change the shape, content, code or other characteristics of the user's information or supply him with new, different or restructured information (8) . The crux of the problem is to decide what charge should be made for renting transmission capacity, and to determine what the intermediaries would be authorized to do on these networks. To give one example : should packet switched data, for the time being the prerogative of postal

---

(8) C. MONVILLE, Bulletin de l'IDATE, Europe des Communications, November 1985.

administrations (in France through the intermediary of the Transpac subsidiary) be open to competition as at present in the United States ?

IV.1.4.2. Implications for the European telecommunications equipment industry

In view of probable developments in postal administrations in the medium term and of technological tendencies European industry appears increasingly vulnerable; confronted with problems of manpower surplus, clinging on to protected or "friendly" markets which are shrinking, it has to establish a strategy of alliances the logic of which is still blurred; agreements with an American partner could lead to subservience, while European agreements are difficult to set up.

The weakness in Europe stems mainly from the lack of integration with the related office automation and data processing sectors : only the Siemens company seems to possess the complete arsenal of technical skills and financial resources to make a possible breakthrough in this market. But faced with stagnating investment on the part of postal administrations and cutthroat competition, the sector is heading towards concentration and the disappearance of some manufacturers : ITT recently announced its withdrawal from the American switching market - a market twice as big as the entire European one. The Alcatel/Thomson merger is a step towards rationalization of French production resources, and the British are tempted to follow suit through financial means, in the case of GEC and Plessey. Philips has already joined forces with AT & T and given up producing its own equipment. The Plessey/Siemens/Alcatel/Italtel agreement signed at the start of 1985 could be the nucleus of a possible regrouping at European level, but that would require a political leg-up comparable to what was done for the major European projects from which Airbus and Ariane emerged (cf. Figure IV.1.e).

For many manufacturers, therefore, the plan is to disinvest from the traditional postal and telecommunications markets in order to develop a range of telematic and office automation products, or even value added services. This recentring, made necessary by market trends, explains the rising number of industrial agreements with firms in related sectors. However, the strategy pursued by European firms is

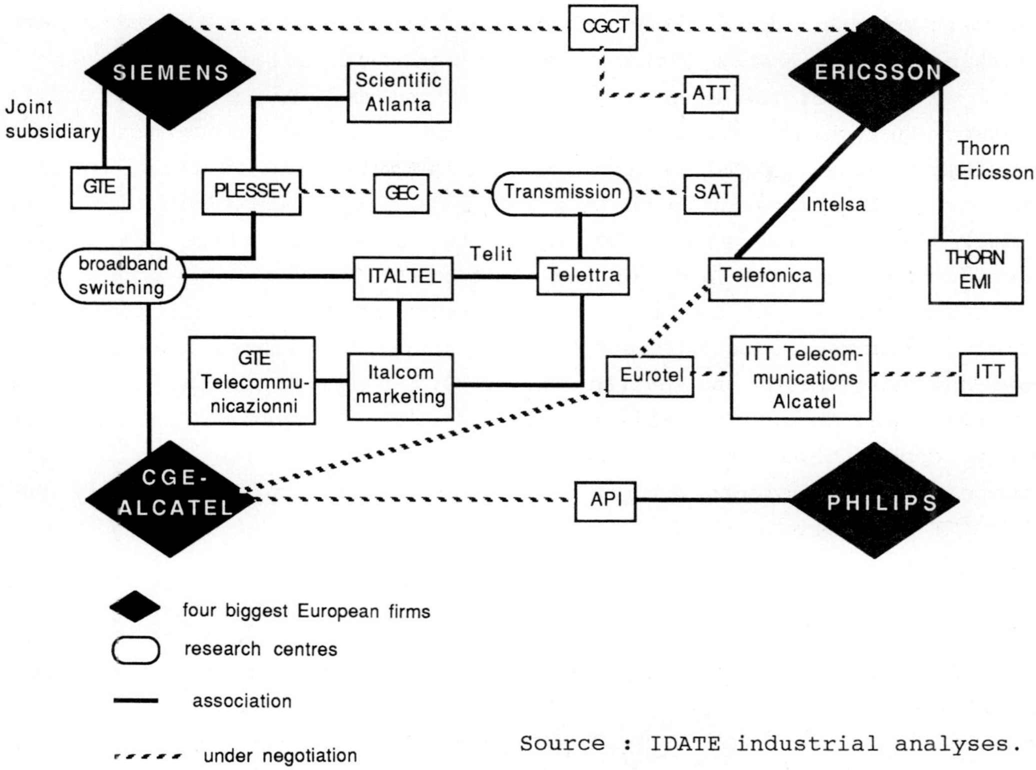

Source : IDATE industrial analyses.

Figure IV.1.e : THE EUROPEAN JIGSAW

different in this respect from that of the American or Japanese (9) : the Europeans give priority to agreements with firms specializing in systems and automated equipment where the tradition of the Old Continent is firmly rooted, while American firms tend to choose vertical integration upstream, especially with component manufacturers. The Japanese, by contrast, are more diversified in their choices, although they give preference to technological agreements over joint ventures or shareholdings. Here again, the European strategy seems vulnerable insofar as it is directed more towards the traditional

---

(9) G. Dang-Nguyen, Communication Policy Research Conference, Windsor, June 1986.

markets of big public clients, whereas the American and Japanese are seeking to meet a more market-oriented demand in different areas.

One of the main telecommunications markets is that of public network switching, which accounts for 31% of world telecommunications equipment sales. In which the European situation is evolving in this sector is characteristic of the problems facing European countries. Their share of the electronic switching market fell from 48% to 41% between the start of 1985 and the start of 1986.

That result may be due to the dispersal of European efforts in the field; Japanese industry has spent 1.6 milliard dollars to develop two switching systems, the United States 3.5 milliard dollars for four systems, and Europe 6.7 milliard dollars for six systems! In order to stay competitive, most European firms have settled for signing cooperation agreements with American or Japanese firms (notably the CGE-ITT agreement) (cf. Figure IV.1.f).

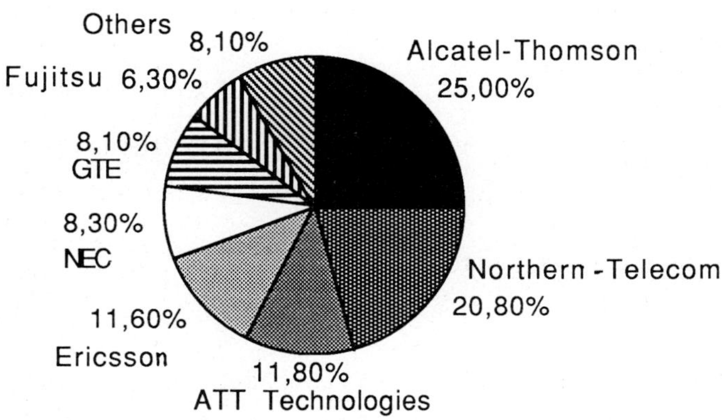

Source : IDATE industrial analyses.

Figure IV.1.f : ELECTRONIC SWITCHING MARKET SHARES
Installed lines - 1st quarter 1986

While American firms appear to be in control of the switching market, it is Japanese groups dominate the market for terminals. Following a period in which specialist manufacturers held sway, traditional office equipment industries have recently entered the market in force, by reason of the effects of integration between electronics and telecommunications.

CHAPTER IV.2

Materials and Energy

Stop press : Superconducting materials

Discussions in the course of the inquiries which led to the writing of this book revealed a broad consensus in the scientific and industrial worlds alike. Prior to December 1986 everyone interested in the development of superconducting materials agreed that the upper temperature limit for the phenomenon of superconduction (electrical conduction free from resistance, free of losses and thus without heat generation) was set by theory in the vicinity of 25° Kelvin (ie -248°C). So the applications envisaged or already in existence for these superconducting materials (mainly metal alloys such as niobium-titanium) were restricted to small, highly specific areas such as high-field electromagnets (with high-resolution RMN as the principal application), ultra-compact computers for very fast calculation (a few prototypes) and high-sensitivity magnetic detectors. It was essential for all these devices to function in a liquid helium bath, and therefore economically impossible to apply these superconductors on a larger scale.

The discovery in late 1986 that new materials based on lanthane-barium or yttrium-barium doped copper oxide were able to show superconducting effects at temperatures in the region of 100°K (-173°C) hit the international scientific community like a bombshell. Apart from the theoretical importance of such a discovery, the fact that liquid helium can be replaced by liquid nitrogen considerably alters the areas of application for these new materials. Electrical power lines, computers and, more generally, all applications involving an electric current, must now be thought out afresh from the technical angle, which could engender an industrial revolution at least as important as the one brought about by the discovery of the transistor.

However, there are still a few problems to be solved - eg high current density and processing of the material - though they are not likely to constitute significant obstacles.

At the time of going to press, we may be witnessing the birth of a hugely consequential technical and economic revolution, once again originating in functional materials.

There are different degrees of complexity in the relationship between materials and energy. The scale rises with the level of industrial sophistication :

- The most elementary relationship (Einstein's fundamental law $E = mc^2$ is relevant here) stipulates that materials and energy are one and the same thing. This relationship is borne out by the use of primary energy materials such as uranium. The use of wood and coal as energy sources is another example of an elementary relationship. The correspondence is also reflected in the advantage to be had from recycling by combustive destruction the materials incorporated in elaborate structures such as spent packaging, demolition scrap and, more generally, all combustible structural materials which have reached the end of their primary use.
- A more complex relationship is involved in either the energy needed to manufacture a material (eg the number of kilowatts needed to produce an aluminium sheet) or the quantity and quality of materials needed to obtain a given energy (an electric generator turbine blade is an example of a structural material, and a refining catalyst is an example of a functional material).
- A relationship one degree more complex between materials and energy can be identified inside an industrial system or article. This was the type of relationship taken into account in studies to find lighter materials in order to save energy in exploiting different modes of transport. Structural materials were mainly involved in that case. But this type of relationship can also be considered in connection with functional materials when, for example, new permanent magnets are designed to improve motor efficiency or in the case of thermal and self-lubricating ceramics which could make the diesel engine particularly competitive.
- Lastly, one final degree of complexity characterizes the simultaneous consideration of relationships between materials, energy and environment. In this global view, such examples as catalytic car exhausts or filters or membranes used in purification processes are good illustrations of a property of materials which enables particular forms of energy to be used because they play a part in pollution abatement processes.

In Parts II and III of this book the main emphasis was on relationships between materials and energy in connection with structural materials (recycled packaging, lighter vehicles, heat insulation of buildings, etc) (10) . This chapter, in which catalysts, ceramics for engines and permanent magnets will be briefly discussed, will focus primarily on relationships between functional materials and energy. That consideration will provide an additional opportunity to highlight the important strategic role played by functional materials.

First of all, however, it is appropriate to remind ourselves of the importance of a problem inherent in all materials in relation to the problem of energy : the need to reason in terms of an energy budget for every product (or technical article) in which materials are included, the energy budget is a measure of the entire cycle, from the inputs (raw materials and energy) needed to manufacture the product until its destruction or recycling.

IV.2.1. ENERGY BUDGET AND MATERIALS

The energy budget can be schematically expressed by the following formula, measuring the mean energy flux ascribable to the product :

$$\frac{E(\text{inputs})+E(\text{process})+E(\text{distribution})+E(\text{final})}{\text{Lifetime}} + E(\text{util.}) + E(\text{maint.})$$

The various parameters are defined as follows :

- E(inputs) represents the energy incorporated in the materials of which the article is composed (intermediate consumption);
- E(process) represents the energy expended in the course of the production process and a fraction of the energy cost of the production plant employed (by analogy with the financial concept of amortization);

(10) When discussing the types of relationship between structural materials and energy, we could of course have given a very large number of examples of their use in obtaining energy (composites for deep sea oil exploration equipment, zirconium cladding in the nuclear industry, epoxy-glass vanes for wind energy, etc).

- E(distribution) represents the energy expended in packaging, transport and storage, either directly or through the materials needed for these purposes;

- The lifetime of the article is a fundamental factor in defining the unit energy cost of the three preceding categories in relation to the unit service rendered by the article;

- E(utilization) represents the energy expenditure systematically associated with the utilization of the article during a given time. The choice of constituent materials will influence this magnitude via such characteristics as weight, ease of handling, multiple use potential, etc;

- E(maintenance) represents the average energy expended during the life of the article in order to maintain it in a normal operating condition, whether the energy is supplied direct or in the form of products consumed for the purpose of maintenance;

- E(final) represents the energy involved in scrapping the article. This is a parameter which can assume positive, zero or negative values. In the first case it represents the energy cost of destruction if the constituent materials are undesirable as such in the environment. The second case covers waste which can be left in the environment or articles which disappear naturally through consumption. The third case is that of waste for which a use can be found : recycling saves at least part of the energy normally required to produce new materials, or else simple decomposition generates energy (heat, fertilizers, etc).

It is theoretically possible to evaluate the contribution each material makes to this global budget. But an immense amount of work would be needed to make such an analysis complete, and could not even be outlined here. However, merely stating the problems does at least give a glimpse of a few fundamental questions which it is right to underline.

- From the energy saving standpoint, the average life of an article is an essential variable to the extent that it affects energy stocks differently (those incorporated in the vehicle, its production process and its final recycling) and the energy fluxes (associated with the utilization and maintenance of the article). Clearly a distinction must be drawn between materials which save the energy utilized during the article's life - however long it may be - and those which, by prolonging its life, automatically reduce the mean fixed energy costs. The first category includes high performance materials such as plastics and composites which help to make vehicles lighter, high conductivity electrical conductors and heat insulating materials, while the second category includes in particular all materials which resist corrosion, erosion, fatigue, etc., as well as those which contribute to the article's overall reliability.

- A close look at every phase in the article's life from the start of its production process onwards makes one realize the importance of the manufacturing process : a great deal of energy is sometimes hidden in this stage, and the total energy content is sensitive to any simplification in manufacture. A material can be efficient when incorporated in the article but ruinous in energy costs at the processing or assembly stage. The crucial factor then is lifetime (and it is long only in the case of consumer durables).
- The overall energy budget must of course also cover the environment. The cost in terms of pollution flux must logically be added to the energy costs of utilization. Likewise, ultimate recycling can either be a net creator or a net consumer of energy : in the latter case, if one considers paying to get rid of an article which has become waste, it is in order to avoid allowing it to accumulate as a potential source of pollution. Thus, when materials are chosen some thought should be given to the fact that they will one day become waste. The problem arises, for example, in the case of composite materials or plastic sandwiches which are difficult to recycle.

In the material-energy dialectic, this budget may have very important consequences when it comes to choosing new material substitution or penetration strategies, or even, as we have seen, choices within manufacturing processes themselves. However, the budget becomes less crucial to those choices at a time when energy constraints are relaxed or, more simply, at a time when energy is abundant.

On the other hand, the possession and utilization of functional materials, which are the keys to the strategic control of energy, must in no circumstances be treated as short-term economic factors : these materials must be treated as permanent development factors. Among them, catalysts have played a prime role almost since the start of the industrial civilization, permanent magnets are a good illustration of the technological evolution of a family of functional materials and alternatives to a "strategic" metal, and functional ceramics open up new prospects for the rational utilization of hydrocarbons in transport. For these various reasons, these three families have been selected to illustrate the problems of materials and energy.

Choosing this particular presentation of course means leaving on one side a number of other fields which it is not possible to deal with exhaustively in such a book as this. Mention should, however, be made of possible developments in materials other than single-crystal silicon in the solar field or of specific materials in tapping nuclear energy and developing high power lasers.

IV.2.2. CATALYSTS

A catalyst accelerates a chemical reaction without being consumed in that reaction. Such accelerations can be as much as 10 orders of magnitude, in other words be multiplied by a factor of ten milliard. The best yardstick of its activity is to evaluate the speed of the chemical reaction, in molecules transformed per second and per gram of catalyst; industry prefers to evaluate in "VVH" ie volume of product treated per unit volume of catalyst per hour.

A selective catalyst may affect only one particular reaction from among all the competing reactions in a reactive mixture. For example, the controlled oxidation of methylene in the presence of oxygen yields, by means of a selective catalyst, ethylene oxide (epoxyde) and not carbonic acid gas and water, though that is the classic combustion greatly favoured by thermodynamics.

Catalysis can be subdivided, according to the physical and chemical nature of the catalytic substance, into five principal branches :

- heterogeneous catalysis, in which the reaction occurs at the interface between a solid (the catalyst) and a gas and/or liquid (chemical reactants);
- homogeneous catalysis which, as its name indicates, occurs in the absence of solids (gas and liquid both being fluids);
- electrocatalysis and photocatalysis where reaction occurs either at any electrode surface traversed by a flow of current or under the effect of light;
- enzyme catalysis which involves natural or artificial substances (enzymes) as catalysts.

While the four last-mentioned branches of catalysis are very important, especially because of the potential they hold for new reactions, the remainder of this section will be concerned mainly with heterogeneous catalysis because (a) it represents 80% of the catalyst market in dollar terms (cf. Table IV.2.B) and (b) it is here that the catalyst is involved as a material. It should however be observed on this latter point that, apart from homogeneous catalysis, the other three branches of catalysis also involve materials - electrode, photosensitive materials, organic materials.

IV.2.2.1. Catalysts : materials or chemicals ?

Catalysts came into being with industrial chemistry and did not really penetrate the energy field until the boom in automobiles, and more particularly the intensive exploitation of oil. Because of their origin, they are still sometimes regarded as chemical products for chemists - sometimes rightly, as we shall see : however, with the advent of techniques of spectroscopic analysis and the search for reproducibility through the microscopic control of their preparation, they are also coming to be regarded more and more as fully fledged functional materials. This distinction between chemical product and material is not unimportant because, if catalysts are regarded as chemicals, they are then studied and developed on the basis of chemical techniques - their reactivity, for example, is studied by screening; this is the method which has proved by far the most effective (claiming at least 90% of the discoveries of all present-day industrial catalysts) and is therefore preferred by industry.

By contrast, the preparation and development of a catalyst considered as a material are often approached deductively from a study of models, using physical techniques - eg spectroscopy for structural analysis; this method has been in fashion over the last fifteen years in research laboratories. While it has had the advantage of explaining quite well the way in which already known catalysts function, generally speaking it has not produced any revolutionary discoveries. At best, it has afforded improvements in catalyst quality (better activity, better resistance to poisoning, etc). Industry has not always been quick to seize on that advantage, believing that the investment in equipment and highly qualified personnel was not justified by the expected benefits. All that is changing now, mainly under pressure from the big private American research groups which have demonstrated the industrial value of understanding how catalysts work, but also under pressure from fundamental research and the scientists involved in it. A proper balance between the two approaches seems most likely to bear economic fruit, although intellectually the deductive approach seems at first sight more "scientific". The big obstacle in the way of this approach is its inability to express all the catalyst characteristics as parameters. The fact is that besides its chemical nature, form, structure, operating temperature, etc. (all of which catalyst properties can be expressed as parameters), the catalyst is "alive" during the reaction. Its surface is rebuilt, poisoned and reactivated constantly under the influence of the chemicals in the reaction. This phenomenon is so inherently complex and its effects on activity and selectivity so dramatic that very often only a pragmatic approach is able to solve the problems that arise. It is as if a computer chip were to change as a function of the throughput and nature of the information traversing it. It is impossible to imagine the complexity of a computer containing several milliards of chips, each changing individually depending on the information processed. But a petroleum catalyst at work is like that : an average catalytic petroleum reforming unit contain 10^{26} platinum atoms, and just as many rhenium or iridium atoms, not to mention the atoms of the catalyst carrier (alumina, for example) which possess both individual and collective catalytic properties.

This example shows how absolutely vital it is to test a new catalyst in increasingly large units - a laboratory micro-pilot plant, intermediary pilot plant and then "life-size" pilot plant - before embarking on an industrial venture. The parameters deduced from small-scale experimentation are not directly transposable to production units because no statistical calculation, even a highly sophisticated one, can today take into account either such a degree of complexity or the problems raised by the heterogeneity of the process (heat transfer, matter transfer, diffusion). This somewhat apocalyptic analysis is not however intended to negate the great potential value of a "materials science" approach, if only to reduce that complexity and perhaps one day bring it wholly under control.

The enormous advantage enjoyed by the big oil or chemical companies as compared to "small" catalyst manufacturers is therefore clear. Who can afford a "life-size" pilot plant apart from these big companies ? (11) . In this branch the material user is inevitably dominant. However, catalyst manufacturing technology, far removed from the central concerns and skills of the big user groups, is often in the hands of, or even developed by, smaller-scale enterprises. Some of these firms have acquired know-how which gives them sufficient power to work with the big groups as partners; thus they are often able to help in spreading new catalysts to other groups when the big groups occupying specific slots in a situation of cut-throat competition are not directly able to pass these materials among themselves.

But the supremacy of the "rich" groups is not general in catalysts as a whole, particularly in catalysts for advanced chemistry and, of course, exhaust catalysts which we shall deal with in a special section.

IV.2.2.2. The challenge to catalysts

While the catalyst market for the whole world in 1986 represents only a very small fraction (three milliard dollars - see Table IV.2.A) as compared to other materials markets, their importance to the GNP of the industrialized countries is considerable. The Pimentel report (12) emphasises that 20% of the GNP of the United States is generated through the use of catalytic processes; and it can be added that a far larger proportion of that GNP is indirectly generated by catalysis, taking into account the industries which incorporate materials or products whose manufacture involves catalytic processes - which in fact would mean practically the whole of industry, or almost 30% of GNP and a sizeable proportion of the primary sector

---

(11) The "life-size" pilot unit owned by ASVAL (Association pour la valorisation des huiles lourdes - Association for the enhancement of heavy oils), although constructed by TOTAL, ELF and the French Petroleum Institute to study reaction conditions and hydro-treatment catalysts for heavy oil fractions, can only operate economically through being partly rented out to other major oil groups.

(12) "Opportunities in Chemistry", report of a group headed by G.C. Pimentel, National Academy Press, Washington D.C., 1985.

Field	Catalysts	Reactions	World	Europe
	Zeolites	Catalytic cracking, petrol and diesel synthesis	440	70
	Cobalt,nickel molybdenum on alumina	Hydrotreatment, desulphuration, denitrogenation	230	60
Energy	Cobalt,nickel molybdenum, tungsten on very acid substrate	Hydrocracking in the presence of hydrogen diesel	100	10
	Platinum on alumina	Catalytic reforming, improvement of octane rating	760*	20*
Pollution abatement	Platinum, rhodium, on alumina or ceramics	Automobile exhaust	760**	160**
	Titanium, aluminium***	Polymerization	714	
	Precious metals	Organic synthesis	225	
	Silver, molybdenum, bismuth	Oxidation, ammoxidation	235	
Chemistry	Copper,nickel, iron, etc.	Methanol synthesis, ammonia, steam reforming	163	
	Nickel,iron, palladium,etc.	Hydrogenation - dehydrogenation	128	
		TOTAL	3095	

* Not including precious metals (recycled).
** 1985.
*** Some included in homogeneous form.

Source : HPI Spending Outlook 1987 and Chemical Week, June 1986.

Table IV.2.A : World and European markets for heterogeneous catalysts (millions of dollars) for 1986.

Field	Reaction	1982 t	1982 $	1982 $%	1987 t	1987 $
	Catalytic cracking	159	235	18%	200	330
	Alkylation, octane rating* by sulphuric and hydrofluoric acid	1814	150	11.5%	2100	170
Energy	Hydrotreatment	9.5	65	5%	12	85
	Hydrocracking	1	40	3%	1	45
	Reforming	2	20	1.5%	2	20
	Sub-total		510	39%		
	Exhausts	-	330	25%	-	285
Pollution abatement	Industry	-	-	0.2%		12
	Sub-total		333	25%		
	Polymerization	29	215	16.5%	38	270
	Organic synthesis	41	80	6%	50	95
	Oxidation, ammoxidation	4.5%	75	5.5%	5	90
Chemistry	Ammonia, methanol synthesis	10.5	50	4%	13	60
	Hydrogenation	4	35	2.5%	4.5	42
	Dehydrogenation	2	10	0.8%	2	13
	Sub-total		465	35.5%		
	TOTAL		1308	100%		1517

* Homogeneous catalysis.

Source : HPI Spending Outlook 1987 and Chemical Week, June 1986.

Table IV.2.B : Catalyst consumption in the United States in 1982, and forecasts for 1987 (millions of dollars and millions of tonnes).

(through fertilizers derived from ammoniac, for example) and of the tertiary sector.

An analysis of Tables IV.2.A and B shows that the American market makes up about half the world market and is structurally quite representative of it. We may therefore take it as a basis for some general economic reflections.

Broadly speaking, the market showed little development between 1982 and 1987, with just a slight increase in terms of value and tonnage. Analysis over a longer period, however, would have shown the growing strength of catalytic cracking, with the discovery and application of zeolitic catalysts in the energy field, the emergence of new catalytic processes (cf. Table IV.2.C) for the synthesis of commodities in the chemical field, and lastly, the birth and development of catalytic processes for the cleaning of exhaust gases.

In fact, these three developments are in part a response to the challenges facing "catalyst science". The need to increase the quantity of energy, the key to our economic development, is probably the principal challenge. Whatever scenario may be adopted to forecast the trend of world energy consumption over the next four or five decades, it will in any event be necessary to optimize all resources, and first of all those in which catalysts play a decisive role - oil and its very heavy cousins (such as bituminous shales and asphaltic sands), but especially coal, reserves of which are about 40 times as great as those of oil.

Two strategies, probably complementary and not competing, can then be envisaged. The likeliest in terms of the catalysts used today is the obtaining of irreplaceable products (petrol and diesel for transport and hydrocarbons for petrochemicals) from large molecules contained in residual distillation fractions, in asphaltic sands and in oil derived from coal or shales. This is already industrially feasible (catalytic cracking and hydrocracking) as long as the molecules are not too big or too poor in hydrogen and provided they do not contain too much sulphur, nitrogen or metal impurities. The catalysts needed for these reactions are, in order of utilization: - Cobalt or nickel and molybdenum sulphides, for hydrotreatments which purify by desulphuration, denitrogenation and demetallization and increase the hydrogen content of products. The main problem which will arise with these catalysts is the growing rarity of the strategic metals of which they are made, and their recycling, which is not now economic, will become indispensable.
- Zeolites, mainly for cracking reactions. The major drawback of these catalysts is their preparation and cost; although as a rule they do not

contain rare metals - being composed only of aluminium, silicon and oxygen - they have a very high cost per tonne. This is even truer of hydrocracking catalysts, mainly because rarer metals are added to them.
- Platinum and a number of additives for catalytic reforming, or sulphuric and hydrofluoric acid for homogeneous phase alkylation, reactions which, among other things, make it possible considerably to increase the octane rating of petrol. Increasing that rating is very important, firstly because of the higher and higher compression ratios used in car engines to increase efficiency - and thus save energy - and secondly because of the gradual elimination of lead so as to enable catalytic exhausts to be used in the interests of environmental protection.

The second strategy in response to the need for non-replaceable products derives from the opposite approach : all heavy fractions and coal are transformed into <u>very small molecules</u>, carbon monoxide CO, hydrogen H_2, and methane CH_4. It may also be necessary to transform methane from natural gas catalytically, into CO + H_2. The oils and products needed in petrochemicals are reconstructed from the basic building blocks of these small molecules. There are two possible approaches : direct synthesis, known as "Fischer-Tropsch" synthesis from the names of the German scientists who discovered these catalysts before the second world war, and indirect synthesis via a methanol stage.

- Direct synthesis, depending on whether one wishes to obtain hydrocarbons or oxygenated products, uses catalysts based on iron or platinum mining metals (Rh, Ru, Pd) respectively. This synthesis is not economic in normal economic conditions; only Germany, with its enormous need for fuel during the war and South Africa today (SASOL process, in an embargo situation, have managed to develop the technique). Huge improvements in selectivity and output are still needed to make it an economically viable process.
- Indirect synthesis uses copper and zinc based catalysts to synthesize methanol. This first reaction is easy and fully controlled. However, the second stage - the transition from methanol to petrol using zeolites (ZMS-5 types, MOBIL process) - is far from competitive in the present economic situation. There is much argument between the defenders and opponents of this technique, but the most commonly quoted figure gives an equivalent of $40 per barrel, whereas traditional oil oscillates between $12 and $20 per barrel. New Zealand is the only country to have started up a unit of this type, because it possesses

large reserves of natural gas and no oil and sees this as the only way of guaranteeing its energy independence and economic equilibrium. Strictly speaking, this is a strategy which needs confirmation. The main technical disadvantage stems from the very high cost of this type of zeolite and its short life span (between 800 and 2,000 hours, depending on the source). Progress in this field would be crucial because it would give another boost to methanol, which is already involved in the synthesis of ammoniac and is at the heart of the synthesis of several commodities such as, for example, acetic acid.

The synthesis of <u>products for chemistry</u> is also a challenge and an incentive to catalysis. Catalysis is expected to play a part here not so much in the synthesis of new products, but mainly in developing catalysts which will make it possible either to save energy and material or to integrate manufacturing stages. A typical example of the first category is the synthesis of ammonia, which is done with iron-based catalysts, the principal characteristics of which were established as long ago as the 1920s by F. Haber's team in the I.G. FARBEN laboratories. New approaches involving nitrides are highly promising. Many examples in the second category have appeared in recent years (cf. Table IV.2.C). The functional conversion of hydrocarbons, usually olefins (ethylene, propylene, butene, etc) into organic acids, alcohols, ketones, epoxides, amines, amides, nitriles, esters, etc is the foundation of synthesis chemistry. It usually takes place in several stages with, of course, cumulative output losses and substantial energy costs. When a selective catalyst makes this synthesis possible in one stage, you have an economic revolution. The ammoxidation reaction which yields acrylonitrile directly from propylene, ammoniac and oxygen, or other amides depending on the nature of the initial hydrocarbon, is the best example of this new type of catalysis.

A new generation of heterogeneous catalysts has also emerged for the purpose of polymerization reactions. Here it is the technological advantages of heterogeneous catalysis over homogeneous catalysis from which industry gains, and particularly the ease with which the catalyst can be separated from the end-product provided it is not incorporated in the mass, as is often the case with homogeneous processes. Thus titanium deposited on magnesium oxide has replaced traditional catalyst of the Ziegler-Natta type for the polymerization of propylene.

Feed stocks	Catalyst	Product	Used to manufacture	1982 USA production (t)
Ethylene	Silver, cesium chloride salts	Ethylene oxide	Polyesters, textiles, lubricants	2,300,000
Propylene NH_3 oxygen	Bismuth molybdate	Acrylo-nitrile	Plastics, fibres, resins	925,000
Ethylene	Chromium, titanium	High density polyethy-lene	Moulded products	2,200,000
Propylene	Titanium, magnesium, oxide	Polypro-pylene	Plastics, fibres, films	1,600,000

Source : PIMENTEL Report (12).

Table IV.2.C : New processes based on heterogeneous catalysis for the synthesis of basic products for chemistry.

The only foreseeable scenario in relation to this second challenge which catalysis has to face is the intensification of these integrations, substitutions and even energy savings through the discovery of new catalysts which must have the essential property of selectivity. Probably the most attractive route is to immobilize on a substrate molecules which are catalytically active in homogeneous catalysis, making the process heterogeneous.

The last challenge which catalysis has to meet is its contribution to the protection of the environment, which has been inevitable by the growth of industry and consequently of industrial waste. But there is one type of waste, car exhaust gases, which has affected people sooner and more directly than others (Los Angeles smog, forest dieback in Central Europe). The media impact has speeded up tremendously the development of catalytic exhausts and it may be that the proposed solutions have been seized on too hastily and will prove not to be very rational in the near future. The development of these catalysts, of concern to Europe first and foremost, is discussed in a special section of this chapter.

IV.3.2.3. Catalytic exhausts

Car exhaust gases contain three principal pollutants : carbon monoxide, CO; nitrogen oxides NOx; and unburnt hydrocarbons HC. To give an order of magnitude for the quantities of these different emissions, the average produced by a European car in 1986 is about 5g/km of CO, 1.5g/km of NOx, and 1g/km of HC (over the same distance the car "consumes" 1,200 g of air, about 1,000 l). These quantities are actually quite tiny when one considers the atmospheric dilution over 1 km, but they become important when traffic density increases (town centres) or when all the traffic in a country is aggregated (13) .

Various technical solutions have been envisaged in order to reduce these emissions. A considerable effort has already been made since 1970 in engine technology in oder to reduce pollution, resulting in -70% for CO, -30% for NO and -60% for HC. Among the different solutions choosen to pursue this improvement, there are two that are more favourably viewed than the others : the lean-burn engine and the catalytic exhaust. A discussion of the merits of the two could be out of place here, but there is a third solution that should not be overlooked : all-diesel, to which we shall return in Section IV.2.4.

The economics of the catalytic solution, the only one which interests us here, depend heavily on the technical aspect which in turn

---

(13) For an average city with 50,000 cars travelling for one hour at 40km per hour in one day, these figures must be multiplied by 200,000, giving one tonne of CO, 300kg of NOx and 200kg of HC per 24 hours. During the same period, daily sulphur dioxide (SO_2) emissions for the same city may reach 120t/day when domestic emissions (oil or coal fired heating) and above all industrial emissions (heating by heavy oil or coal) are taken into account.

Annual European consumption of normal and supergrade petrol is 100 million tonnes. If average consumption is 0.08 kg per kilometre (10 l/100 km), that represents 1,250 milliard km travelled, giving total emission levels of 6 million tonnes CO, 1.9 million tonnes NOx, and 1.2 million tonnes HC. The amount of air over Europe is approximately 150 million cubic kilometres, or 2.10^{14} million tonnes. Assuming, in the extreme case, that this air is immobile and not renewed, that gives dilutions of 3.10^{-8} ppm (parts per million) of CO, 1.10^{-9} ppm of NOx and 5.10^{-10} ppm of HC. From these figures it will be obvious that only local concentrations need to be taken into account.

depends to some extent on the legislative situation, in other words, norms laid down by states (14) .

The norms laid down for the coming years by the two groups of European states, the EEC and the Stockholm group, differ from the American and Japanese standards which have already been firmly applied since 1983. Without going into detail, the two main consequences are (i) the technical solution eventually adopted in Europe will be partly different so that the experience acquired by the Americans and Japanese will have less impact in the battle for the new European market, and (ii) this technical solution seems to pose the problem of rhodium and platinum resources in an even more acute way.

The design of the catalytic system (only one catalytic converter or several, recirculation of gases etc) has been fluctuating for some time, but opinion now seems to be coming down in favour of the "three-way" system. The catalyst consists of a carrier, a honeycomb or equivalent ceramic, or a metal alloy containing chromium, iron and aluminium, the "monolith", a carrier coating, the "wash-coat", made of alumina doped principally with rare earths like cerium and, lastly, the metallic active phase impregnated on the wash-coat, containing platinum and rhodium (cf. Table IV.2.D).

COST BUIL-UP FOR A TYPICAL EUROPEAN THREE-WAY CATALYST

1.3 litre ceramic carrier with 1.24 g/l Pt + Rh (ratio 5)
Price : $47 per unit (large quantities)

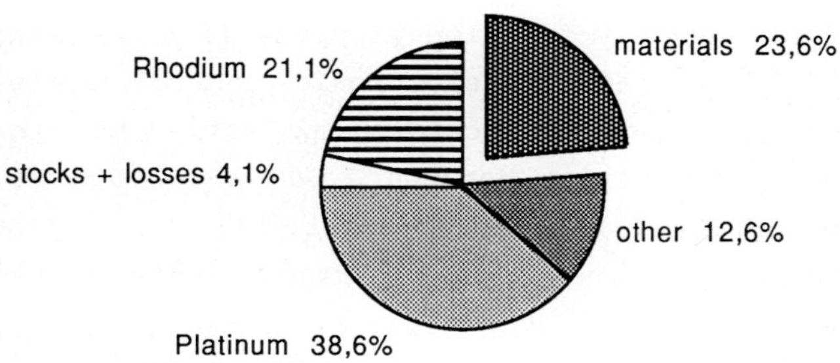

Source : (14)

Figure IV.2.a

(14) "The market for car exhaust catalysts in Western Europe - A review of trends and development", W. GROENENDAAL, CAPOC, Brussels, September 1986.

COMPOSITION

Carrier	Cordierite monolith with 400 passages per square inch and a wall tickness of 0.15 mm.
Wash-coat	20% wt. pseudo - boehmite promoted with a.o. lanthanides, to improve the high temperature stability and the adhesion to the carrier.
Metals	Pt + Rh : 35 - 40 gr/cu ft (1.24 - 1.41 gr/l), Pt/Rh wt. ratio : 5.
Bulk density	0.45 gr/l.

OPERATING CONDITIONS

Temperature	300 - 900°C
Space velocity	100 000 - 200 000 l/l.h

Catalyst to engine cylinder volume ratio 0.8 - 1.5

PERFORMANCE

Controlled within : = 0.99 + - 0.06
Conversions in % :
Fresh; HC : above 80% CO and NOx : above 70%
Uncontrolled within : = 1.05 + - 0.2
Conversion in % :
Fresh; HC : min. 50%, av. 70%; CO min. 20%, av. 55%; NOx min. 10%, av. 50%.

Source : (14).

Table IV.2.D : European "three-way" catalyst for a catalytic exhaust.

This "three-way" catalyst can only function properly if the engine is equipped with an oxygen sensor and an electronic control system which keeps the air/fuel ratio constant. According to estimates, the price of this electronic equipment varies between 2,000 and 4,000 FF, and the unit price of the catalyst itself is about 500 FF. Figure IV.2.a shows the cost percentages of the different constituents of the catalyst.

Europe's position overall is satisfactory in relation to Japan, and even advantages in relation to the United States, where electronic control is concerned. The German firm Bosch is a world leader in this field. Inside Europe, competition among the major French, Italian, British and German (and soon Spanish) car manufacturers suggest a less rosy picture.

The European situation is less favourable when it comes to the carrier, which accounts for about 24% of the price of the catalyst. The solution adopted worldwide for more than 90% of catalytic converters is the cordierite (ceramic) monolith, whose manufacture is in the hands of two companies, Corning in the United States and NGK in Japan, although Corning pioneered this technology. These two companies have invested in Europe in two factories with a capacity of 6.5 million units per annum (the European demand in 1986 was about two million units; in 1990 it will be about three million units and in 1994 between 5 and 9 million units). The metal alloy monolith is principally "European" (Federal Republic of Germany) and is of course the alternative which may loosen Corning's monopoly. In addition, its high-temperature performance and the thiness of its walls are major plus points for Europe's "small" high-performance engines. Present production capacity is 0.5 million units per annum in Europe. However, because it is still expensive (two or three times the price of ceramics), only top-of-the-range cars are being fitted with it (eg Mercedes and Porsche). A research effort into manufacturing techniques could eventually make it competitive across the board.

The situation is very balanced where the firms engaged in impregnation of the active phase and the "wash-coat", as it were the real catalyst manufacturers, are concerned. The American market for monolithic catalysts (15) is shared between four companies : Degussa (Federal Republic of Germany), Engelhard (United States), Johnson Matthey JMC (Great Britain) and Allied Signal (United States). Seven local companies are present on the Japanese market, three of them subsidiaries of car manufacturers. Lastly, of course, in Europe, Degussa and JMC appear to be ready, Engelhard is building a factory and Allied Signal has announced that it is to build a factory in France. In the Federal Republic of Germany, Heraeus and Doduco supply catalytic converters for cars which do not have them factory fitted.

The fact that this type of catalysis does not require vast investment in pilot plants, unlike the other branches of heterogeneous catalysis, means that the firms able to develop and market this type of catalytic converter can be relatively small. The assets which a manufacturer must possess fall into three broad categories : either he has know-how in impregnation techniques (Allied Signal, Engelhard, etc), or he has preferential access to the precious metals market

---

(15) General Motors fits 30% of its automobiles with a carrier made of alumina pellets, a technology which is now almost out of date.

(JMC), or he belongs to a highly integrated car manufacturing groups (Japan). The battle in Europe will probably be played out between these three tendencies and, in particular, confronts the managers of the big European car manufacturing groups with an important strategic choice. On the industrial level, the emergence of markets such as Australia, Korea and in the more distant future Brazil and Mexico, adds an extra dimension to that choice.

Last but not least among the economic aspects, the large amounts of rhodium and platinum needed could in the long run bring everyone together! These two metals account for 60% of the price of the catalyst (incidentally giving a fantastic advantage to firms which have preferential access to these metals) (cf. Figure IV.2.a above) since, because of the lead which European petrol contains, twice as much of these rare metals is needed as in the United States. Over and above the cost, however, which is never prohibitive up to a certain point, it is in fact the very limited physical resources which have to be taken into consideration. Figure IV.2.b shows western world demand (including Japan) for rhodium (7.09 tonnes) and platinum (79.7 tonnes) in 1985. The share taken by catalytic exhausts is 54% for rhodium and 31% for platinum respectively. In order to meet EEC norms Europe will need 8.85 tonnes of platinum each year by 1994, which is more than 10% of present consumption, and 1.8 tonnes of rhodium which is <u>more than 25% of present consumption</u>.

The ratio of recoverable reserves throughout the world to cumulative demand in the western world (1979-2000) is 9,200t/1,800t, or about 5, which is not unduly alarming in the case of platinum as compared for example with silver (ratio = 0.5); it is much more so in the case of rhodium, although there is great uncertainty over the value of recoverable reserves, the figure varying between 100 and 800 tonnes, in other words a ratio varying from 0.5 to 4 for a cumulative demand of 200 tonnes. The uncertainty is maintained by the speculation conducted in this metal by various interest groups. The most worrying aspect where access to reserves of these two metals is concerned is the geographical situation of those holding the reserves - South Africa and the USSR. Canada, which also has reserves, is part of the consumer pool for catalytic converters. In 1985, 85% of all platinum used in the western world came from South Africa!

This rather gloomy picture calls for two remarks : firstly, is the three-way catalytic system chosen to meet the challenge of car exhaust pollution a wise one ? Secondly, if we stay with that choice, we must at all costs develop a system for the recuperation and recycling of the precious metals; this will be rather uneconomic at first, but will become economic as the ratio of reserves to demand falls. It will also be necessary to control speculation in order to give this industry a firm foundation.

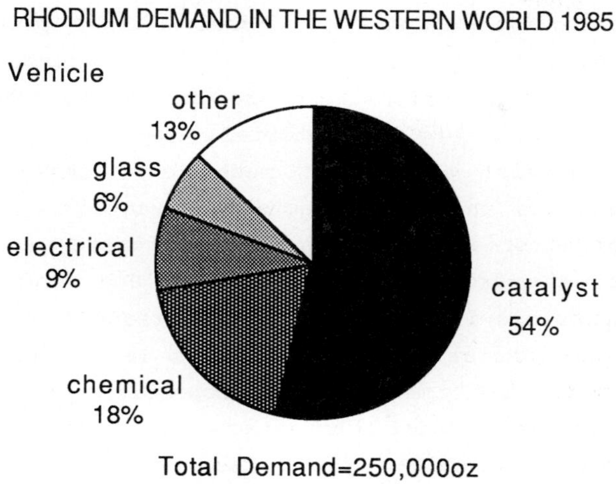

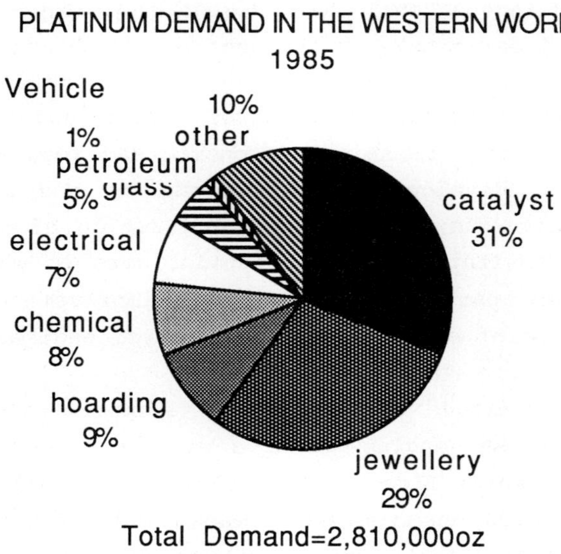

Source : EVANS et al, CAPOC, Brussels, September 1986.

Figure IV.2.b

IV.2.3. DIESEL AND CERAMICS

Why should we be dealing in the same breath with a traditional mode of locomotion and a technically promising category of materials ? If the application of ceramics to diesel engines proves technically feasible and economically viable, it will of course give diesel a decisive boost; if solutions can be found to the problems of reliability (fragility of the material) and industrial mastery of processing (a material at present much less convenient than metal casting), diesel will enjoy many decisive advantages. The introduction of this type of material, which in its application to motor vehicles plays a role at the frontiers of structural and functional materials would make it possible in particular, to increase thermodynamic output by reason of its insulating properties, to limit friction, wear and corrosion through its mechanical properties and low chemical reactivity, to achieve weight advantages and generally speaking to improve performance greatly without resorting to rare or strategic elements as special metal alloys do.

It is quite conceivable that the subject of the diesel engine could be discussed independently as one mode of energy transformation - followed by that of ceramics - in the context of structural, functional or intermediary materials. Progress in diesel technology and its competitive position in relation to petrol engines and gas turbines are bound up with many other technical, economic and regulatory aspects - new engine designs, developments in the quality and price of fuels, pollution norms, etc; furthermore, the application of new ceramics to thermal systems constitutes a vast potential area of which the diesel engine is only one particular part. Many improvements afforded by ceramics ought to be of advantage both to petrol engines and to diesel engines.

However, an overall technical and economic analysis of the prospects for new diesel engines using a variety of ceramics show that this development scenario fits in particularly well with a large number of topical energy and environment issues. Many factors are coming together to make the diesel engine, with a new ceramic dimension, a model of energy application for the future.

IV.2.3.1. The position of diesel

Let us begin with the importance of the diesel engine in the overall economic picture. The public is mainly aware of an application where this engine does not dominate, namely, private overland transport. The picture varies greatly from one country to another, even within Europe, but the situation can be summed up by saying that, except in Germany, diesel still secondary where light vehicles are concerned. On the other hand, it plays an important role in commercial transport, in heavy goods vehicles and on the railways. It represents 90% of the maritime transport market. Finally, we should not forget the considerable importance of fixed diesel power plants used as compressors in industry, electricity generators, etc.

As regards its position in relation to the competition, let us observe first of all that diesel is at present dominant in powerful energy systems, where changing operating speeds are not a crucial problem. It is quite efficient in this specific sector. Its other big advantage over most of its competitors is that it can accomodate a whole range of fuel qualities : for vehicles "diesel" fuel is used, and in industrial applications we find a whole variety of fuels including heavy fuel oils, and in the near future perhaps synthetic products derived from coal, or even coal dust suspensed in an oil product (micro-slurry). This description shows that diesel technology is not unrelated to the energy problems discussed in the section on catalysts.

Apart from the petrol engine (and possibly the Stirling engine) the main present and potential competitors are : (i) the <u>electric motor</u> and (ii) the <u>gas turbine</u>. The former is not likely to make any impact on the <u>transport</u> market in the near future, until such time as fundamental advances have been made - and these often entail materials, for batteries (organic polymers?), to lighten magnetic engine components etc. In <u>fixed applications</u>, substitution is already well under way in countries pursuing a strategy of nuclear generated electricity (or electricity in general) : if the cost per kilowatt-hour is not too high, the electric motor is an attractive replacement for steam or diesel in process industry compressors. The impact on the proportion of motive power consumed in the basic chemicals or agri-foodstuffs sector may be considerable. Replacement by electricity often leads to significant energy savings (examples are heat pumps and mechanical recompression of steam). But the electric motor's breakthrough in the types of plant mentioned has the effect overall of substituting KWh for tonnes of oil : it does not challenge the diesel engine in particular for shares of the market. By contrast, the gas turbine represents a new direct competitor, by reason of the recent technical improvements achieved in air transport applications and of progress in industrialization, opening up the possibility of further diversification towards overland transport and even fixed applications (where costs become a decisive factor in competition).

In motor vehicles, provided the overall balance of oil refining is not unduly upset, diesel ought to make a substantial breakthrough because it is more robust and thus able to deal with a moderate downgrading of fuels, as well as being cleaner and more economical. However, the contradictory development of the input and output structure in refining (ratio of heavy to light fractions) could result in the diesel fraction being sacrificed. If the qualitative downgrading of this fuel comes up against the development of unduly strict norms applicable to diesel engines, motorists will no longer be able to stay with it (either technically or at acceptable cost as compared with the petrol engine competition). The disappearance of diesel in this type of application - along the lines of what has happened in the United States, for example - is therefore a possible scenario, as indeed is its opposite - the expansion of diesel. Everything will depend on the way in which norms move and on the refining industry's ability to safeguard diesel oil quality.

IV.2.3.2. Technologies developing and coming together

The coming together of diesel and ceramics technologies could have an impact all the more marked because of the time it has taken to come about. We shall begin by showing the extent of the progress made independently in these two areas, before going on examine the important synergy which may well occur between the two technologies.

Diesel engine designers have not been idle in recent years. Designs have been progressively modified and the form of component parts optimized. Combustion control has also done much to improve engine performance, as has the widespread application of turbochargers and increasingly complex systems designed to recover the energy lost in the exhaust. The profession has always managed to innovate sufficiently to keep up the rate of adjustement required by the market -

particularly following the oil shocks. On the other hand, the present challenge is on a different scale : moving beyond the present performance levels achieved would mean radically rethinking the design of the engine and the materials used. In addition, the consumption and pollution norms required are becoming draconian. Ceramics appear to be an essential step in coping with this challenge.

Ceramics, for their part, have made good progress in the engine sector. While this technology is still quite foreign to the manufacturers of truck engines (with the exception of Cummins) it has long been a subject of research and indeed of applications in helicopters (gas turbines) and jet engines (nozzle cones). Rocket nose cones are an area which lies outside engine technology, but the problems are similar - resistance to erosion and high temperatures. Applications in nuclear energy and chemical processes also provide interesting models which can be transposed to engines - for the properties developed, such as corrosion resistance - and for processing techniques such as coating, moulding of solid parts etc.

It was only to be expected that on-board gas turbines should have benefitted from ceramics technology earlier than industrial compressor diesels. However, this is not simply the difference between a "luxury" industry which demands high performance and an ordinary industry for which every dollar counts. The thermodynamic formulae for the gas turbine and for the piston engine are not the same, and as a result of this difference any increase in combustion temperature substantially improves the efficiency of the turbine and somewhat less that of the engine. Bringing ceramics in to replace metals is therefore not a luxury in the case of a helicopter. Manufacturers are no longer necessarily engaged in such research in oder to improve the military performance of a prototype but, more tritely, in oder to give their aircraft (whether civilian or military) greater fuel economy. The question is whether the same thing will happen soon with piston engines.

The answer to that question is complicated because a great many factors are at work simultaneously. An increase in combustion temperature would certainly yield an improvement in efficiency, but only provided the engine were redesigned : expert studies have shown that mere replacement of the material part for part is not enough to make the effort of changing worthwhile. In addition, ceramics would offer other advantages into the bargain which it would be absurd to ignore : heat insulation which might even make it possible to do away with cooling systems, corrosion resistance which would enable different grade fuels to be used, longer engine life etc. Of course we are talking here in the conditional tense, since major technical problems still have to be resolved before reliable parts which are easy to process industrially become available.

IV.2.3.3. Diesel and ceramics : the main lines of technological development

In its present average configuration, the diesel engine converts 40% of the fuel's chemical energy into mechanical energy : 30% is lost as heat in the exhaust gases and 30% in the cooling system. The exhaust calories are the easiest to recover, since they reach the silencer at 450°C; this is what the turbocharger does. But any increase in combustion temperature results in hotter gases and improves the potential for recovery. Complex systems can be envisaged, notably the "Rankine cycle". Water is used for cooling, and the lost calories are at a temperature of 100°C and consequently of little thermodynamic value. Various arrangements can be imagined, provided the admissible temperature is increased. With a fluid at 150°C or 200°C it is conceivable that these calories could be added to the exhaust gases. A completely redesigned engine could produce gases leaving the cylinders at 700°C and achieve a total saving of 20%. Since ordinary metals are limited to 500°C or 600°C, there would have to be a change to high-temperature alloys. Some progress has been achieved though the introduction of nickel alloys, but the ultimate objective remains the elimination of the cooling system (adiabatic engine), and this necessarily means using ceramics.

Generally speaking, in order to cross a significant threshold towards improving efficiency it will be necessary, as we have seen, to resort to materials that are no longer traditional structural materials (even high-performance ones) but materials which also present important functional properties (especially thermal and anti-corrosion ones).

There are two possible medium-term approaches : to replace certain parts by solid ceramics or to apply ceramic coatings on exposed metal surfaces. The search for special metallic alloys is yet a third approach envisaged, but we shall not consider it here except to say that it has the disadvantage of relying on rare elements.

In any event the parts requiring modification are : pistons, piston rings, cylinder heads and cylinder head gaskets, valves, exhausts, and turbocharger components. The ceramics envisaged for coatings are silicon nitride or aluminium-magnesium-lithium ceramic glasses. One technique often talked about is ceramic plasma spray on zirconium. The main problem to be solved is coating adherence, and for this reason the ultimate solution might be the moulding of solid ceramic parts. In Japan, Nissan has fitted a vehicle with a turbocharger with a rotor made of massive silicon nitride. Similarly, in the United States Cummins is investigating heavy goods diesels fitted with solid ceramic piston heads, big-end bearings and cylinder linings.

We have seen that engineering ceramics are multifunctional materials, and indeed that their advantage lies in the fact that they solve several problems at once. However, as value analysis teaches us in a great many fields, it is important when developing products to know exactly what function is performed in each case, in what conditions, and how much one is prepared to pay for it. For this reason, although many ceramics possess good thermal, chemical and mechanical properties, research is divided into two main areas :

- where the prime requirement is to prevent corrosion and wear, interest focusses on zirconium, yttrium and magnesium oxides or on such composites as NiCrAl alloys.
- where the aim is to reduce heat loss, particular interest is taken in carbides (of chromium and molybdenum) which can be used as plasma sprays.

To conclude, let us return to the competition between types of engines. Even though progress in materials and electronic control are likely to boost all the competitors, the specific types of use can be tentatively classified as follows :

- for light vehicles, the petrol engine will remain a formidable competitor, especially with advances in electronics. But its competitiveness could be damaged by the problem of oil product quality, especially if the situation is complicated by more stringent pollution norms (because of the precious metal cost of catalytic exhaust systems).
- above 500kW power, gas turbine will become very attractive because they benefit considerably from developments in materials (new metals, ceramics).
- over the whole power range, the electric motor (which has made great progress in terms of maximum power and variable speed) may become a

serious competitor (i) for fixed applications depending on trends in the relative price of fossil/electrical and (ii) for vehicles if significant advances are made in batteries; but that is a much longer-term prospect.

- pending a possible breakthrough by new alternative technologies such as powerful lightweight batteries or fuel cell batteries, diesel will continue to dominate the vehicle market in the important slot which separates the petrol engine from the gas turbine and, assuming that fossil energy becomes scarcer and poorer in quality, that market sector could expand at the expense of the petrol engine. This strategy, allied to a considerable research effort into ceramics, is apparently the one the Japanese have opted for.

Finally, a reminder that technological developments can lead to separate branches coming together : one of the lines of research pursued in seeking to achieve better fuel combustion is to equip the diesel engine with a high-temperature precombustion chamber, thus obtaining a kind of petrol/diesel hybrid which combines the conventional engine's good response to variable loads and the ordinary qualities of diesel. Here again, progress involves the use of ceramics.

IV.2.4. PERMANENT MAGNETS

The principal applications for permanent magnets are as (a) magnetic memory carriers (cf. Chapter IV.1) and (b) electro-mechanical appliances including some electric motors, loudspeakers and generators (dynamos). It is this last category which is of concern to us in this chapter on materials and energy (cf. Figure IV.2.e) (16) .

---

(16) W. BARAN, p. 191, in "Nd-Fe Permanent Magnets, their Present and Future Applications", Ed. by I.V. Mitchell, CEC. Elsevier Applied Sciences Publishers.

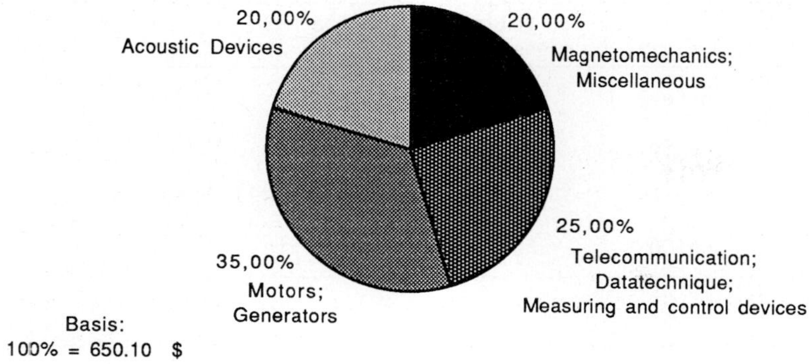

Source : (16).

Figure IV.2.e : Applications of permanent magnets in 1982

The market for these magnets in the western world in 1982 is about 115,000 t or 650 million dollars. The market is entirely dominated by hard ferrites (iron oxide based ceramics) with 93% of the tonnage and 55% of the price, followed by AlNiCo (aluminium-nickel-cobalt) alloys, with 7% of the tonnage and 33% of the price, and lastly by cobalt-samarium alloys with 0.3% of the tonnage and 11% of the price. These figures alone show the enormous advantage which ferrites have over all other materials in terms of price/weight ratio, which is 3.3 $/kg. However, AlNiCo or cobalt-samarium alloys, despite a very unfavourable price/weight ratio (27 $/kg and 210 $/kg) hold market shares in 1982 which are unassailable by their big rivals by reason of specific technical properties :

- remanence, a measure of the magnetic field which the magnet can supply and thus of the electrical output of the motor in which it is mounted;
- coercivity, which determines the limit beyond which the magnet loses it magnetism, and therefore its properties (coercivity depends on temperature);
- volume magnetization, a measure of the magnetic field supplied per unit volume of the magnet (cf. Tables IV.2.E and IV.2.F).

	Magnetic characteristics Remanence	Coercivity	Volumic magnetization	Thermal stability	Price
Good	Nd-Fe-B	Nd-Fe-B	Nd-Fe-B	ALNICO	Ferrite
↓	ALNICO	Co Sm	Co Sm	Co Sm	Nd-Fe-B
	Co Sm	Ferrite	ALNICO	Nd-Fe-B	ALNICO
Poor	Ferrite	ALNICO	Ferrite	Ferrite	Co Sm

Table IV.2.E : Comparative advantages of families of permanent magnets.

Nature	Composition	Remanence (Tesla)	Ambient temperature Coercivity (Tesla)
Ferrite	Sr $Fe_{12}O_{19}$ Strontium + iron	0.42	0.28
Alloy	AlNiCo	1.1	-
Alloy	Co_5Sm Cobalt + Samarium	0.9	0.8
Alloy	Sm $(Fe,Co,Cu,Zr)_{7.5}$ Iron + copper + zirconium + cobalt + samarium	1.05	1.0
Alloy	$Nd_{15}Fe_{77}B_8$ Neodymium + iron + boron	1.2	1.1
Alloy	Nd Fe B	1.32	-

Sources : H. ZIJLSTRA, Journal de Physique, Colloque n°6, CG 1985 and (16).

Table IV.2.F

AlNiCo alloys are particularly valued for their high-temperature utilization (about 300°C), in a range where all other materials have lost their magnetization, as are cobalt-samarium alloys in small lightweight motors requiring high unit power or in dynamic loudspeakers.

This market was fairly stable in 1982, but in 1983 there appeared the iron-neodymium-boron magnets developed, apparently simultaneously, by Sumitomo (Japan) and General Motors (USA).

While these new magnets do not pose a threat to ferrites initially, AlNiCo and CoSm alloys are directly replaceable by Fe-Nd-B, for this new alloy has magnetic properties which are at least as good or slightly superior, the only drawback being demagnetization at low temperature (approximately 120°C). But it has the enormous advantage of not containing any strategic metals such as cobalt or even, to some extent, samarium. Neodymium is a fairly abundant element, though that does not mean that present supply capacity could meet a sudden demand; its applications prior to this discovery were virtualy negligible. Lastly, the manufacture of this material is no more technically difficult than that of samarium-cobalt alloys. For all these reasons the unit price seems likely to be relatively low, thus giving a price/weight ratio which might even compete with ferrites in certain applications (15 to 20 $/kg).

The above considerations suggest that this new alloy is going to make a powerful impact on the market for permanent magnets operating at relatively low temperature.

The only real economic obstacle to its development is the shortage of neodymium production capacity, though that problem, will probably be quite quickly solved once the technical options in respect of the product form which the magnet industry wants have been presented to the manufacturers.

The critical area in which this alloy might come to occupy a leading position is the manufacture of robots with articulated arms, where the need for electric motors which are both powerful (though without needing high torque) and lightweight (being located on the ends of the arms) is obvious. Medical NMR applications are also conceivable (liquid helium cooled superconductor magnets are needed nowadays to reach magnetic fields which can be obtained with remanences of 1.2 Testa). On the other hand, starter motors for cars, often suggested because of the low operating temperatures, seem to offer little interest because in this particular case it is useful to have high torque (high power at slow rotational speeds, as in the case of a locomotive) and the "standard" electric motor without a permanent magnet is unassailable here.

Finally, there is one remarkable point to be made about this particular functional material. It is the only example where Europe has seized the opportunity in time to develop a strategic material for the industry of the future, through concerted action (CEAM), under the auspices of the EEC. This group brings together materials experts, magnet manufacturers, magnet users and raw material suppliers (rare earths). 53 laboratories divided into four groups, coordinated by CNRS-Grenoble (France), Trinity College, Dublin (Ireland), the University of Birmingham (Great Britain) and T.U. Berlin (Federal Republic of Germany), are being funded to study and develop magnets based on neodymium-iron-boron).

This project deserves to be highlighted and set against the catastrophic dilatoriness which has bedevilled the development of materials for information technology.

The reader will find an exhaustive technical economic and industrial study of this new material in reference 16.

CHAPTER IV.3

Biomaterials and Biotechnology

The use of materials for technologies applied to living organisms has in the past been empirical in the highest degree. To find replacements able to perform the functions of defective tissues or organs in the human body, recourse was had to the broad categories of available materials (wood, metals, etc) which were sometimes unable properly to meet the required specifications, either because the implant-tissue interface gave rise to poorly controlled chemical and electrical processes, or - most important - because the requirement of biocompatibility was not satisfied.

Technical advances in biomedical materials are currently revolutionizing the health field. It is not simply a matter of replacing some materials by other, more efficient, ones. In conjunction with the development of new materials for biomedical applications, new medical techniques are being employed, new therapeutic possibilities inconceivable a short while ago are being opened up, and closer collaboration is coming about between medical specialities which in the past have been strictly segregated.

But the links between new materials and biotechnology are no longer limited simply to biomedical materials. The development of veritable molecular electronics, of "biological computers", is probably one of the most promising prospects for the decades ahead. Even though research today is still at the laboratory stage and even though the early applications will probably be very modest ones, the use of biomaterials to perform a new function - processing information - is under way. And the first markets seem likely to be in the field of biosenscrs, where there are innumerable applications.

Lastly, the new materials are playing a key role in another area of biotechnology - in industrial processes involving bacteria, where the membranes used in a number of critical phases of biotechnological processes (separation, purification, concentration, etc) represent the most imрortant markets.

IV.3.1. THE DEVELOPMENT OF MATERIALS FOR BIOMEDICAL USE (17)

Bicmaterials are the foundation of what is sometimes considered to be the major therapeutic revolution of the last third of the twentieth century. Initially they were kept for critical situations, but they are now employed to satisfy needs perceived by the patient in terms of rehabilitation, comfort, amenity and appearance. The problems that they pose lie at the crossroads between the life sciences, the engineering sciences and the legal, political and socio-economic sciences. Biomaterials science, which is to surgery what pharmacology is to medicine, thus brings together specialists who were not often in the

---

(17) This passage is taken from the paper presented by Dominique MUSTER, Laboratoire de Recherche, Chaire de Stomatologie et de Chirurgie Maxillo-faciale, CHR, BP 426, F-67091 STRASBOURG Cedex, on 21 June 1985 to the meeting of the group of experts to prepare innovation courses in the biomaterials sector. Dominique MUSTER, "Utilisation des biomatériaux en chirurgies osseuse et dentaire, aspects médicaux légaux" ("Utilization of biomaterials in bone and dental surgery, legal medical aspects"), 1983; J.P. MASSUE/D. MUSTER, "Action biomatériaux : transferts recherche-industrie en Europe"), May 1985.

habit of talking to each other. While the main purpose of this transdisciplinary applied research is still to improve the rebuilding of organs and tissues it also enables the effect of medicines to be targetted and, lastly, adds to our fundamental knowledge of the interactions between living structures and chemical compounds.

In the first place, researchers looked only for materials with suitable mechanical properties which did not provoke catastrophic biological reactions in the short term. Nowadays biocompatibility is defined not simply as an absence of unfavourable reactions, but as control over the body's reactions to the implantation of a biomaterial throughout the period during which the two are in contact, so that these reactions are not detrimental either to the organism or to the implant.

We fear the pollution or contamination of our external environment in situations where the risk can only be evaluated retroactively and statiscally. But introducing biomaterials into the body brings our internal environment into contact with foreign bodies (internal pollution). The implantation, every years, of millions of materials represents experimentation under more controlled conditions than a great many studies on the biological effects of chemical agents in our external environment, with the additional advantage that the experiments are carried out on the animal species that interests us most - our own.

The success of a biomaterial depends mainly on three major factors : the implant's physical, chemical and biological properties as a whole, the condition of the recipient, and the skill of the surgeon who carries out the implant and monitors its behaviour.

Once inside the body, materials implanted will have to operate under biological constraints. They will be subjected simultaneously to chemical attack by the biological environment and to physical attack by the mechanical environment at the implant site. Chemical attack will take the form of corrosion of metals and deterioration of plastics, and physical attack will deform the material, both combining to make it fragile and consequently liable to breakdown in a constantly mobile environment. Corrosion and deterioration of biomaterials can have harmful consequences of two kinds : firstly, the implant may become increasingly fragile, and secondly, there may be toxic or allergic reactions inside the organism through the products it liberates when reacting to materials passing into the body's internal environment.

Nowadays, most biomaterials applications employ industrial substances which were originally developed by industry for non-medical use (particularly the aeronautics and space industries in the case of advanced materials such as titanium alloys or carbon-carbon composites, or by the nuclear industry in the case of a material such as pyrolitic carbon, which has high strength and is blood-compatible). While the functional aspect of the performance of materials inside the human body can be predicted with some reliability, it is still difficult to assess the biological performance of implants.

The enthusiasm aroused by the results of experiments on animals is sometimes dampened by clinical testing. The scale of the testing currently required to demonstrate an acceptable level of biocompatibility for a new material takes so much time (at present five years) and is so costly (more than five million francs for an ASTM type protocol) that the range of implant materials is growing very slowly, hence the need for a reasonable approval procedure so as not to hamper innovation and the desirability of mutually recognized proving tests in Europe.

The development of less expensive, more efficient and more reliable equipment is unavoidable if biomaterials are to keep healths costs down. In France, for example, the Advisory Committee on Hospital Equipment recently stressed this point and the need for purchasers to be properly trained. Account has to be taken of considerations deriving from industrial research, product development, quality control, safety measures and regulations.

Where the liability aspect is concerned, there is as yet little case-law in this field. It is a question both of medical liability and of product liability, because the device itself can be a factor beyond the control of even the most skilled practitioner.

The biomaterials used to manufacture implants for human beings represent an important economic prize in Europe : the market for them is growing by 15 to 20% a year and is expected to catch up with medicines by the year 2000. Europe has to import the bulk of these products, though Germany is an exception, particularly where "soft" materials are concerned (Bayer). The principal types of material which enter into biocompatible materials are the following :

- for cardiovascular prostheses (where biocompatibility with blood is the paramount characteristic) : silicons, polyurethanes (envelope and diaphragm), pyrolitic carbon (valves) or carbon in vitreous form.
- for artificial blood (where only a few vital functions of human blood can be reconstituted) : polymers such as polyvinylpyrolidone or perfluorocarbons.

- for hard tissues, as replacements for metal prostheses, a great deal of hope is being placed in "bioglasses" (which have the advantage of chemically bonding with bone) in which the silica which is normally a component of glass is replaced by natural bone components - calcium, phosphorus and sodium - and in ceramics (calcium phosphate ceramics for maxillary applications, glass ceramics for odontology) or composite materials (tooth fillings, crowns, etc). The 3M Company (USA) holds a leading position in this field.
- for ligaments and tendons : solid elastic fibres (kevlar, dacron, etc).
- for artificial skin : various polymer combinations.

Current annual figures for a country like France are over two million traditional dental prostheses, about 50,000 total hip replacements, 5000 knee replacements, 15,000 heart valves, 15,000 arterial prostheses, 20,000 heart pacemakers, 10,000 lens implants (to correct cataract), 2,000 dental implants, and so on. These figures are certainly well down on those in the United States, where the number of biomaterial implants is estimated at several million each year. It is as well for the industrialists concerned (metallurgists, plasturgists, ceramists) to realize the value of this economic sector and find out what the medical fraternity want so that they can win new market sectors in the manufacture of biomaterials in Europe.

The biomaterials field is in a traditional situation between the artisanal stage and integrated research in the product development effort. It should be a particularly favourable field for small and medium enterprises, whose size and organizational flexibility fit them particularly well for the manufacture and marketing of biomaterials. However, it seems that these firms may perhaps have difficulties in producing the biomaterials under consideration. Thought will have to be given to bringing together the SMEs in question and industrial groups able to prepare the basic biomaterials. The small tonnages characteristic of this sector can be made economic by the high added value of the high-technology products which it makes. In this way, the biomedical industry must secure its place in industrial redeployment, and European cooperation is needed in order to support it.

Implant	Total/ yr ('85)	$/ Implant	'85 Market	Total/ yr('95)	$ Implant	'95 Market
Interocular Lenses	1.0	$ 400	$ 400	$ 1.5	$ 750	$ 1.125
Total Hips	0.3	600	180	0.4	700	280
Total Knees	0.15	400	60	0.2	500	100
Ankles, Elbows, Shoulders	0.05	500	25	0.7	600	420
Finger Joints	0.4	50	20	0.6	100	60
Mammary Prostheses	0.4	100	40	0.5	120	60
Tooth Implants	0.3	500	150	1.0	300	300
Ridge Augmentation; Periodontal Materials	0.2	20	4	2.0	40	80
Cardiovascular (heart pacers, valves, bypasses, arterial prostheses)	1.0	250	250	1.5	300	450
Middle Ear Prostheses	0.03	150	4.5	0.035	200	7
Cochlear Prostheses	0.0005	20000	10	0.01	10000	100
Orthopedic Fixation devices	1.0	50	50	1.2	60	72
General (sutures, skin grafts, catherers)	2.0	20	40	3.0	25	75
TOTAL (U.S.)	**6.8 million**		**$ 1.233**	**11 million**		**$ 3.129**
TOTAL (World Wide)	**12.0 million**		**$ 2.800**	**25 million**		**$ 6.000**

(Source : HENCH and WILSON "Biomedical materials, MRS Symposia Proceedings, Vol. 55, p.66, Pittsburgh, 1986).

Table IV.3.A : Estimated biomaterials market (millions of dollars)

IV.3.2. BIOSENSORS AND PRODUCTS FOR ANALYSIS

Bicsensors measure organic substances in liquids and gases by combining a biological device with electronics to translate biological interactions into electrical signals for interpretation (18) Starting with biological recognition of the elements to be detected in a reaction mixture or substrate (eg blood), the elements being recognized by enzymes, antibodies or microbe or animal cells, the reaction products are selected by a membrane, following which the biological interactions are translated into a coherent signal. it is here that the function of electronic sensors or transmitters comes into play. There are several types of sensor in competition with each other, among them ionic sensors of the ISFET type (Ion Sensitive Field Effect Transistor) which detect ions and amplify the signal in a silicon substrate, or optical sensors.

The advantage of using biosensors in the various fields in question is their better selectivity (it is easier to make a biosensor which is sensitive to just one product than to make a mineral sensor to do the same job), as well as sometimes a very substantial time saving. Measurement capacities in industry can easily be multiplied by a factor or five or more without additional constraints, and intervention to remove poor samples can take place far more quickly and efficiently. Moreover, these techniques are generally much less expensive than traditional methods.

The main areas of application are as follows :

- the biomedical field (measurement of blood sugar level in diabetics, extracorporal measurement of blood cholesterol, monitoring the presence of anticoagulants to prevent thromboses and strokes, followed by hormones in blood, chronological and quantitative monitoring of an injected drug, etc. (19) . In this field, biosensors can be a key factor in the decentralization of medical checks done either by general practitioners or by the patients themselves;
- the foodstuffs industry (purification of sugar or milk);
- the pharmaceutical industry (monitoring production of vitamins or antibiotics);
- the environment and pollution (monitoring of soil acidity, liquid effluent, industrial gases, etc). This is an area that ought be of great interest to public authorities and local government. It is in any case relevant to defence : the English firm Thorn EMI, for example, has

---

(18) "L'essor des biocapteurs" ("Boom in biosensors"), P. GENVRIN, Sciences et Techniques, December 1986.

(19) According to the "Biosensors", Study CPE Bulletin, December 1984.

just won a contract (for 23 million pounds sterling) from the British Ministry of Defence to produce a neurotoxic gas detector (Naiad system). With the prospects of a possible increase in the development of chemical and biological weapons, biosensors are of paramount strategic importance.

All these devices, situated at the interface between biology and electronics, are of strategic importance for another reason : they offer a huge, practical area of learning by experience in a field which is still sometimes thought of as futuristic, but which is attracting growing interest - that of "biological electronics".

IV.3.3. THE USE OF MATERIALS IN BIOTECHNOLOGICAL PROCESSES

These materials consist mainly of membranes which can operate at several critical process stages - separation, purification, concentration, sterilization, etc. In their various applications (via the three main techniques of ultrafiltration, microfiltration and inverse osmosis), membranes rival the traditional operations of evaporation, centrifuging, concentration, spin drying or distillation, which normally consume energy and time and which often entail destructive heat treatment. There is a large potential market for membranes. It is widely thought that it could well triple over the next ten years, particularly in agri-foodstuffs (milk, wine, etc), the pharmaceutical or biochemical industries (blood and blood derivatives, etc) or in the production of ultra-pure water for electronics. The effort going into membrane research is in proportion to these potential markets (Japan, for example, is spending 420 million dollars over five years on a programme of research into all membrane techniques).

Membrane development relies heavily on the development of new materials. Following the first generation based on cellulose acetate, and then the second based on natural or synthetic polymers, a third generation using ceramic or graphite carriers could well become established (20) .

---

(20) Another promising line of development is that of liquid membranes for the purposes of separating gases (a very important one strategically).

A great many of the ultrafiltration processes used in the dairy industry now rely on a mineral membrane (which, unlike polymer-based membranes, can withstand very high temperatures, up to 500°C), consisting of a layer of porous carbon coated with a zirconium oxide film (21) . Apart from the membrane itself (which, as it were, acts as component), the role of the modules which serve as membrane carriers (and act as the system) is also fundamental. The shape and nature of the materials used as modules are strictly conditioned by the function which the membrane is required to perform (so there are modules in the form of hollow fibres for seawater desalination by inverse osmosis, in tubular form for processing high-viscosity products, in flat form for the ultrafiltration of machining emulsions, etc).

The utilization of membranes in industry raises the problem of integrating new techniques in traditional production processes. Most of the elements observed in other materials fields are to be found in this sector : the development of membranes for wine making is a characteristic example here. More and more demanding international norms are currently forcing the wine industry to improve the stability, sterilization and clarification of wines. The development of tangential microfiltration membranes offers a very attractive solution for this purpose : by integrating the sterilization and clarification phases in a single-stage industrial process, tangential microfiltration yields a whole series of economic advantages : savings in production time, less risk of oxidation, less product loss, elimination of the cost of consumable items, the possibility of processing musts as well as wines, the possibility of regenerating the membrane, etc. Moreover, there are long-term hopes that the use of membrane techniques will make it possible, by modifying the alcoholic content of wine, or even by making it possible to produce alcohol-free wine, to open up new markets to production, and particularly European production (perhaps opening up markets in Islamic countries, for example).

---

(21) This type of membrane is a remarkable instance of technology transfer, having been originally developed for the nuclear industry.

PART V

STRATEGIES
Elements in an Industrial Policy and a Policy for Employment and Skills

It has taken a long time for people to become aware of the consequences for society arising out of the current changes in materials. It is true that for the most part these upheavals are hidden, and come about deep down, in the actual heart of the technical system, in the mastery of the microscopic, in the reconception of the application of material, or in the working out of new industrial processes. Yet superficially the impression of continuity persists. In the technical articles which are part of our everyday lives - motorcars, buildings or the most banal objects - the progress that has been achieved in materials is barely perceptible. It is hard to imagine, for instance, under the paintwork on the latest model of a car, the ingenious manner in which the different materials co-existing have been assembled, just as it is hard to grasp that the elaboration of something with so traditional a shape as a bottle demands today the most complicated optimization calculations and the use of complex materials. Objects regarded as new - computers, pocket calculators, CD players, etc - suggest the progress of information technologies far more than of materials technology.

This skin-deep continuity is found again when we examine the statistics. There is no marked break in the growth curve of technical plastics in vehicles, or the upward trend of the proportion of optical fibres as against copper cable in telecommunications.

In this way, and because the newest structural materials generally copy earlier forms when they come to be used for the first time, because the most strategic functional materials are sometimes invisible and only account for a small proportion of the volume and value of technical articles, but also because industrial habits and traditions help slow down the rate of innovations (because old methods "react" to new methods, or more simply because time is needed for testing and trying out something new), the importance of changes in materials, and their role in economic expansion are not always realized.

The appearance of continuity is particularly dangerous when a strategy comes to be worked out, both for industrialists and the authorities, since they run the major risk, by dint of taking it for granted, of being rapidly overtaken by competition or of intervening too late - which is what has happened in the steel industry. This is why, more than in any other field, it seems necessary to reflect on the future of materials by tracing development scenarios, in order to help in decision-taking, to bridge gaps and to limit the price that has to be paid for current transitions, with particular reference to job losses.

The arguments developed in the previous chapters suggest, for the future of materials, two major development scenarios : the first relates to the "growing importance of the phenomenon of variety" and the second to the "irreversible ascent of plastics". These scenarios are indicated in detail in Chapter V.1. Each prompts a specific reflection concerning industrial policy, and policy for employment and skills. Even if there is inevitably an excessive contrast effect in the presentation of the two scenarios, their comprehension makes it possible to clarify considerably current issues and obstacles, and to shed light on the strategic elements for the European Community, presented from Chapter V.2. onwards.

The consequences of these two scenarios concern particularly the nexus of problems posed by structural materials. They must not, however, cause us to overlook the eminently strategic role of functional materials as a key to society. The European Community cannot place a block on these materials to the advantage of structural materials where, thanks to its traditionally strong industries (chemicals, transport industry, public works), it occupies a position that is perfectly competitive vis-à-vis the United States and Japan. The arguments set out in Part V indicate the importance which the European Community must attach to its training system, so as to make it possible to accentuate fundamental research, which is a source of progress and makes for the mastery of functional materials, and so as to permit constant refresher training for its engineers, technicians and labour force, the essential condition for a mastery of continually changing materials technology.

CHAPTER V.1

Description of Two Development Scenarios for Materials

V.1.1. FIRST SCENARIO : "GROWING IMPORTANCE OF THE PHENOMENON OF VARIETY"

In accordance with this scenario, the industrial system as a whole becomes capable of functioning in a permanent system based on variety. It presupposes that there has been a final break with the previous regime, based on the standardization of products and that the industrial system finally succeeds in managing efficiently a form of production of goods and services that is being diversified all the time, with the aid of a range of technical solutions that is constantly being renewed with a continuity expanding spectrum of possibilities.

The arguments developed earlier in this report have shown that the role of materials in this hypothesis (in conjunction with information technologies) appears decisive for the purpose of ensuring that the new production regime is viable. Continual progress in the conception and application of new materials means that increasingly numerous technical solutions can be devised, while the mastery of processes for the perfection and utilization of materials notably through integration processes - makes production processes viable, notwithstanding the increase in complexity.

It is worth commenting that this hypothesis of the permanence of the variety-based regime does not mean the absence of any company structuration - far from it. A variety-based economy is only possible if there is a "fixed" factor around which the different forms of production are arranged and renewed. The "fixed" factor for a company is in this case its organization network in the broad sense, embracing the internal production network and the "external" network, after the process, nearer to the market place making it possible to obtain information about demand and the environment, and the network before the process for the structuring of research. The proper operation of the interaction between these three networks is becoming a more and more decisive factor in a company's competitiveness.

V.1.1.1. Importance of the variety-based regime for materials

For materials, the permanence of the variety based regime a priori gives pride of place to the concept of "multimaterials" or "complex" materials resulting from the alliance of several kinds of materials. Composites, ceramics, super-alloys, multi-layer materials, etc, are naturally all different forms of response to this pattern. But it is clear that no material is ever condemned in a permanent variety-based system. When a material makes way in a given application for a new kind of material, there is immediately a reaction from the material that has made way and research into it is intensified (reaction of silicon to the emergence of gallium arsenide, for instance). Variety provokes a continual stimulation of progress in materials, and in particular steadily increasing progress in the mastery of the microscopic structure.

This advance in the mastery of the microscopic structure of materials is in fact a powerful factor for the continual replacement of one material by another. To some extent, therefore, it is safe to say that the concept of "composites" is a purely <u>transitory</u> concept in the history of materials development. It is a kind of provisional "short-cut" as a means of offsetting the absence of full-scale mastery of the microscopic structure of material, since with the composite we adopt a macroscopic structure so as to counter defects which are microscopic. From this viewpoint, ceramics may appear by comparison with composites to be situated at a higher level in microscopic mastery, and so one of the families of materials that are likely, among others, to displace composites. The main difference between "ceramics" and "composites" lies, in effect, in the fact that ceramics are single-phase and thus eliminate all problems connected with the interface between the different phases. (We are not talking here about ceramic-ceramic composites or sandwich ceramics).

It is important to point out, furthermore, that if variety concerns structural materials in an initial stage, in the long term it applies to all materials, including functional materials. It is possible to see a constantly growing diversity of basic materials and components in the sphere of electronics, for instance. In semiconductors, silicon is destined to be less and less "on its own", and the gradual recourse to III-V or II-VI compounds (which can lend themselves to an infinite number of combinations) makes for a considerable degree of variety.

Lastly, one of the major advantages of the permanent regime based on variety is that it liberates us, to a very considerable degree, from the concept of "critical" or "strategic" material. The temporary fear of a shortage of a given material is markedly allayed by the existence of replacement solutions. For one thing a large proportion of materials is based on elements which are present in abundance on the surface of the earth (carbon and silica, among others). In one way, the variety-based regime hypothesis could correspond to an expensive energy scenario, with restricted access to raw materials. It is no coincidence that the premisses of an industrial system based on variety made their appearance after the great oil shocks, with the greatest impact on heavy materials industries such as the steel industry and the cement industry.

V.1.1.2. Importance of the variety-based regime for skills

In the first scenario, the permanence of a variety-based regime implies an extremely emphatic hypothesis concerning <u>skills</u>. The industrial system cannot work, in fact, in this kind of context unless it presents perfect mobility enabling a <u>rapid</u> change-over to be made from the use of one material to the use of another, depending on the state of parameters relating to demand, the environment or research; this ability demands a far-reaching evolution in skills towards a unified technical skill vis-à-vis materials and the process whereby they are applied.

Perfect mobility, then, presupposes the existence of a <u>training</u> system that teaches the bases of a <u>unified science of materials</u> including, in particular, chemical and metallurgical skills. But the incessant evolution of knowledge, the perpetual questioning of settled principles, the achievement of higher and higher levels in the complexity of production necessitate the general recognition of an educational principle that is geared to the variety-based regime :

alternating and continuous forms of training in which periods of study alternate with periods of work on the shopfloor. If a variety-based system is to work, then, an extremely high educational level and continual refresher training are needed. By way of corollary, it is necessary to maintain a pool of capable training and research staff on which to draw for this kind of training; such competence can only be maintained at the required level through advanced fundamental research coupled with teaching, a role which by tradition has fallen to the universities.

The new skills required for this scenario presuppose the mastery of new tools necessary for the management of a higher level of complexity : the utilization of expert systems, the management of local networks, full integration of the different production functions and generalized recourse to data banks.

V.1.1.3. Importance of the variety-based regime for industrial structures

The concept of a "multi-materials" firm is of course the predominant industrial concept in the case of a variety-based regime. This multi-materials structure makes it possible to prevent a set of processes or skills from disappearing in the event of a sudden transfer of downstream demand from one material to another. It also makes it possible to manage the substitution process by foreseeing it, that is to say ensuring technological continuity instead of the reverse.

A multi-materials firm in a variety-based regime dominates both the knowledge and application of "primary" materials (steel, plastics, aluminium, glass, etc) and those of "secondary" materials (composites, sandwiches, etc). However, the profile of different multi-materials firms (the nature of their principal customers, for instance) may appear very different depending on the technological route followed by those firms from their initial specialization. The different firms will in fact evolved towards a multi-materials structure by gradually developing outward from their initial strong point. A firm originally specializing in glass, for instance, might progressively build up multi-materials expertise by taking on board, in an initial stage, a plastics firm and then a firm producing composites, and so on. The fact that all firms are tending to become multi-materials firms does not in any way mean that industrial profiles become uniform. In this sense, the variety-based regime gives every materials manufacturer his "chance" at the outset (be he chemist, metallurgist, etc).

The main feature of relations between utilizers and producers of materials is the concept of industrial partnership. This notion corresponds logically to the greater complexity of the industrial system, which rules out fixed and lasting relationship between industrial firms. It also corresponds to the necessary mobility of skills and abilities.

V.1.2. SECOND SCENARIO : "THE IRREVERSIBLE ASCENT OF PLASTICS"

To some extent, this second scenario corresponds to a simple extrapolation of the present situation. At the end of the 80s, plastics (in the broad sense, it would be better to say organic materials) seem to be slowly, but irreversibly, continuing to take the place of other materials. Since the early days of the utilization of plastics, there has been virtually no example of plastics being ousted from applications to which they had acceded. This phenomenon is particularly evident in the case of structural materials (in transport, packaging, buildings, etc), but is little extending even to functional materials and particularly those for electronics .

Behind this straightforward extrapolation there lies in reality a major hypothesis concerning the evolution of the industrial system : a variety-based regime is basically unstable and over-complex to manage. The variety observed since the mid-70s was only a transitional regime between two industrial epochs, marking the change-over from a "steel and standardization" civilization to a "plastics and tailor-made production" civilization.

V.1.2.1. Advantage of plastics

The advantage of plastics is that they offer a technologically acceptable and industrially viable solution to the requirements of demand and production. Plastics constitute, in the first place fundamentally, the type of materials in which the mastery of the microscopic structure of material can be reproduced on a large scale (a large batch of plastics possesses, at the polymerization stage, the same microscopic quality at each point, while defects to be found here and there are generally due to ageing).

Plastics also have the advantage in production processes of considerable flexibility of application, in particular lending themselves when necessary to the most complex forms, and lastly of responding to the requirements of customized demand. Their faculty for imitating other materials, but also the scope they provide for conceiving fresh technical low-cost solutions, their fundamental property of resistance to corrosion and lastly their infinite range over the entire industrial scale, from the most current commodity product to the most elaborate technological plastic, are all assets which give them clear advantages.

To some extent, the logic of plastics makes it possible to "encapsulate" - from the production of standardized commodity plastics to the elaboration of a very high-performance composite - all the stages and degrees of elaboration in materials production. It is a logic which (in contrast with the variety logic) preserves a considerable proportion of the production organization intact. The "cost" of the changeover from the present system to a "plastics" scenario is perhaps less radical and less onerous to manage (in terms of jobs and skills, in terms of technological culture, in terms of materials, etc) than is changeover to a variety-based regime.

Lastly, in the face of demand, plastics have the advantage of functionalization (function being taken here in the strategic sense, not in the technical sense applying throughout the rest of the text in referring to functional materials). By ringing the changes on their extremely vast range it is possible to devise plastics which respond to several separate functions at the same time (in the field of insulation, plastics can be made which combine the properties of lightness, ease of application and heat insulation).

The ascent of plastics means that the other materials retain their long-term share of markets only in highly specific applications where they are unrivalled (high temperature, for example).

The disadvantage of the plastics "solution", on the other hand, is its marked dependence on raw materials availability. The recent oil crisis was admittedly not on ordeal for the plastics industry, as might have been feared, because it became apparent on the one hand that the best applications of oil in the event of a crisis were its utilization for freight and the chemical industry, and on the other that chemical firms have evolved towards high value-added production. It was shown also that the energy content of plastics was frequently lower than that of other materials. It is not certain, however, that in the event of very extensive development of plastics in the industrial system, a further oil crisis could be tolerated without difficulty. In other words, the "plastics" scenario is a scenario geared to maintaining the availability of access to oil, and more generally the absence of an energy crisis. It must be pointed out, however, that an oil crisis in a few decades will be felt more as a raw materials crisis (affecting plastics and other petrochemical products) than as an energy crisis (hitting aluminium or magnesium harder), thanks to the new sources of electrical energy offered today by nuclear fission, and in the longer term by nuclear fusion. (It is worth pointing out at this stage that a third scenario could have been added - with a very low probability of realization, at least between now and the end of the century - that of mastery of nuclear fusion, giving a "civilization" of aluminium or magnesium considerable scope for development).

V.1.2.2. "Plastics" scenario and skills

In terms of skills, the irreversible ascent of plastics corresponds to the gradual replacement of metallurgical know-how by chemical know-how. Chemical know-how is to be interpreted here in the broad sense, and covers those who make plastics as well as those process or apply them.

This irreversible replacement by plastics may appear to be a very profound revolution in the division of labour and in the nature of skills. Metallurgy established itself as the art of developing materials from their raw materials, then incorporating them in industrial articles by gradually "de-forming" them in the mass. The trades of blacksmith, machine-tool operator, milling machine operator, etc, follow each other in the metallurgical logic as so many "basic" skills essential for achieving the shapes needed. Around these basic skills a great many "additional" skills are required : welding, surface treatment, among others, so as to add the properties required by demand which were not present in the material in the form it presented at the start.

The logic of the plastics solution proposes a different structure of skills. It attaches definite importance to the materials designer who can already, at the stage when the material is being made, integrate some of the specifications demanded, and to the mould-maker who makes sure that the right shape is achieved. Between these two extreme elements in production comes a succession of other specific skills. One important stage is the formulation of products obtained, ie preparing them for the processing they are to undergo bearing in mind the uses sought. The raw plastic has to be mixed with additives, fillers, reinforcing agents and other plastic raw materials. From that point onwards, however, the third phase of preparation, the processing of plastics, which consists of processing and shaping the plastic material, is no longer specifically a matter for the chemist, but rather a skill of engineering origin. It is at this level that an excessively sudden break due to the changeover could be avoided, since very many firms using materials are gradually incorporating in their production units processing lines for plastics (which to begin with are often "in addition" to existing lines for metals, and subsequently replace them once know-how is acquired).

V.1.2.3. "Plastics" scenario and evolution of industrial structures

Behind the hypothesis of a "plastics" scenario there is of course that of the gradual domination by the chemical industry of all other materials-producing activities. The power of the chemical industry might not be confined to the mere production of plastics. A great many chemical concerns have already absorbed raw materials processing units. The need to satisfy the requirements of functionalization, that is to say be perfectly familiar with the conditions in which the end user's needs have to be met may encourage the chemical industry to try to go still further and control the application sectors which use materials. Chemical firms are as it were becoming the rivals of their own customers.

The solution for those customers is to incorporate chemical know-how in their own firms. It is worth noting that this strategy is already being applied on a broad scale, as many groups in the computer and automotive sector, for instance, already have substantial chemical laboratories and subsidiaries in the chemicals sector.

V.1.3. COMMENTS ON THE TWO SCENARIOS. CERTAIN PREDOMINANT TENDENCIES IN MATERIALS

The way in which the scenarios are presented is inevitably crude and provocative (see Table V.1.A). It may appear particularly so in the field of materials, since today it is still the steel industry which sells the most materials, where the traditional forms of skills still predominate. It may appear so all the more because, to a large extent, the training system in Europe perpetuates teaching traditions by segregating the different special categories of materials.

	SCENARIO 1	SCENARIO 2
	"GROWING IMPORTANCE OF PHENOMENON OF VARIETY IN MATERIALS"	"IRREVERSIBLE ASCENT OF PLASTICS"
Effects on materials	* Marked trend towards the concept of multimaterials (ceramics, composites, sandwiches, etc) * Although no material is definitively discarded	* Gradual domination by plastics in broad sense (including organic-matrix composites, etc) * Other materials command only restricted outlets
Effects on industrial structures	* Trend towards multi-materials firms * Industrial relations marked by partnership	* Domination of chemical firms producing materials having assimilated chemical know-how
Effects on skills	* Multi-disciplinary training based on unified knowledge of materials	* Domination of chemical know-how
Effects in terms of economic strategy	* Scenario compatible with a high-cost energy hypothesis and problems of access to strategic materials	* Scenario of cheap energy and easy access to supplies

Table V.1.A : Two development scenarios for materials.

Even if the chances are that steel will retain substantial applications in the industrial system, it is clear that changes in the technological system as a whole require thought to be given to the transition to other predominant patterns, and it is in this sense that the reference to scenarios may be useful for strategic reflection. The harbingers of one or other scenario are already to be seen, in fact: multi-materials firms, industrial partnerships, steady growth of the proportion of plastics in all materials, etc. But without a reference model for the future, the present situation is extremely difficult to interpret, since it corresponds to a point when "all the options still seem open".

Already, however, some predominant tendencies are appearing sufficiently clearly for them to be immediately isolated before any strategic analysis :

- The evolution of society towards a service economy, reflected in a decline of the physical quantity of mass-produced goods. In more simple terms, the byeword is more than ever "less energy, less materials and more information".
- Whatever the scenario, materials are going to play a threefold role in the industrial system : respond to more and more varied quantitative and qualitative requirements of demand, promote simplification of the production process and keep to designs and processes that are sufficiently straightforward, with no excessive break with the past.
- Structural materials are having to take on constantly more <u>functions</u>. The boundary between structural and functional materials is therefore tending gradually to disappear. The technological routes followed by these materials, which hitherto differed from each other, could well come together. The boundary between these types of materials was already relatively vague for materials for high temperatures and heat insulation, where the structure and function aspects are closely bound up. On the other hand, the integration in structural materials of functional elements relating to information could be an important future direction (concept of material rich in information notably in such sectors as the automobile industry, where electronics are going to grow considerably). The safety imperative which has to be satisfied by large structures (aircraft) is moving in the same direction, and means that pride of place is given, among structural materials, to greater functionality in terms of information (tendency towards "structure-function-information" composites).

It is to be noted, in fact, that in checks on technical products (such as vehicles), there is a very marked distinction in the perception at a given moment of the performance achieved by the product. It is possible to be informed in real time of the state of functions peculiar to it (consumption, temperature, etc) but not on the state of its structure (wear and tear on materials, microscopic flaws, degree of corrosion, etc). The need to devise materials that can without difficulty provide information on the state of their structure thus appears particularly significant. Already, in a random sample of applications, attempts are being made to satisfy this need (indicator panels are included in the packaging of certain agri-food products which change colour as the product ages, and as it undergoes "heat treatment"; again, in important yacht races, sensors are inserted in the masts of the most modern vessels providing navigators with regular information on wind force, at all mast heights). It is easy to imagine, in these applications, that biomaterials may be first-rate elements which help solve the problems.

- The question of relations between chemists and metal-workers is all too real. The rivalry between them is particularly marked in the two main categories of new materials, composites and ceramics. The figure V.1.a, which concerns the different forms of composites, indicates clearly what is involved, namely process control (still open) of the production of ceramics in the realization of composites.

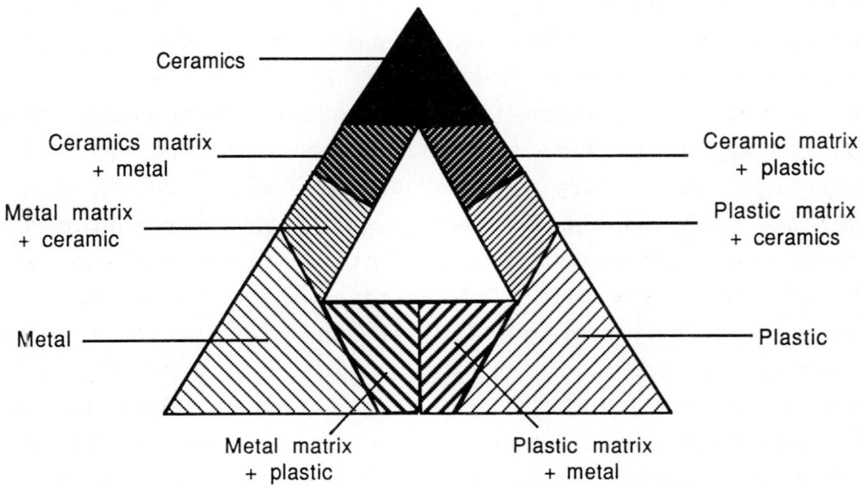

Source : Les Echos, supplement, 9 October 1985.

Figure V.1.a : Different forms of composites.

It is now possible, with the aid of the two scenarios sketched in this chapter and the over-riding tendencies just referred to, to indicate for the sphere of materials the outlook for industrial strategies, the role of the authorities and the outlook for employment and skills , situating all these reflections from a European viewpoint.

CHAPTER V.2

Industrial Strategies in Materials : European Situation

V.2.1. GENERAL SITUATION

The situation of European industry in the field of materials can be very roughly summarized as follows :

- Europe occupies a position which is entirely competitive with the rest of the world in the conception, production and application of <u>structural materials</u>. This favourable position is based to a large extent on the might of the European <u>chemical industry</u> (notably on the three German groups Bayer, Hoechst and BASF and the other major European firms - see Table V.2.A - but also on an overall European "chemical culture" of high quality), which confirms its considerable strategic role for Europe.

This favourable position is also based on the existence of European materials producing groups, which in their different areas occupy prime places throughout the world : Péchiney, aluminium; Wacker, silicon; Saint-Gobain and Pilkington, glass and derivatives, etc. Lastly, the European position is also accounted for by the presence of major groups utilizing structural materials (motor, aeronautical, building and public works sectors, etc) having a strong competitive position in the world.

Position	Firms (country)	1985 turnover ($M)	Net 1985 profit ($M)	As % of turnover
1	Bayer (FRG)	15598	428	2.7
2	Du Pont (USA)	15046	875	5.8
3	Hoechst (FRG)	14510	408	2.8
4	ICI (UK)	14241	740	4.8
5	BASF (FRG)	13141	339	2.2
6	Dow (USA)	11537	58	0.5
7	Eastman Kodak (USA)	10631	332	3.1
8	Schell Chem. (UK,NL)	9174	300	3.3
9	Union Carbide (USA)	9003	- 581	n.s.
10	EXXON Chem. (USA)	7900	249	3.2
11	Ciba-Geigy (Suisse)	7404	598	8.1
12	Monsanto (USA)	6747	- 98	n.s.
13	Rhone-Poulenc (F)	6243	257	4.1
14	Elf (F)	5630	19	0.3
15	AKZO (NL)	5422	254	4.7
16	Mitsubishi (J)	5112	98	1.9
.				
.				
.				

1) Excluding energy (oil, gas, coal).
2) Net operating profit : for chemical activities (excluding "corporate" expenditure)
3) Net profit : all activities combined
4) The chemicals sector of the ELF group includes ATOCHEM and subsidiaries, Sanofi, Habitat and "sulphur" activity
5) In terms of total consolidated turnover

ns = non-significant
nb : exchange rate : mean 1985 rates vis-à-vis dollar

Source : Informations Chimie n°277 - November 1986.

Table V.2.A : The 16 leading world chemical firms in 1985.

- On the other hand, Europe depends very heavily on the United States and Japan when it comes to functional materials. This dependence is evident in electronics, magnetic materials, engineering ceramics with high temperature resistance, in numerous biomaterials. Europe does admittedly present several bright spots (for example, the German Wacker, referred to above in connection with production of silicon, or Rhône-Poulenc for rare earths, or many other companies for catalysts) but these are too dispersed to prevent a general dependence at European level, the effects of which are extremely damaging for the European Community. Europe is still obliged, for instance, to import a considerable proportion of the electronic components it needs from the United States or Japan, with disastrous effects on its trade balance. Not holding the "keys" to the technological system, Europe is heavily penalized when it comes to developing major industrial lines based on the mastery of basic functional materials (information, for instance), or else it has to pay the price (as in the field of optical fibres, where the American Corning has "blocked" most of the patents). The consequences of this situation are also measurable in terms of jobs. For instance, the jobs needed for the constitution of the great industrial lines connected with functional materials have been for the most part created in the United States or Japan, at the expense of job creation in the countries of Europe.

This very rough, very brief description of the situation of European industry in materials needless to say must be qualified, and this is something that will be enlarged on in the following sections. Whatever the verdict on particular situations, however, there is no avoiding the central finding. If we accept the idea that materials represent a fundamental element in an industrial system, in the same way as information technologies, the European situation does not appear to be satisfactory overall. It gives evidence of too many areas of dependence, too many lost opportunities, too much dispersal and failure to predict the future aright.

The following analysis of the industrial situation of Europe considers Europe's dependence in strategic materials, the situation of European industry in structural materials and the situation of European industry in functional materials, respectively.

V.2.2. EUROPE'S DEPENDENCE IN STRATEGIC MATERIALS

The last decade has been marked by a profound easing of the constraints governing Europe's supplies of commodities (oil, gas, etc) and strategic materials (gold, silver, etc). This appreciable easing of constraints is not only due to cyclical factors. It is also the result of strenuous efforts in Europe to rationalize the utilization of energy, to lighten structural materials, to improve the conditions of utilization of scarce materials, to diversify its sources of supply. It is clear that growing variety in materials has also been a decisive factor in allaying fears of a shortage of supplies. The concept of "critical" material or "strategic" material indispensable for the economy and for defence remains, admittedly, but it is no longer so keen as it was a few years ago. Of course, as indicated in Chapter

IV.2, the supply of chromium, manganese, platinum, rhodium, antimony, cobalt, molybdenum, nickel, titanium or niobium is still threatened, but the prospect of finding substitutes rapidly helps to ease the constraints concerning these metals. It would be far better in fact to refer to "critical products" rather than "critical materials", since a supply breakdown in one other of these materials would hit the products concerned extremely hard, at least momentarily. In any case, what appears today to be strategic is "grey matter". Carbon is not strategic, but carbon fibre is. This example illustrates well that a material can be strategic without being physically scarce.

If the easing of constraints vis-à-vis materials and commodities has had undeniably positive effects for Europe, as for other western countries, it is liable to pose a grave problem if it continues for certain developing countries which must hold reserves of metals or oil if they are to secure their development. The collapse in the price of oil throughout the first half of the 80s, but also the collapse of certain markets in base metals (tin, for example), threaten numerous developing countries that produce raw materials. Europe, by guaranteeing its independence, committing itself to new materials (ceramics, composites) whose basic material often consists of very plentiful elements (silicon, carbon, etc), like the other developed countries, is liable to embark on a process where a new system of relations with developing countries must be clearly established. One factor making for a solution may be aid to gradual evolution towards higher added value, as in the example of setting up basic chemical units in the countries of the Middle East, with their abundant oil sources.

Even if the problems of Europe's dependence for basic raw materials have become less acute, they are liable to play an important role in the next decade in the direction followed by the technological system. The study of different scenarios has shown that a hypothesis of extensive dependence would accentuate the transition to a variety-based regime, whereas a hypothesis of lesser dependence tends to give pride of place to the "plastics" scenario.

V.2.3. INDUSTRIAL SITUATION AND STRATEGIES IN STRUCTURAL MATERIALS

In any examination of the industrial situation and strategies in structural materials, a distinction has to be made, for the purpose of analysis, between materials producers, materials processors and companies using materials. The distinction must of course be handled with caution, since the phenomena of strategic integration among the three categories of industry are among the decisive issues.

V.2.3.1. Producers of materials

From time immemorial Europe has played a major part in the production of materials. While the European steel industry has been and still is one of the most efficient in the world, the European chemical industry possesses a might which there is no gainsaying, and in other areas (aluminium, glass, cement, etc) Europe on the whole leaves no doubt its competitiveness.

However, all the major industries producing materials have been affected by a sizeable problem which concerns all the developed countries : the decline in per capita consumption of materials. In Europe as elsewhere there has arisen the serious (and continuing) problem of excess capacity. The situation is particularly grave in the steel industry but it is present as well in the field of aluminium or commodity plastics.

These problems have affected the United States or Japan equally, but the specificity of the European situation is marked by a relatively poor mobility of industrial structures, and the failure to switch to functional materials which in overall terms, after a period of adaptation, have taken over. It is necessary at this stage, once more, to refer first of all in this connection to the old familiar problem, which is that whenever problems of excess capacities in basic industries arise in Europe, solutions are harder to find than elsewhere, because of the compartmentalization of production units and the dispersion of strategies over the territory of Europe. Where it is enough to close down one steelworks in the United States, consideration has to be given to the closure of between two and three in Europe.

The other reason seems more serious, in the sense that it relates to a manifest strategic sluggishness on the part of a considerable proportion of European companies when it comes to identifying future opportunities and to diversifying, the archetypal example being the steel industry. In the United States, as in Japan, steel companies have opted for a strategy of diversification towards "multi-materials". The favourite targets are advanced ceramics, carbon fibre and basic materials in electronics. In Japan, for instance, Kawasaki Steel has associated itself with Péchiney for the production of permanent magnets, has absorbed the American firm NBK to produce silicon wafers, and has concluded a joint venture agreement with another American company to produce semiconductors. Japan Nippon Steel has a subsidiary, Nittetsu Electronic, which has just announced the construction of a large format silicon wafer factory, and another subsidiary, Nippon Steel & Chemical, which is developing carbon fibre. Again, Nippon Kokan has concluded a licensing agreement with General Electronic for the production of polycristalline silicon, and is collaborating with Martin Marieta in the manufacture of titanium or aluminium alloys for the aerospace industry.

This kind of diversification is not confined to steel firms. In aluminium, a firm like Alcoa in the United States has undertaken a diversification strategy on a very large scale, by acquiring holdings in composites, ceramics, complex alloys and plastics. Alcoa is probably one of the firms showing the most strategic determination to become a genuine "multi-materials" firm. In a recent MRS report a top Alcoa executive wrote that one thing was certain, that the epoch that still prevailed in the early part of the 80s, marked by a prosperous aluminium industry which was predominantly conservative and concentrated on a single line, was over once and for all. The future belonged to an advanced multi-materials industry, facing considerable risks and definitely high-tech.

By comparison with these strategies, the attitude of European firms, and particularly steel firms (1) , appears relatively immobile.

---

(1) This comment can be qualified with the recognition that some European steel firms (ARBED, COCKERIL-SAMBRE, USINOR, SACILOR, etc) have declared that they are interested in plastics. In order to combat the problems of corrosion and achieve lighter structures, metal-plastic sandwiches and plastic coatings are envisaged. Again, advantage is taken of the sound insulating qualities of plastics in metal-plastic-metal sandwiches for motor vehicles. It is worth pointing out here that the diversification coming about (slowly) in plastics is in some ways being "forced" so as to improve basic steel materials, and is not the result of a real "multi-materials" strategy. One exception, however, is the firm KLOCKNER, which was world leader in the production of polyvinyl.

European steel firms point to financial reasons. They have to invest massively, in particular, to finance modernization operations and notably the redesign of factories (continuous casting, horizontal casting, automation, etc) so as to satisfy a multiplicity of objectives (reproducibility of qualities in narrower ranges, greater dimensional regularity of products, greater dimensional variety, etc). Reasons of this kind are doubtless plausible, but American and Japanese firms respond to such arguments with the comment that the funds to be committed to new materials are relatively slender. The level of R & D expenditure devoted to diversification in new materials represents a bare 0.5% of their turnover. Nonetheless, these investments are regarded as strategic. Two hypotheses may be put forward : either the industrial system is moving towards a permanent variety-based regime (scenario 1) and a "multi-materials" diversification is the right long-term response, or other scenarios win the day (plastics scenario, for instance), but in that case it will always be attractive to resell activities offering very high growth prospects (minimum 15% for engineering ceramics) over at least a decade.

Among the other European producers, the same hesitation may be seen between improving existing production units and avoiding over-production, on the one hand, and on the other choice of a multi-materials strategy. But the attitude is generally more dynamic, possibly because the crisis has taken less toll of financial capacities.

Péchiney is probably one of the most significant examples of a great European materials firms (again centred on aluminium) which has opted for a "mixed" strategy, seeking on the one hand to make its base metal activities as profitable as possible, and on the other to deploy a multi-materials strategy.

In aluminium, the group noted a lasting over-capacity situation and the existence of over-high energy cost (one-third of production cost) in France, and decided to reduce capacity (the oldest units in France will progressively be closed down) and to recentre its aluminium production in countries where energy is cheap, such as Australia and Canada. (It must be noted that in the latter country, however, the groups has had to make over one-half of its share in the ultra-modern factory at Bécancour, given the imperative need to reduce capacity, to the American firm Reynolds).

This strategy for the defence of the group's aluminium sector has obliged Péchiney to move further downstream and to integrate processors closer to final applications (notably Cegedur, in soft rolled products), not to say final utilizers (Cebal, in packaging). Péchiney has also moved, with the development of aluminium-lithium for the aircraft industry, in the direction of high-technology aluminium.

The group is also actively developing a "multi-materials" strategy, concentrating particularly on new metals and materials (zirconium, titanium, molybdenum, tungsten, carbon fibre, engineering ceramics, etc). Having decided to dispose of its chemical activities when the most important thing was its recovery, the group is clearly a materials enterprising integrating above all metallurgical know-how.

For the moment, then, the group has decided to keep a foot in both camps, or more precisely to build up for itself a portfolio of options for the future, defending its overall position in aluminium and developing a multi-materials range. The strategy adopted by Péchiney is not the same as that adopted by Japanese aluminurgists (who under the impetus of MITI, the Ministry of International Trade and Industry, have defined an overall strategy : abandon the bulk of primary aluminium production but retain a small core of competitive electrolysis, invest in developing countries, such as Indonesia, so as to have the certainty of long-term contracts, and rely otherwise on the spot market), or by American aluminurgists, whose strategy is firmly multi-materials.

The real difficulty for Péchiney (and for many firms producing materials) is to manage this twofold strategy - not only in financial terms, but also in terms of human beings. How is the group to adjust its structures and the mentalities of its teams to the changes in hand ? How, in concrete terms, is a synergy to be established between departments which are each other's natural rivals ? How can preparations be made for doing battle against the chemical firms which in the long run will be formidable rivals (in ceramics, for instance) ?

Chemical firms producing plastics have come up against the same problems, in the production of commodity plastics as the steelmen and other producers of materials : how to bring down <u>over-capacities</u>. Many units were closed down in this way in Europe, while the production of commodity plastics was expanding in "new" countries such as the Middle East countries (Saudi Arabia, notably, althought in 1985 the country accounted for only 0.8% of total European imports) (see Figure V.2.a).

The other solution in the field of plastics was to move further downstream of produce plastics with a higher added value. It is important to distinguish between these two strategies for the chemical industries, as the integration of companies situated closer to the market place (processors) is restricted in some European countries by anti-trust rules. In Germany, for instance, Dynamit Nobel alone produces and processes its polymers. The "major" producers, Hoechst, BASF and Bayer, own a number of processing enterprises but only on a restricted scale, for the sake of complying with anti-trust legislation. This is why the other "plastics" strategy is to design and produce plastics with a high added value, responding to the specifications set by demand. This strategy prompts many chemical firms to respond more and more to "customized" requirements and to become familiar in ever increasing detail with what the main clients situated further down the line want; this applies particularly, it would seem, to ICI.

POLYETHYLENE

(all grades)

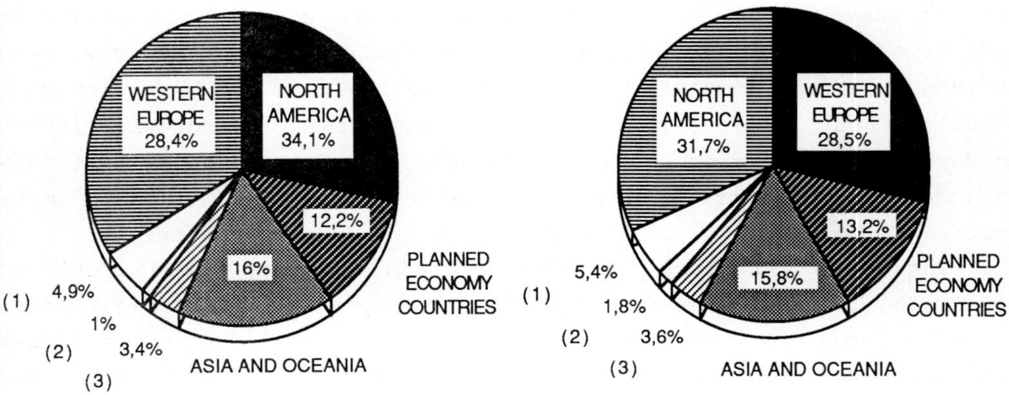

(1) CENTRAL AND SOUTH AMERICA : 4.9%	(1) CENTRAL AND SOUTH AMERICA : 5.4%
(2) AFRICA : 1	(2) AFRICA : 1.8%
(3) ARABO-PERSIAN GULF AND MIDDLE EAST : 3.4%	(3) ARABO-PERSIAN GULF AND MIDDLE EAST : 3.6%
Distribution of capacities by geographic zones in 1,000 mt/per annum : 27,375 in August 1986	Distribution in the next few years in the light of projects already announced in 1,000 mt/per annum : 29,726

Source : Informations Chimie n° 277 - November 1986

Figure V.2.a

Another type of strategy is for chemical firms to envisage "multi-materials" diversification. The most frequent way in which they do this is by controlling companies producing composites, since the "composite" philosophy is not far removed from "plastics" know-how; other kinds of diversification are to be noted, however, notably in ceramics and materials for information technology. Needless to say, the sphere of catalysts belongs in part, historically, to the chemical industry.

BASF, for instance, has decided to secure a controlling share in the American group Celanese Corporation (fibre-reinforced composite materials) and is also developing in carbon fibre (agreement with Toho Rayon), glass fibre and special adhesive films; Rhône-Poulenc has acquired Ceraver (ceramics); Hoechst has acquired Infotec (informatics), and so on. It is to be noted here that the strategy of American chemical firms seems still more agressive, particularly vis-à-vis the electronics sector : du Pont has absorbed Solid State Dielectrics (ceramic-based electronic components), Allied has absorbed Eltra (electrical products), Bunker Ramo (electronic components) and Semi-Alloys (components for the manufacture of high-reliability semiconductors, etc). Generally speaking, however, the soundness of the European chemical industry, and particularly the German chemical industry, is a pledge of competitiveness for the Europe of materials. One instance, among others, testifying to this was a recent Japanese fact-finding mission to Europe, in an attempt to establish the bases for co-operation between Europe and Japan in the sphere of plastics, whose members recognized the full extent of European know-how.

V.2.3.2. Processors of structural materials

The situation of those engaged in processing materials is tricky. Placed as they are between those who design and produce materials and those who use them, processing firms - often small or medium enterprises - have to make allowance for the far-reaching changes affecting the design and production of materials, and in demand. Tailor-made design, technical integration procedures, the tendency of materials producers to move further downstream, attempts of firms utilizing materials to take over control, are all manifestations of the difficulties facing this group of industries. The problems are not specifically European, but they affect in Europe a universe of small and medium enterprises representing a considerable number of jobs, and liable to find themselves coming up more and more against current changes. In reality, those engaged in processing guarantee the "connection" between materials production and utilization. They grew up in a stable standardization regime which made for the creation of a network of numerous small and medium enterprises (the network has played the part of a real "reservoir" of jobs for Europe), based on know-how predominantly concerned with engineering. These companies generally operate as sub-contractors vis-à-vis users.

Irrespective of the scenario envisaged, the foreseeable consequences for processing companies appear critical :

- A variety-based scenario, pre-supposing in particular powerful technological integration phenomena (studied in Chapter II.5), is liable to eliminate a substantial proportion of processing industries. This scenario entails a profound change in the user firms' sub-contracting network, which might well do away with probably many more materials processors than it would help creates (although it is possible to imagine the creation of multi-materials processors). On the other hand, it is clear that at the same time this scenario gives an advantage to the development of a network of enterprises providing <u>industrial "services"</u> for producer and user sectors. The permanent variety-based regime implies a change in the actual <u>nature</u> of sub-contracting.
- An irreversible ascent of plastics scenario would probably have less radical consequences, but would mean extensive adaptation of a part of the sub-contracting network currently operating, replacing metal-working know-how by know-how relating to the processing of plastics (this trend is in fact already apparent in the case of certain processors). One of the main problems of processors, in this hypothesis, is how to finance the change. It is certain that the change will be easier for processors which are subsidiaries of major producers or users. The "plastics" scenario, then, necessitates closer inspection of the development pattern of plastics processors (Table V.2.B). All the analyses carried out in this report suggest that the plastics processors which are going to develop are the "mould-makers", whose main problem will be how to stand up to the user groups' attempts to integrate them in their production system.

1. <u>General structures of PPMs</u>

The purpose of Processors of Plastics Materials (PPMs) is the manufacture of semi-finished products in plastic materials. The PPMs are made up of three groups : independent processors, processing subsidiaries and user or integrated processors. A further possible distinction is the classification of PPMs into those which effect a "preliminary processing operation", producing intermediate goods (granules, fibres, etc) and those effecting a "second processing operation", producing final consumer goods (including finishing operations). Many PPMs use discontinuous processes for initial processing, while their second processing operation represents a substantial part of their activity.

<u>PPMs in Europe</u>

Characteristics	FRG	France	GB	Italy	Netherlands	Belgium	Europe (9-memb)	Spain (85)(1)
N°of firms with 30+ payroll	1549	905	1151	902	181	138	4969	2307
Payroll	164866	98107	138650	74052	14203	12675	517840	59821
GVACF (2) (M Ecus)	2704.4	1408.6	1398.5	867.4	267.7	259.4	7181.0	-
Turnover (M Ecus)	7048.4	3770.3	3497.3	2483.5	800.5	629.0	189649.7	11,4 BIL.FF
Total turnover proportion, EEC (9-member)	37.2%	19.9	18.4	13.1	4.2	3.3	100.0	-
Tonnage processed 1985(1)	6.500 M	2.536 M('83)	2.0 M		0.545 M	0.640 M	-	1.21 M ('84)
Population, M 1985 (1)	61.2	54	56	56.4	14.4	9.9	-	38.5
Per capita consumption, kg, 1985 (1)	106,2	47 ('83)	-	50	-	50	-	38,1 ('84)

<u>Source</u> : DAFSA, 1983.

(1) EUTRA PLAST, Brussels 1985 (2) GVACF = Gross Value Added to Cost of Factors.

The three main structural features of PPMs in Europe are as follows : the predominance of small and medium enterprises (SMEs) with a payroll of between 1 and 49, a very small number of large and highly capitalized companies (payroll of 500+) and the tendency towards substantial industrial integration of the PPM in the other sectors using plastic parts.

Geographically, the PPM represents the main activity in certain parts of Europe (76% of concerns in Alsace operate as PPMs).

It will be noted that the German PPM sector is of considerable importance in Europe (1,549 establishments; 37.2% of the total proportion of turnover in the 9-member EEC).

2. Comparison of structure of PPMs in Germany and France

By way of example, the comparison between the structures of PPMs in Germany and those in France is studied below on the basis of four main factors :

The number of concerns is slightly up (decline from 1980 onwards). In terms of payroll numbers, extensive recruitement in the 70s (earlier substantial decline from 1978 onwards). Big increase in turnover. The PPM industry represents a fairly considerable rate of concentration (highly capitalized concerns). On the whole, the level of concentration in the PPM industry in France (3%) appears higher than in Germany (1%), see table below.

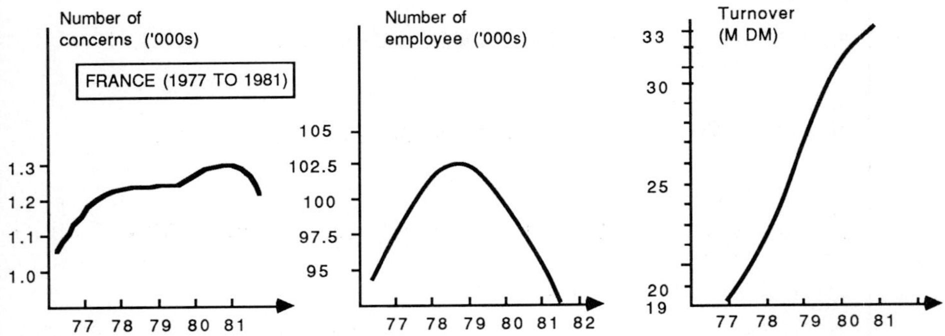

Number of concerns rising substantially (slight stagnation in 1980). Rise in number of employees following extensive recruitement in the 70s, decline in 1980. Turnover rising sharply (slight stagnation in 1980). The PPM industry is fairly concentrated but the degree of concentration between 1976 and 1980 was not high.

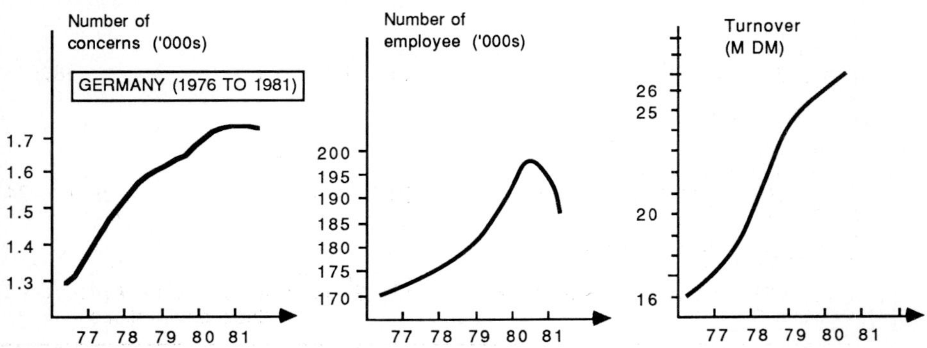

Table : Degree of concentration in PPM industry in Europe

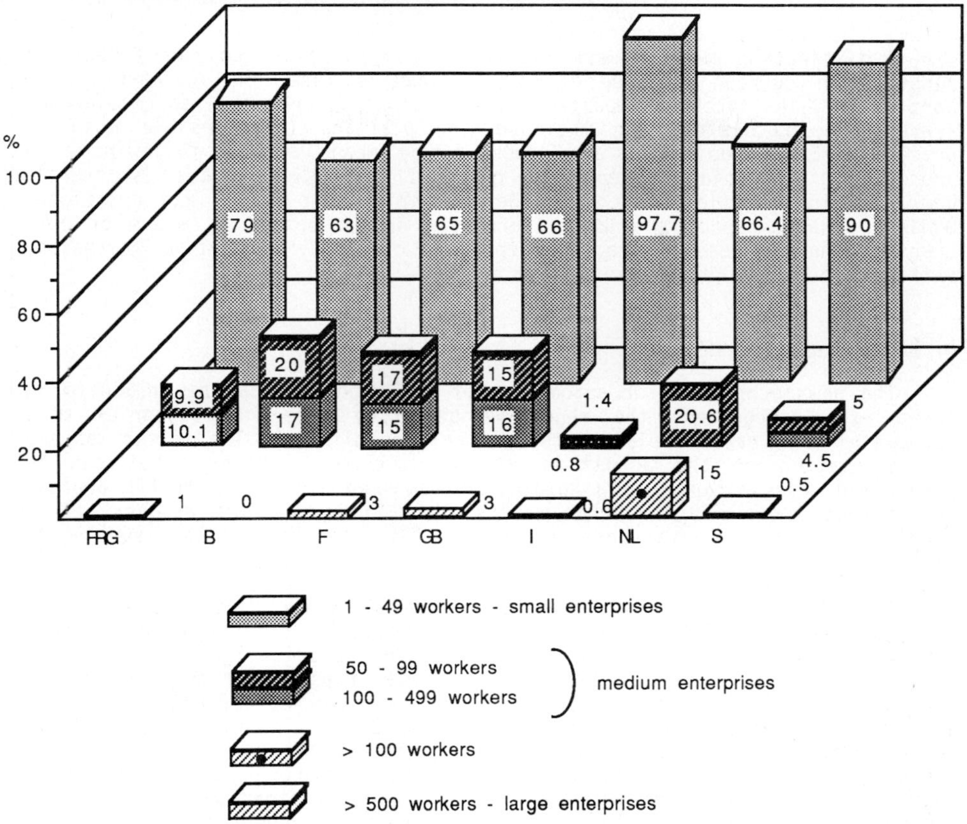

3. PPMs' main industrial partners in Europe

There are four main industrial partners of European PPMs : (1) producers of plastics materials (major oil and chemical groups, regenerators, compounders, producers of glass, mineral and organic fibres; (2) mould constructors, mould makers); (3) manufacturers of processing equipment (machinery, peripheral equipment automats, robots, CAO/CAM, etc); (4) user industries (medium and large enterprises - motor manufacturers, household electrical appliances, sport and recreational equipment, medical products, etc).

Outline : Industrial logic of PPMs

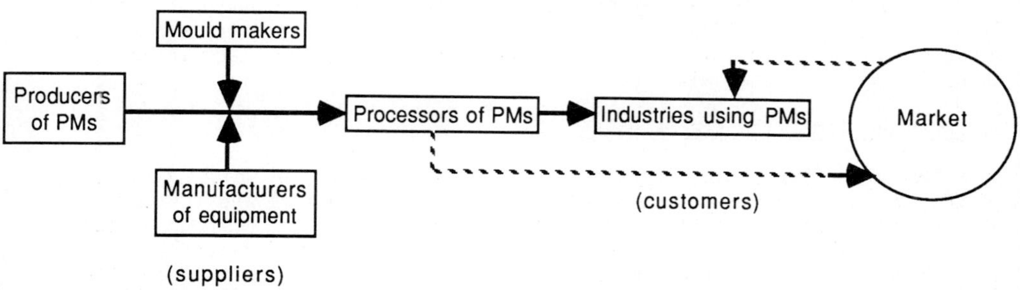

4. Main processing methods and new automation techniques of PPMs in Europe

A distinction must be made, above all, in plastics (polymers), between plastics in current use and technical plastics (used in engineering, in particular). A further important distinction is between thermoplastics (reversible processing) and thermosetting plastics (irreversible chemical processing). Thermoplastics are processed by means of the following five processes : injection (telephones, girders), extrusion (tubes, shutters), blow moulding (bottles, flacons), thermoforming (boat hulls) and rotational moulding (large-capacity tanks). Thermosets are processed by means of both transfer and compression processes (printed circuits, decorative laminates, switches).

5. New automation techniques

The new techniques for automation of PPM processes can be indicated in three main categories, the objective being to make PPM factories more economic (lower raw material costs, better streamlining). The categories are (1) automation of the actual processing (use of machinery that can be programmed, for instance); (2) automation and robotization of the PPM process environment (manipulators and robots for susequent operations; parts handling, automated transport systems); (3) computerization (supply to processing machinery generated by computer, computer assisted mould construction, CAO/CAM).

In PPMs, production was initially automated in the first stages of the process. The automation of processes has a substantial effect on the economic structure of European PPMs : (1) improved productivity (renewal of manufactured products - more technology, shorter runs, lower level of direct manpower, savings in intermediate consumption, greater use of high technology - control systems); (2) better quality (improvement of products quality, better process control (execution of work, flexibility of equipment); (3) higher profit level (economies in direct labour are relatively slender : machinery, tools and materials are the heads which dominate the company's gross margin (to the tune of over 70%) with the main proportion accounted for by the cost of materials, the overall efficiency of production (lasting profitability) depends on the reliability and flexibility of the whole production system) (according to CPE Etudes 1983).

6. Strategies of small and medium industries in PPMs in Europe

The European PPM industry covers more than 25,000 small and medium concerns, a considerable number being firms working in sub-contracting networks of the biggest PPM concerns or industries in other sectors (automotive). It is possible to distinguish, by way of example, three principal strategies employed by sub-contractors in the European motor industry (cf. CPE Etudes, 1983) : (1) industrial diversification strategy (new products and sector so as to avoid the constraints of sub-contracting); (2) technological diversification strategy (in the framework of sub-contracting); (3) financial diversification strategy (acquiring new companies not connected with the PPM).

7. European PPMs : the stakes

The main issue for European PPMs is the speed at which they can restructure their production systems (control of cost price) and the outcome of international competition, which has becomes very keen (United States, Japan, small and medium industries). One of the most frequent responses is to specialize mainly in activities with a high added value. The present predominance of small and medium industries may prove to be a major obstacle for the sector, since some firms are not really receptive to new production techniques. To date they prefer to keep to "reliable" and traditional production structures. This strategy may be dangerous, as industries in other sectors further domwnstream are increasingly interested in PPM activity (incorporation of large workshops in the case of the motor manufacturing industry and household appliances, for instance). This sector is thus liable to lose its own industrial autonomy in the current industrial integration process. One of the solutions for the future could be the general spread of industrial partnerships (collaboration between specialists firms).

8. Steady sectors for European PPMs

Potential markets for PPMs in Europe are mainly the following eight industrial areas : (1) Transport (structural parts, in the motor industry, for example); (2) Transfer of information (fibres, components); (3) Electrical (household appliances); (4) Building (vast potential); Agri-food (plasticulture); (6) Corrosion and pollution control; (7) Health (small parts, high added value); (8) Sport and recreation.

Technoplastics and components are going to compete mainly with traditional materials : steel (automotive industry, bodywork, oil drilling pipes), light alloys (housings, petrol pumps, carburettors), other metals (taps, plumbing) and wood (handling : pallets). Their industrial properties are based on the advantages of stability (in time) and lightness (with the exception of certain metal alloys), good performance and adaptability, low energy content.

Table V.2.B : Processors of plastics in Europe.

V.2.3.3. Users of structural materials

The main sectors in which structural materials are used have been studied at length in Part III, so that at this stage we shall merely give a reminder that the analyses carried out showed, for these industries, the absolute necessity of dominating the development of information technologies applied to production (CAD, CAPM, expert systems, etc) and at the same time new techniques for producing and applying materials. This two-fold constraint is a stage through which it is essential to pass for the sake of achieving international competitiveness. Europe has learnt its lesson, and to its cost, by losing out to Japan and the United States, over the last 10 years and more, relatively large parts of the market in the sale of industrial articles (automobiles, for instance), although its skills in the field of materials are not in question. But cumulative shortcomings and present-day production management have cast a shadow over the general excellence which it displays in the design, production and application of structural materials.

This statement of course needs qualification : it is true that a country like Germany, with a firmly entrenched tradition in dealing with production management problems and the advantage of a body of high-level know-how in materials, seems to be well placed for rapidly effecting the adjustements necessitated by international - particularly Japanese - competition. It is also true that civil aviation, where European co-operation (principally between France and Germany) exploits the different partner's technological and economic capacities, is opening up a world market for itself with the Airbus programme.

It is probable that the power of materials users situated further downstream, having close contact with demand requirements, will expand. This power would become particularly extensive in the event of a variety-based regime becoming permanent, since the logic of the system is based on the management of profound interactions between demand and production. Conversely, if the scenario is that of "plastics", chemical firms will probably be tempted to take their integration still further (in a functionalization logic) down the line and control some of the users for the sake of access to all the knowledge they possess about demand.

V.2.4. SITUATION AND STRATEGY OF EUROPEAN INDUSTRY VIS-À-VIS FUNCTIONAL MATERIALS

The overall diagnosis of the situation of European industry in functional materials is alarming. Attention is naturally caught by the fact of having lost out when it comes to materials for electronic data processing. Europe is competitive in a few isolated strongholds (production of rare earths, material for infra-red, catalysts, etc) but generally it is clear as the 80s draw to a close that Europe is trailing behind the technologies developed in the United States and Japan for the very great majority of functional materials, and that this situation produces a relationship of marked dependence on the outside. The solutions sought for making good the backlog may, incidentally, not facilitate the creation of a "technological Europe", as the most tempting and the easiest solution is forming and alliance with a Japanese or American firm. The small number of European firms with a good record here still tend to give priority to investing in the United States, or even Japan, where the great future markets for functional materials are being determined.

V.2.4.1. Industrial situation and strategies in information technologies

The world semiconductors industry was for a long time entirely dominated by American companies, which benefitted from the fundamental discoveries of their research laboratories and public orders (military and aerospace). Anti-trust legislation in the United States, by excluding IBM and Western Electric from this market, opened the doors to dynamic small and medium enterprises. Competition was intense, helping to perfect products and bring down prices. The pattern in Japan was entirely different : the driving force behind the expansion of the sector was consumer electronics. The main integrated electronics groups (Hitachi, Toshiba, Mitsubishi, NEC) in this way, for their own production of downstream goods, tapped the demand for semiconductors (which they first imported and then gradually produced themselves). Their endeavours enjoyed the very active support of the Ministry of International Trade and Industry (MITI), notably in the VLSI (very large scale integration) project between 1976 and 1979. The result of the Japanese effort is spectacular : Japan is today the leader in high-density memories, using MOS and CMOS technologies, and in cheap microprocessors; it is rapidly catching the United States up in the making of high integration circuits. Only <u>custom</u> circuits, very high speed integration circuits (VHSIC) and top-of-the-range microprocessors are still dominated by the United States. The Japanese strategy - of attacking the American market first by low tech in high tech", ie by bottom-of-the-range products in bread and butter business and then shifting quickly to more sophisticated top-of-the-range products - has thus paid off (see Figure V.2.b).

In this battle of the titans, Europe has consistently lagged behind. Few new enterprises have been created and the production of semiconductors has remained confined to a few isolated major groups (Philips and Siemens), which have had immense difficulties in adapting their production structure and strategies to the evolution of microelectronics. As a result, six out of the 10 biggest world producers of semiconductors are American, and the other four are Japanese. Europe, in contrast with Japan, has as it were lost out by accepting American supremacy and not taking up the challenge of competition, and frequently being content to conclude agreements with top-level American firms. The economic penalty for this dependence is measured, among other things, in the production of everyday consumer goods, which incorporate components - often to the tune of over 10% of their value - which must be bought from abroad. It is directly in Europe's trade balance, then, that the failure to keep up in the field of semiconductors takes its toll. It can be estimated that in 1985 close on two-thirds of European electronic components were imported (American and Japanese firms accounting for 48% and 13% respectively of supplies to the European market).

PURCHASING PACKAGE

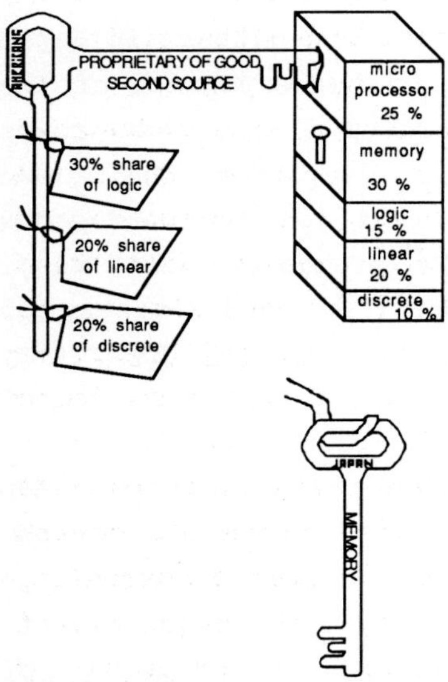

WIN FACTORS FOR ICs

PRODUCT	MOS MEMORY	MOS MPU AND PERIPHERAL	BIPOLAR MEMORY	BIPOLAR LOGIC	LINEAR
NECESSARY ATTRUBUTES	PRODUCTION EXPERTISE - COST -DENSITY - AVAILA-BILITY	APPLICATIONS SUPPORT DEVICE STANDARDS PERFORMANCE	DEVICE PERFORMANCE	PRICE DELIVERY COMPLETE LOGIC FAMILIES	APPLICATION SUPPORT COST FUNCTIONA-LITY
LEADERS	JAPANESE	AMERICANS - INTEL - ZILOG - MOTOROLA - COMMODORE	AMERICANS	AMERICANS - TI - MOTOROLA - FAIRCHILD - SIGNETICS	AMERICANS & EUROPEANS - NATIONAL - SGS - TI - FAIRCHILD

Source : Mackintosh International, "An analysis of the micro-electronic industry for medium and long term strategy purposes".

Figure V.2.b

This dependence has even contrived, at certain stages, to place the defence of some European countries under the American umbrella.

This somewhat bald comment needs to be qualified, of course. At the furthest point up the electronics line, it is a European firm (Wacker in Germany) which is world leader in the production of silicon wafers. Again, it is at the other extremity of the electronics line that Europe is relatively competitive in the utilization of electronic system, through the development of numerous dynamic software companies (Europe also occupies a number of competitive positions in professional electronics).

The fact remains that Europe is conspicuous by its absence from the "heart" of the sector manufacturing electronic components. Several reasons can be put forward to account for this state of affairs. Europe has been in some way caught between two fires. <u>On the other hand</u>, the United States, thanks to private or public research budgets (IBM's research budget exceeds Belgium's total R & D budget) has had access to a considerable financial "critical mass", which was sufficient to produce spectacular break throughs in fundamental research.

Thanks to the execellence of university teams, the flexibility of relations between universities and industry and financial mobility, the efforts undertaken have enabled the Americans to develop the industrial applications arising out of their innovations. In a way, the conventional pattern traced by the economists (investment in fundamental research ⟶ scientific discovery ⟶ innovation ⟶ industrial development) has worked in the United States - but only because the country had the means, <u>both financial and human</u>. We shall see later on that the pattern is at least in part being questioned.

On the other hand, Japan's strategy has been quite different. As we have seen, it has imitated innovations from the United States and adapted them in ordinary consumer goods, such as consumer electronics. But this strategy would not have been enough had Japanese companies not at the same time affected another kind of innovation : in the <u>manner of producing</u>. The development of new production methods, popularized by the principle of the "five zeros" (zero breakdown, zero stock, zero waste, zero paper, zero delay) is a revolution in the organization of production which has enabled Japanese companies to place their products on the major markets thanks to two vital assets : quality, and the ability to produce in the shortest possible time.

Situated half-way between these two firmly based strategies, Europe has not known which to adopt. First of all, it perhaps made the wrong diagnosis : for a very long time European authorities and industrialists believed in the main that the economic crisis was only an energy crisis, and tackled it as if it were. They failed to see that the bases of the technological system were being questioned. Europe has been constantly wavering between the strategy of sectors and the strategy of outlets. Many European industrialists believed for a time that they had simply failed to occupy one slot, in micro-electronics. Only a few realized that it was the bridgehead for the principal strategic route of the latter part of the 20th century that was being established. Conversely, European industrialists imagined that they had the upper hand with a whole line in optics, but conventional optics have today become just one slot in the photonics sector.

A further explanation which should be proffered is that public expenditure effected in Europe to stimulate the electronics sector, directly or indirectly, did not produce much knock-on effect. The unsuccessful large-scale public plans in certain countries (Plan Calcul in France, for example) are familiar, but it is also a fact that the major programmes, despite their considerable indirect economic effects (2) have fallen short of expectations for the development of the electronics sector and particularly the development of silicon-based components.

If we look at the case of the European space programmes, then, while it has been shown that they have had considerable spill-over in the case of structural materials (composites), it is also clear that their impact on the development of silicon technology functional materials has been only slight. When the European programmes were drawn up, Europe's dependence here on American technology was more or less admitted, and this implicit recognition meant that the American domination held good throughout the period of development. On the other hand, it has to be recognized that the space programmes (and military programmes) have had a definite impact on the development of a European capacity in gallium arsenide. This capacity is likely to be a hefty asset should the information technologies move in the direction of high-speed logic.

One of the most serious consequences of European dependence in components is the fact that in very many cases Europe is excluded from industrial competition when it comes to laying down the <u>main</u>

---

(2) See BETA study on the indirect economic effects of the European Space Agency contracts, Strasbourg, 1980.

international standards. Here too, competition is between the United States and Japan.

Aware of its unfavourable position, Europe is trying to arm itself, through an energy policy, with the means of winning back its competitiveness. The setting up of the ESPRIT programme and the efforts of major European groups (which have contrived to put a foot in the door of the United States and Japan) are encouraging factors. But if the undertaking is to succeed, industrial co-operation will doubtless have to be still further reinforced at European level.

- Realization of a scenario based on materials variety would not necessarily be a negative factor for Europe in the case of functional materials for electronics. In this case, the scenario in question would in fact mean growing diversification in semiconductors and a trend towards "made to measure" components (which is already a current trend for very complex circuits). This evolution would help challenge the monopoly in silicon-based components which Europe has failed to overcome, making way for the development of components using numerous other basic materials, such as III-V compounds for which Europe still appears to be competitive.
- The "irreversible ascent of plastics" scenario has its own counterpart in the field of materials for electronics : the development of electroconductor polymers and, more generally, the heightened development of plastics in electronics. Already information technology (and telecommunications) represent a market with an annual growth rate of some 15% (polymers used in the manufacture of printed circuits, magnetic tape films, audio and video cassettes, etc), which according to du Pont de Nemours could reach approximately $10 milliard by the end of the century. It is quite clear that prospects of this kind have already encouraged the chemical industry, in Europe as elsewhere, to acquire substantial holdings in electronics.

V.2.4.2. Industrial situation and strategies in telecommunications technologies

Strategies in this sphere have been examined in Chapter IV.1, but it is necessary to stress once more the importance of the connection between information technology and telecommunications. This technical evolution (with materials playing a decisive part) is crucial to

Europe. Roughly speaking, since the connection is relevant on the one hand to an area where Europe is dominated (information technology) and, on the other, to an area where it is competitive (telecommunications), it is easy to imagine that two outcomes are possible. Although is is too early to trace the direction of developments, there are grounds for disquiet in the collapse of the strategies of the different European telecommunications groups.

CHAPTER V.3

Role of the Authorities in Europe in the Field of Materials

The arguments developed so far converge towards a general recommendation which is mandatory for the European authorities : means must be found, whatever the price, to enable Europe to catch up in the field of functional materials and to maintain its advance in the design, production and application of structural materials.

Europe's future competitiveness hinges on this global objective. American and Japanese domination in functional materials has in fact already meant the loss of too many potential jobs (in electronics or other advanced technologies) which might have been created in Europe but have in fact been created elsewhere. American and Japanese domination has caused an excessive brain drain, with people leaving Europe for the great American research centres, both private and public, spelling too many wasted opportunities in the technologies which came into being in Europe but have been developed in other countries.

With every day that passes, the competitive position occupied by Europe in structural materials (thanks mainly to the might of its chemical industry) is threatened more and more by the headway realized in other countries. The production of technical objects is no longer solely dependent on mere skill in setting up the best structures. If a form of production is to be organized that is both efficient and flexible, that skill must be accompanied by an ever growing mastery of information technologies.

Over and above this prime objective - competitiveness - the role of materials in society is perhaps far more profound. Mastery of their development may lead to the liberation of new creative forces, a guarantee of effective defence, a response to increasingly stringent social constraints such as safety requirements, or a response to the exigencies of developing other human activities.

It is clear, if responses are to be found to these objectives, that the action of the authorities is decisive. The governments of other countries or groups of countries do not hesitate to intervene on a massive scale in order to support the development of materials in their respective countries. They do so either directly, through huge subsidies to fundamental research, or more indirectly by means of policies involving large-scale programmes (space, nuclear or most of all military programmes, such as SDI), or by means of policies that are less apparent but just as effective, with the establishment of norms on import duties. The idea that a "laissez faire" policy is in operating among Europe's competitors in the materials field, then, is entirely erroneous.

Lastly, it is clear that the competitive level of a country is closely bound up with the education and occupational training system it has adopted, and here ambitious policies have been carried out by countries like Japan. Downstream from the educational system it is to be noted that in Japan, the United States or the Federal Republic of Germany, for instance, the social standing conferred by careers in technology as compared with that enjoyed by white collar workers is not a neutral factor, and that those countries which have an advance in the field of materials have contrived to give their engineers, technicians and workforce an enviable position in society. In none of those countries do oil stained fingers carry any stigma!

The need for intervention on the part of the authorities is all the greater in the materials field because companies producing, processing and using materials have entered a transitional phase. There are clear signs of a gradual abandonment of the system of standardization which was a feature of the previous production system. And, because the shape of the future regime is still uncertain (even if it is possible to juggle with scenarios) and because "degrees of freedom" are today very considerable, appropriate action on the part of the authorities can help reduce uncertainties and generate competitive strategies.

Many people in positions of public responsibility have clearly become aware of what is at stake in Europe. After pursuing an essentially defensive strategy in connection with materials (through measures to "rescue" the steel industry, notably), the countries of Europe, either individually, or most of all at Community level, have recently set themselves up with ambitious programmes for maintaining and improving Europe's competitiveness. The impact of these programmes will of course reflect the favourable climate of the European scientific and industrial "environment". It is hard, for instance, to imagine carrying out successfully, and in depth, an ambitious European materials policy if the conditions for disseminating innovatory policies are not at the same time fulfilled, or if the conditions in which a full-scale European market operates are not improved. To take but one example, it is certain that the initial developments of a functional material are today tried out on a world-wide market, or at least on the scale of the American market.

In the specific case of materials, the problem of harmonizing a European policy is complicated by the existence of very considerable differences in the degree of European countries' competitiveness. More specifically, virtually all the company or university heads we met in the course of this survey were emphatic about Germany's advance compared with the other European countries in a very large majority of spheres relating to materials. Some European countries do not in fact have any real materials industry, particularly at the level of producing the main basic materials. Admittedly, this judgment needs to be qualified, since undisputed centres of competitiveness are to be found in European countries other than Germany - glass, aluminium, carbon composites, rare earths (France), materials for infra-red, biomaterials and organic composites (United Kingdom), etc, but the might of German industry (particularly the chemical industry), its organizational capacity and system of occupational training have helped create an exceptional environment in Germany. Finally, Germany is perhaps the first country in Europe to have successfully forecast the scale on which certain new materials like engineering ceramics would expand. It is a state of affairs that cannot be denied or brushed aside. It is incidentally referred to constantly by foreign industrialists and academics. It shows, in any event, that it is up to the European authorities to select, when evolving their strategies, the means whereby they are going to space out their efforts "geographically". There are two principal snares which they must avoid : the first is a dispersal of effort which could produce fruitless results, or, instead, the creation of over-capacities. The second snare is policy that would widen the gap between one country and the others still further. It was no part of the authors' remit to pronounce on this point, but it is one worthy of emphasis. On the other hand, the conclusions set out in Chapter V.4 prompt us in any case to highlight one factor in the German system : the principle of linked work and study, which apparently corresponds well to the requirements of future shifts in the European materials production system.

These comments form the basis for the following examination of the role of European authorities, which follows certain lines, to wit : research and training policy and policy on norms and regulations.

V.3.1. IMPORTANCE OF RESEARCH AND THE ROLE OF THE AUTHORITIES

Competition between the United States, Japan and Europe in the field of materials research is extremely keen. The United States and Japan stepped up their efforts considerably at the end of the 70s. The countries of Europe, their eyes having been opened by high-level diagnoses (including the Collyear Report in the United Kingdom, the Causse Report in France and the BMFT Report in the Federal Republic of Germany), manifestly reacted. The scope and ambition of the major European programmes which have just been given concrete form (outlined below) are the best evidence of Europe's determination to make good a still disquieting gap.

Competition is particularly intense in materials with the highest expected growth rates for the coming decade (technological ceramics, composites, technological plastics and materials for electronics). But the scale of budgetary funds earmarked for processing (powder metallurgy, high-precision casting, superplastic, high pressure and isostatic forming, methods of bonding, non-destructive testing, etc) demonstrates effectively that what is at stake is not merely the so-called "new" materials, but covers the whole range of both materials and processes.

V.3.1.1. United States-Japan-EEC competition in research applied to the field of materials

It is not always simple here to trace a boundary between fundamental research and applied research. The following survey relates above all to applied research.

Situation in the United States

The scale of intervention by the American authorities in support of materials research is massive. In all, the Federal Government spends approximately $1 milliard every year in this field (allowing for military expenditure). According to the United States Congress, materials availability is essential for the security of the nation, for economic well-being and for industrial output (3) . As can be seen from Table V.3 below, responsibility for materials programmes is shouldered by a large number of ministerial departments or agencies : Department of Energy, Department of Defense, National Science Foundation, NASA, Bureau of Mines, National Bureau of Standards, etc. These massives sums are of course swelled by the efforts realized by private research, on a scale which is certainly higher than that of public research. The United States is still today the only country able to set aside a "critical" mass of ressources in materials, both through major projects (Apollo, Shuttle, SDI, etc) and through financing of networks of high-tech laboratories.

The result is that the United States today occupies a dominant position in virtually the whole array of materials. The United States in fact has the capacity to advance on the whole "research front" in materials, which is a huge asset, particularly in a variety-based scenario. While the different programmes naturally tend to emphasize the "newest" materials (ceramics, composites, amorphous metals, thin layers in electronics and new catalysts), the different directions followed by research on the same material help considerably towards enriching knowledge, especially since mobility in this sphere of

---

(3) National Materials and Minerals Policy Act 1980.

research is considerable in the United States. (It is worth pointing out, however, that while internal mobility is still high, increasingly tight restrictions apply to exchanges between research workers from the United States and other countries, including Europe).

Support from the American authorities is not confined just to research into materials or programmes. The University of Berkeley, for instance, has just acquired an ultra-modern centre for testing materials.

It is important to mention a new aspect in the research trend in the United States, however. In 1986 several very big private research centres (EXXON, BELL LAB, among others), abruptly parted company with a very large number of top-level research workers. There has been much speculation about the reasons prompting groups like EXXON and ATT to do without the services of the brightest stars in the materials firmament, to whom materials science owes virtually all the discoveries of the last 10 or 15 years.

Group in-fighting is not to be excluded (rivalry, power, etc) but it seems all the same that the Challenger disaster suddenly gave top executives (not to mention the rest of the US population) a glimpse of the fragility of a science, and particularly a science of materials, on which they had perhaps pinned too many hopes, tending to forget - as a result of resounding successes in space, for instance - that an industrial development policy is not based only on research. Japan was showing them that production organization, even with a relatively low research content, produced success; and a minor design defect, a synthetic rubber gasket, in a few seconds toppled a grandiose space policy. More than the technical aspect, it was most of all poor organization of production, design and management which was disclosed by the accident.

Amount	DOE (1)	DOD	NSF (2)	NASA	BOM	NBS
R&D funding	368	296	100	75	32	12

(1) 1982 (2) 1984
DOE : Department of Energy.
DOD : Department of Defense.
NSF : National Science Foundation.
NASA : National Space and Aeronautic Administration.
BOM : Bureau of Mines.
NBS : National Bureau of Standards.
Source : General Accounting Office report GAO/RCED 85.63, September 9, 1985.

Table V.3.A : Funding of materials R & D by US departments and public institutions in 1983 ($ millions).

Situation in Japan

Intensive research into new materials is a relatively recent phenomenon in Japan. However, in the mid-70s MITI took matters in hand and decided that materials represented one of the country's over-riding technological priorities. The MITI drive in 1981 quickened its pace with the establishment of a programme concentrating research endeavours on a small number of materials (fine ceramics, carbon fibre, electroconductive polymers, high crystalline polymers, semiconductors) although it was fairly distinctly centred on engineering ceramics. The

Fine Ceramics
To develop ceramics with improved toughness, strength, anti-corrosion and anti-abrasion properties.

Materials for Membrane Technology
To improve separation technology and energy efficiency in the chemical industry.

Electro-conductive Polymers
To achieve new electrical or electronic properties.

High Crystalline Polymers
To develop new polymers having comparable mechanical properties to metals for structural applications.

High Performance Crystalline Controlled Alloys
"Single crystal" superalloys, superplastic forming, strip casting of steel.

Composite Materials
High performance metal and polymeric matrix composites.

Bioreactors
Cell Mass-Culture.
Re-combinant DNA Applications.

Superlattice Elements
Materials and processes, superlattice elements in semi-conductor thin films.

Three Dimensional Circuit Elements
Multi-layer VLSI.

Enforced Environment-Resistant Element
Element, packaging and integration technology for severe environments.

Source : Collyear Report, Department of Trade and Industry, 1985.

Table V.3.B : Japanese programmes.

R & D effort will be coordinated over a period of 10 years, and MITI is trying in particular to speed up the process whereby university and industrial research workers are brought closer together. Total R & D spending can be estimated at some $3 million per annum (see Table V.3.B).

The co-ordination of Japanese efforts in materials extends well beyond research into "new materials". The decision by Japan to abandon its activities in primary aluminium is thus an intrinsic part of an overall materials plan.

The results achieved over the past decade are in any event highly impressive. In the space of a few years, Japan has become world leader in the production of fine ceramics for electronics engines, and second world producer after the United States of semiconductors and carbon fibre.

One of the reasons for a number of successes chalked up by Japan is perhaps the ability of Japanese companies to integrate at speed the most recent developments in materials and industrial management of production and application processes.

Situation in Europe

For a very long time Europe played a leading role in materials research, notably in the production of "traditional" materials (steel, cement, glass, aluminium, etc). It would perhaps be going too far to say today that Europe has been overtaken, but there is no doubt that it has allowed the United States and Japan to steal a march on it in research into most of the materials with the highest expected growth rates between now and the end of the century (such as amorphous metals, superplastic alloys, certain new engineering polymers, carbon fibre, ultra-pure engineering ceramics, high-performance magnetic materials, etc). In overall terms, however, fundamental research in Europe upstream from this applied research is roughly at the same level as that in the United States and well ahead of that in Japan.

There are perhaps many reasons why Europe lags behind (such as dispersion of activities, lack of foresight, want of strategic coherence, divide between research and industry, etc, factors which are studied at length elsewhere), but it is clear that inertia due to tradition is one that merits emphasis. Research carried out in Europe by the steelmen to "counter" the inroads made by aluminium drinks cans from the United States is a very significant instance of the traditionalist reaction here. In many ways, that reaction seems perfectly sound and economically efficacious, but it is certain that in no case can it be the sole instrument for halting competition from other countries.

The realization that Europe has considerable leeway to make up has prompted almost every European country to draw up action programmes for materials. Under the stimulus of BMFT, the Germans for instance in 1981 adopted a 10-year plan representing some $385 million; the United Kingdom, in the wake of the Collyear Report, France, following the recommendations of the materials taskforce, and other European countries have drawn up programmes intended to make good the accumulated shortcomings. Each country's efforts have been supported on a Europe-wide scale by ambitious programmes which respect national efforts and which aim to give Europe the "critical mass" it lacks, to establish a denser network of relations between the different European research workers, and to tighten the bonds between research and industry.

The importance of materials is already explicit in the first Community R & D programmes. Part of the following programmes is devoted to research into materials : ESPRIT (VLSI integrated circuits), RACE (optical fibres), BRITE (development of materials at the pre-competitive stage), BIOTECHNOLOGIE (biomaterials), etc. Materials also occupy an important place in the EUREKA programme (see Table V.3.C).

It is most of all through the EURAM programme, however, that the EEC has acquired a full-scale Community programme specifically concerned with materials. The EURAM programme (European Research on Advanced Materials) is a programme spread over a number of years, the period 1986 - 1989, devoted to materials research and development, with a budget of 30 million ECUs.

The principal objective of the programme is the development of advanced materials, upstream from manufacturing industries, coupling basic research with engineering work. It is designed to create, develop and apply new materials, and to secure the evolution of existing materials to a higher level of sophistication and a competitive production cost. The programme covers three main research fields : metallic materials, advanced ceramics and composites (see Table V.3.D).

Nature of project	Projects approved end 1986 (Stockholm)	Cost (M ECU)	Duration (years)
	(nb D = Germany, CH = Switzerland, E = Spain, S = Sweden, F = France, N = Norway, Netherlands, A = Austria, I = Italy, GB = GB)		
EU 3 Amorphous silicon	Solens (F) - MBB (D)	53	5
EU 5 Filtration membranes	Lyonnaise des Eaux (F) Dégrémont (F) - De Danske Sakker Fabriken (DK)	56	6
EU 13 CARMAT 2000 Development of solid automobile structures, ultra-light fibres	Peugeot (F) - BASF (D) Battenfeld Extension Technic (D) - Bayer A.G. (D) - Cristaleria Esp.(E) Ecole des Mines (F) - Elf (F) - Pechiney Alu (F) - Saint-Gobain Vitrage (F) Usinor (F) - Vetrotex (F) DSN (NL) - ICI (GB)	60	5
EU 25 Aluminium in place of chromium in treatment of leather	(Spain) - (Germany) (GB)	25	3
EU 29 Development of new ceramics for car engines	Desmarquet (F) - Péchiney Alu (F) - Peugeot (F) - Fiat Holding (I)	15	5
EU 33 Ceramics in gas turbines	SEP (F) - Hispano Suiza (F) - Snecma (F) - Alfa Romeo (I) Volvo Slygmotor (S)	16	5
EU 40 Construction technologies (infrastructures and networks for major building developments)	Giblin (D) - SAE (F) - Società Costruzioni Generalisti (I) Bovis Construction LTD (GB) - Charcon Tunnel (GB)	9,2	5
EU 42 Light-weight materials for transport systems	VAW (D) - Institut für Angewandte Stahlentechnik (D) KSR (D) - Péchiney Alu (F)	15	4
EU 47 Ceramics for diesel engines	Mann (D) - SEP (F)	14	5
EU 52 Disposable sensors for the medical field	(Netherlands) - (Switzerland) - (Belgium) (Germany)	4	5
EU 96 Super-conducting coils	Université de Genève (CH) Société Elin-Union (A) - Metall-Werke Plansee (A) - Tecknishe Universität Wien (A)	8	3

	Spectro A.G. (CH)		
EU 102 EPROM (Integrated Circuit Non-Volatile Memory). Memories with a capacity of 16 M bits	(France) - (Italy)	400	5
EU 113 Development of lasers (solid state included)	(France) - (GB)	180	6
EU 117 Fibre-reinforced plastics, glass fibre, composites	(Finland) - (Belgium)	2 to 3	2
EU 127 JESSI (Joint European Sub-micron Silicon). Sub-micron technology	(Germany) - (GB) - (France) - (Netherlands) (Italy)	2.4 to 2.8	10
EU 132 Transmission by optical fibre systems	(GB) - (Sweden) - (Portugal)	8 to 16	2 to 2.5
EU 138 Coatings for advanced technologies	Coat A.B. (S) - Lebolt Heraeus (D) - Flort (S) - Volvo (S) - Imcol (GB) Pilkington (GB)	4.5	4
EU 139 Methods of forecasting properties of (injection moulded thermoplastic articles	Royal Technology Institute (N) - Centre for Industrial Research (S)	1.6	5
EU 155 International co-operative research for laser applications	(Germany) -	7.5	3
EU 159 EURODYN (Programme demonstrating technology of gas turbine engine)	(France) - (Netherlands)	33.8	4
EU 160 Development of mineral membranes and procedures for separating fermentation products (antibiotics)	(France) - (Italy)	14.7	4
EU 162 Electron beam treatment : applications in processing industry (micro-emulsions, polymer granules)	(Sweden) - (Finland)	3.3	3
EU 163 EUROVISE (display systems)	(Ireland) - (GB)	1.6	3.5
EU 164	(Finland) - (GB)	0.5	-

Microencapsulation, development of new sensors, pharmaceutical industries			
Projects pending			
Thin films and deposit on materials (EU 66) (electronic, lubrification)	Lausanne Ecole Polytechnique and other Swiss bodies	1.1	2
New process for polymer production (EU 74) (fermentation, lactic acid)	Spanish companies	-	-
Production line for mass-market integrated sensors	Metravibe (F) - CSEM (CH)	-	-
Flexible automated microlithograph line, for integrated circuits	MATRA (F) - CNET (F) - Cambridge Instruments (GB) Frauenhofer Institut (D) SGS (I), etc...	-	-
Inspection and automatic testing of integrated circuits	Electronique Serge Dassault (F) - CSEM (CH) BATELLE (CH)	5	3
Advanced microprocessors, GaAs integrated circuits, microwave components, high-density memories, flat screens, sensors)	Thomson (F) - GEB (GB) - Philips (Netherlands) - Siemens (D)		-
Flexible automated factory for manufacture of electronic equipment	Eurosoft (F) - CSEA (I) - Inisel (E)	28.8	5
Non-invasive medical diagnosis equipment (biosensors and IA)	(France) - (DK) (Spain)	-	-
Custom-built integrated circuits (ES2)	(Luxembourg) - (Germany) - (GB) - (Belgium) - (Finland) (Sweden) - (Switzerland) (Austria)	105	3
Materials and new assembly technologies for transport	Pêchiney (F) - VAW (D)	-	4
New materials for semi-trailers	Berlin (F) - Vetrotex (F) - Solvay (B) GKN (GB) - Fruehauf (D), etc...	-	-
Medium-power ceramized gas turbines	SEP (F) - Hispano Suiza (F) - Volvo (S) - Alfa Romeo (I)	16	5
Materials and computer-assisted design and manufacture	Aêrospatiale (F) MBB (D)	-	-

Broad-band telecommunications	CIT-Alcatel (F) - Italtel (I) - Plessey (GB) etc...	-	-
Full automation of chip assembly	Matra (F) - SGS (I)	-	-
Computer-assisted design and manufacture of GaAs integrated circuits	Thomson (F) - GEC (GB)	58.5	3
Thyristors for high-power application in rail traction	Thomson (F) - GEC (GB)	18.5	2

Table V.3.C : Materials in the EUREKA programme.

Development of metallic materials

Topic 1 - Aluminium alloys

Aluminium and its alloys are already well developed yet they still offer much potential for development - especially in transport (aerospace, automobile) where the ratio of mechanical properties/specific weight plays an essential role.

Development could follow these three lines :

1.1 Further development of conventional aluminium alloys to achieve greater performance and reliability

- Improving the mechanical properties such as fatigue strength, fracture toughness, resistance to abrasion and weldability;
- improvement of purity of the basic alloy constituents;
- in-service safety and reliability should be also improved in order to reduce inspection, repaire and maintenance costs;
- improvement of corrosion resistance by surface treatment;
- development of non-destructive testing methods (NDT) for all alloys.

1.2 Development of new powder-metallurgy aluminium alloys

- Systematic study of the powder metallurgy of various compositions of new alloys. The rapid solidification route will be used in all of its forms;
- further work on Al-Li alloy and its derivatives;
- targets to be aimed at : ultra-fine (micrometre sized) particles, temperature-hardening and stabilization of alloys by addition of insoluble dispersion hardeners;
- study and scaling up of powder shaping : compaction,

moulding, sintering, extrusion, rolling, etc.

1.3 Development of superplastic Al alloys

- Since superplastic is obtained via ultra-fine grain structures, it will be useful to study ultra-fine Al powder by using the rapid solidification technique under an inert atmosphere;
- definition and characterization of new superplastic alloys, study of their practical application;
- study of the casting of components by the fluid-phase technique (rheocasting);
- investigation of relationship between high purity constituents and the mechanical alloy properties;
- improvement of joining properties (weldability and diffusion bonding) of Al alloys.

Topic 2 - Magnesium alloys

2.1 Development of new Mg alloys having improved characteristics

- The aim here is to achieve a net improvement in the mechanical strength and chemical (corrosion) resistance of Mg alloys. The creep threshold would also have to be increased to 200°C;
- adaptation of new Mg alloys to modern casting techniques (rheocasting or thixo-casting);
- work on ultra-high purity Mg;
- upgrading of Mg alloys by controlled addition of hardening agents (e.g. ZrO_2, TiO_2);
- adaptation of the production engineering techniques for new Mg alloys (forging, extrusion, rolling, etc).

2.2 Improvements to magnesium alloys coatings

- Development of simple, effective methods of coating magnesium components which are compatible with semi-continuous and continuous production processes. Elimination of chromium salt treatment;
- development of coatings and surface treatment processes, and associated automated equipment;
- study on reducing the fire risk during magnesium component production;
- studies on thermal damage to Mg components (fire, heat).

2.3 Development of a new range of rapid solidification (RSR) magnesium alloys

- Opening up of new applications for magnesium by the rapid solidification route and powder metallurgy. The products developed must be compatible with the mass production of components;
- improvement of shaping processes : pressing, casting, extrusion;
- improvement to product reliability and NDT methods;

- increase in the operating temperature threshold of magnesium components.

Topic 3 - Titanium alloys

3.1 Simplification of titanium alloy preparation via the direct reduction of mixed oxides

- systematic study of the high-temperature calcium treatment of mixed oxides in order to obtain titanium alloy powder (parameters : kinetic temperature reactions, grain size distribution, product purity, variations in composition);
- study of suitable methods and scaling-up technology for the shaping of these alloy powders, (compacting, sintering, extruction, rolling, etc). Cost reduction.

3.2 Metallurgy of high-performance titanium powders and alloys

- Study of ways of producing titanium components at a lower cost using powder metallurgy and new high-performance titanium alloys;
- study of new more cost-effective titanium alloys containing oxide dispersants in order to improve their characteristics (dispersion hardening);
- development of rapid solidification routes for the production of titanium alloy powders.

3.3 Technology of semi-finished products and fabrication of titanium-alloy components

- Systematic examination of the fundamental aspects of forming titanium alloy "semis". Both physical and metallurgical aspects should be taken into account;
- study of superplastic titanium alloy forming in order to produce near-net-shape components;
- study of methods of reducing the weight and cost of producting titanium alloy components without impairing their reliability;
- study of the physical strength and chemical resistance of titanium alloy components (maximum operating temperature, oxidation); work on stiffness improvement;
- systematic study of corrosion, and, in particular, sea-water corrosion, of titanium alloy offshore equipment.

Topic 4 - Electrical contact materials

- Systematics study of new materials for use as electrical contacts to replace the metals normally used, such as silver, platinium, noble metals, etc.;
- development of new ways of treating the surface of electrical contact materials in order to protect these from erosion by electric arcing;
- development of new porous-structure materials (tungsten

carbide) infiltrated with silver or copper alloys and the study of their electrical functions.

Topic 5 - Magnetic materials

- General and systematic study of new high-performance magnetic materials in order to replace Co and samarium with more freely available and cost-effective metals;
- study and technology of magnetic alloy powder metallurgy and other fabrication processes belonging to the Fe-Nd-B family and other rare earth iron rich compounds with the addition of other elements in order to optimize their magnetic characteristics;
- study and practical application of well characterized permanent-magnet components for specific industrial applications. Parameters for study : grain size distribution of powder, powder metallurgy, RSR, sintering, magnetization curves, direction and uniformity of magnetic field, component geometry, machinability, temperature, etc.;
- analysis of new permanent magnets from an economic and market-impact viewpoint.

Topic 6 - Materials for surface coatings, for machining, tooling, cutting and machining applications

- The overall aim of this R&D should be the development of new coatings, to improve technological processes and to replace strategic metals;
- development of new ceramic materials, (oxides, borides, nitrides, carbides) for surface coatings and their application technology;
- research, development and technology of new ceramic cutting materials (including ceramics) for high-performance machining;
- development and optimization of surface treatments and coating methods (example : ion implantation, PVD CVD, lasers, etc.);
- development of new steel alloys having higher Si, Al and Ni contents.

Topic 7 - Development of improved thin-walled castings

- Improving the mechanical characteristics of spheriodalgraphite cast iron for thin-wall components (cast iron shells having improved rigidity and lower vibration levels for engines);
- development of casting techniques for thin-well components (cast iron, aluminium alloys and other materials). Improvement of moulds with refractory linings for bath production;
- study of new cast alloy structures, grain fineness, segregation, micro-porosity.

Development of engineering ceramics

Topic 8 - Optimization of engineering ceramics for use in IC engines

- Development of advanced ceramics and their production technology using closely specified SiC, Si_4N_3, ZrO_2, Si-Al-On, etc. powders;
- study of ceramic powder characteristics as a function of component quality, reproducibility and reliability;
- design of the scaling up processes and automation of batch production methods for high-quality ceramics, (for example : in I.C. engines);
- research into new of compacting ceramic powders and quality control;
- optimization of the complex relationship between powder, pressure, temperature, density and mechanical quality.

Topic 9 - Study of metal/ceramic interface : cermets

- Systematic, comparative study of ceramics, metal or alloy expansion rates with their association in mind;
- research into adequate ceramic/metal bonds;
- production of cerments by powder metallurgy;
- conceptual and engineering studies relating to the mass production of components.

Topic 10 - Study of ceramic composites with fibres or whiskers as reinforcement for industrial applications

- General study and development of reinforced ceramic composites able to withstand very high stresses (mechanical shock, tensile stress, temperature variations);
- optimization of powders and fibres in line with their uses in IC engines and other applications;
- systematic study of methods of compacting high-density components. Ooptimization of the relationship between pressure, temperature, density, mechanical strength, powder composition, structure and fibre orientation;
- optimization of batch production costs for high-performance ceramic components; study of the relationship between the cost and mechanical properties of engine components.

Topic 11 - Basic study of the high-temperature behaviour of engineering ceramics

A study of the basic aspects of ceramic behaviour under the operating conditions inside an IC engine (adiabtic diesel). The effect of temperature will be a decisive factor. Experiments will be conducted at maximum temperatures in order to plot operating limits. The temperature will also be varied very widely in order to stimulate engine acceleration states.

- Systematic study of all of the basic aspects of ceramic/ceramic friction and abrasion as a function of temperature (max. 1500°C);
- conceptual study of an effective, simple system of measuring high temperature ceramic friction;
- development of a solid or liquid high-temperature lubricant that can be incorporated into the ceramic;
- examination of suitable methods of treating ceramics in order to improve their friction coefficient;
- improvement of the oxidation and corrosion resistance of engineering ceramics, and particularly their resistance to combustion products or mineral salt used on roads (for example, gas turbines).

Development of composite materials

Topic 12 - Organic-matrix composites

12.1 Optimization of thermoplastic organic-matrix composites and their stiffeners

- Study of the basic properties of ductile thermoplastic matrices (polysters, polysulphides, etc.) in which advanced fibres such as carbon, aramids, ceramics, etc. are embedded. Development of a family of self-reinforcing products;
- widening the range of operating temperatures of this family of composite materials (250°C during steady-state conditions and 400°C during peak conditions);
- further development of hot forming and reaction injection moulding (RIM);
- improvement of mechanical and thermal properties of thermoplastic composites;
- basic study of matrix/fibre interface and of the structural characteristics of the composites.

12.2 Development of thermosetting organic-matrix composites

- Systematic and basic study of the properties and matrix/fibre combinations of thermosetting composites (polymides, epoxy, etc.);
- investigation of matrix/fibre interfaces as a function of temperature. Fibres to be studies : carbon, boron, aramids, ceramics, etc.;
- study of semi-finished products as a function of maximum permissible operating temperature;
- tests on hybrid composites - mixtures of two types of fibre.

12.3 Development of fabrication methods and technologies for organic composites

- Adaptation of organic composites to mass production methods (motor vehicle or aerospace);

- study of the physical and chemical parameters relating to the joining and assembly of heterogeneous composite materials;
- development and control of the production of larger, more sophisticated components;
- development of non-destructive testing methods for in-service monitoring purposes.

12.4 Development of design methods for composites structures

- Provision of a methodology and design rules for organic-matrix composite structures;
- development of methods of analysing composite behaviour under mechanical stress;
- design and development of energy-absorbing composites;
- study of organic-matrix composite reliability, improvements to detection and fault identification techniques. Study of fault propagation mechanism in matrix and fibres.

Topic 13 - Metallic-matrix composites

13.1 Composites with an aluminium-alloy matrix

- Development of the basic constituents of these composites and optimization of their composition;
- substantial improvement of the physical, chemical and mechanical characteristics of these composites (modulus of elasticity, fatigue strength, increase in heat resistance threshold);
- systematic investigation of the following matrix/fibre interfaces : Al/B, Al/SiC, Al/Al_2O_3, Al/C, Al/Ti or Al/steel.

13.2 Development of fabrication methods and technology of raw compounds using Al composites

- optimization of the quality/price factor;
- optimization of fibre size and alignment within the Al matrix as a function of the mechanical properties desired. Study of fabrication problems;
- development of quality control and non-destructive testing (NDT) methods.

13.3 Metallic-matrix composites : Magnesium alloys

- Development of a new generation of magnesium-matrix composite materials;
- study of the development of methods of producting semi-finished or raw components;
- study of the optimization of both the quality/price and density/strength ratios;
- improvements to the operating temperature threshold and to fire and corrosion resistance.

Topic 14 - Other advanced materials for specific applications

For obvious reasons the list of topics covering advanced materials given above cannot be considered as exhaustive since this field is widening all the time. Therefore some development work should also be devoted to specific borderline materials with an interesting industrial potential. These will be treated on a case by case basis.

The following research lines can be given as an indication :

- **memory alloys;**
- **materials with high energy absorption;**
- **highly corrosion resistant marine materials for offshore structural applications;**
- **composite materials containing amorphous matrices;**
- **composite materials with glassy or vitreous matrices;**
- **composite materials with elastic matrices for energy-shock-noise absorption.**

Table V.3.D : EURAM programmes.

The importance of public research funding must not make us lose sight of the role of private research in industry. This recalls, in particular, the importance of research carried out by the European chemical industry, which has retained almost intact the European competition potential in plastics. This research is at present centring on composites and ceramics.

Ceramics perhaps represent the materials field in which competition is keenest at the moment, and where intervention by a public body - MITI in Japan - is the most manifest for the purpose of co-ordinating a very large-scale strategy in research and development. This is examined below.

Engineering ceramics : an example of massive intervention by the authorities in research and development

Advanced ceramics in Japan

Intervention by MITI in the field of advanced ceramics, with the agreement of the very great majority of the Japanese industrial groups concerned, takes several forms :

- Training of a considerable number of research workers and technicians by comparison with the other major countries (as demonstrated in Table V.3.E).

Countries	Scientists and engineers	Techniciens
Japan	2 000	2 000
USA	1 000	1 000 to 2 000
Europe	500	not estimated
Canada	50 to 100	not estimated

Source : Advanced Ceramics, Technology Situation Report by MKM Consultants International.

Table V.3.E

- Establishment of particularly close links in this field between universities, industries and the authorities.
- Support for an association of industrialists, the Japan Fine Ceramics Association, which includes over 170 companies involved in the development of advanced ceramics. MITI support for research and development in industries is reflected by ambitious, clearly-targetted programmes. For example, four companies (Kyocera, Toshiba, NGK and Asahi Glass) are developing under the MITI umbrella advanced ceramics for turbines, reciprocating engines and rotating systems (see Table V.3.F).

WHAT IS THE CURRENT SITUATION IN ENGINE CERAMIZATION ?

The second international colloquy on the use of ceramics in heat-powered machines (Lübeck, April 1986) and the third international colloquy on zirconia (Tokyo, September 1986) were opportunities to take stock of the present situation concerning ceramization of engine parts. (The growing use of ceramics in sensors to monitor the functioning of an engine - for instance, the composition of exhaust gases - was also touched on, but does not come within the subject matter of this article).

The first "mass" market application date back to 1981, when the Japanese motor manufacturer Isuzu began to integrate in diesel engines silicon nitride preheating plugs and then, in 1983, silicon nitride diesel pre-combustion chamber bottom sections in the Aska model. These parts, developed by the Japanese ceramist Kyocera, make it possible to reduce ignition time and the noise caused on cold starting. Toyota followed this example in 1984, giving its Crown model a silicon nitride diesel pre-combustion chamber bottom section, developed by the Japanese ceramics firm NGK. In the same year, Porsche fitted its Turbo 944S model with aluminium titanate Al_2TiO exhaust pipes, developed by the German ceramics firm Hoechst-Ceramtech. These make it possible to reduce heat constraints in the cylinder head, to improve the performance of the catalytic anti-pollution system on starting and at low speed, and performance of the turbocharger. The model also has turbocharger parts in partially stabilized zirconia (developed by the German company Feldmühle).

In 1985, Japanese engine specialists fitted the Nissan Fairlady model with a silicon nitride turbocharger rotor developed by NGK. This ceramized rotor cuts accelerator response time, improves supercharging at low revs and reduces turbocharger mass. The engine can also operate at a higher temperature. Lastly, in 1986 Mazda fitted engines with a whole silicon nitride diesel precombustion chamber, also developed by NGK. The Ford company purchased the part, which makes it possible to reduce the discharge of combustion particles for fitting to some of its Escort and Links models.

Ceramization is not confined, however, to hot engine parts. It also applies to parts subject to friction. Since 1984, for instance, Mitsubishi has been testing silicon nitride rocker arm seats on its Galant model. Developed by NGK, their wear is reduced and durability at high speed is increased. What is more, engine drive torque at low revs is improved thanks to these parts.

Only partial information is available concerning military applications, the most familiar being those attempted by the Cummins company, which is investigating in the United States a reciprocating engine in which zirconia coatings are used as one way of improving heat insulation.

These, then, are the ceramic parts actually in use at the moment in engines. All the series are limited - some tens of engines per day - for top-range models. All parts are tested before being assembled.

Most engine specialists have experimental programmes, among them Renault and Peugeot in France, Volkswagen (reciprocating engines) and Mercedes (turbine engine) in the Federal Republic of Germany, Volvo in Sweden, Ford (turbine engine) in the United States. Studies are also being carried out, notably in France by the CEMT (Centre d'Etudes des Machines Thermiques) on ships' engines and electrical power turbines. But there is no doubt that it is in Japan that the programmes are the most numerous, and spelt out with the most publicity!

It is worth mentioning by way of conclusion the European programmes EURAM (European Research Programme on Advanced Materials), several of whose themes relate to engine ceramization, and EUREKA. The aim of the first is to co-ordinate European efforts in the development and study of new materials. The second includes, among the 40 projects in which French companies are participating, three five-year projects concerning engine ceramization, "Development of new ceramics for car engines", which it will cost 15 million ECUs to fund, "Ceramics in gas turbines" (16 million ECUs) and lastly "Ceramics for diesel engines" (14 million ECUs).

Source : La Recherche, January 1987.

Table V.3.F : Ceramics and engines

- The research programmes are sufficiently flexible to allow firms to develop "horizontally", however. For example, Kyoto Ceramics, the biggest producer of ceramics for integrated circuits in the world, and one of the main producers of ceramic capacitors, is seeking a leading position in structural ceramics by adapting its technologies. The advantages of this kind of integration lie in the technological advance built up in the production, characterization and non-destructive testing of powders. This advance also applies to other utilizations of materials.

- Research and development programmes are bound up with a strategy for the early marketing of advanced ceramic products. The strategy is not without its risks (loss of financial investment if the performance of the product falls short of users' expectations). It must also be noted that in this country, these ceramics utilized in high-technology applications are introduced as a matter of course into products such as scissors, fish hooks, etc. This confers considerable advantages, because it enables producers to develop a market at a very early stage and to move more rapidly along the learning curve.

The veritable challenge posed by Japan, through the impetus of MITI, is the acquisition of a considerable stake in the advanced ceramics market. Japan already holds close on 50% of the market (which can put at about $9 milliard at present, at world level, and over $20 milliard between now and the end of the century, including a relatively substantial proportion representing ceramics for electronics - between 60% and 80% of the whole).

Japan's interest in advanced ceramics can be accounted for in various ways, most of which have already been referred to in several sections of this work :

- the expected growth rate for these materials is at least 10% per annum up to the end of the century;
- advanced ceramics occupy a strategic role in two sectors where there is a very considerable <u>market</u> : motor manufacturing and electronics.
- advanced ceramics are likely to oust metals in very many applications.

This has been seen, for instance, in the manufacture of nuclear fuels, where, after hard-hitting competition, advanced ceramics (oxides, carbides or nitrides of fissile metal) have won out (thanks, in particular, to their powers of heat resistance) against the most competitive metal alloys (alloys of uranium, uranium-plutonium, or plutonium stabilized by other metals).

Applications	United States ($ million p.a)	Japan	Western Europe
Electronic applications : multilayer capacitors	10 - 25	R&D budgets not known. 10 times more patents and engineers than in USA	Not known
Gas detectors	1 - 2	R&D budgets not known. 3 times more technical reports and 10 times more patents than in USA	Probably in line with the American effort
Structural ceramics: ceramics for engines	35 - 40	>$50 million pa	Less than the USA : FRG and Sweden lead in Europe
Cutting tools	Probably not more than 1	20 times more patents than in USA,1973-82	Probably more than in USA
Optics, integrated optics	10	Slightly less than budgets set aside by USA	About one-half of US effort : France and UK lead in Europe

Source : Charles River Associates, 1984.

Table V.3.G : Comparison of R & D efforts in different countries in the field of advanced ceramics.

- the know-how needed for the purpose of controlling the development of advanced ceramics necessitates a combination of know-how reflecting a higher level of complexity and embodying knowledge of a great many materials and processes.
- For a country like Japan, not really enjoying the support of a powerful chemical industry, presenting a challenge on ceramics is one way of building up a considerable specialization in those materials with a promising future which can least easily be replaced by plastics. It is worth noting that the ceramic route through powder metallurgy, making it possible to produce "made-to-measure" parts, adopts the same kind of solution as plastics in order to meet demand. In a way, one of the major questions for the future of materials might well be the

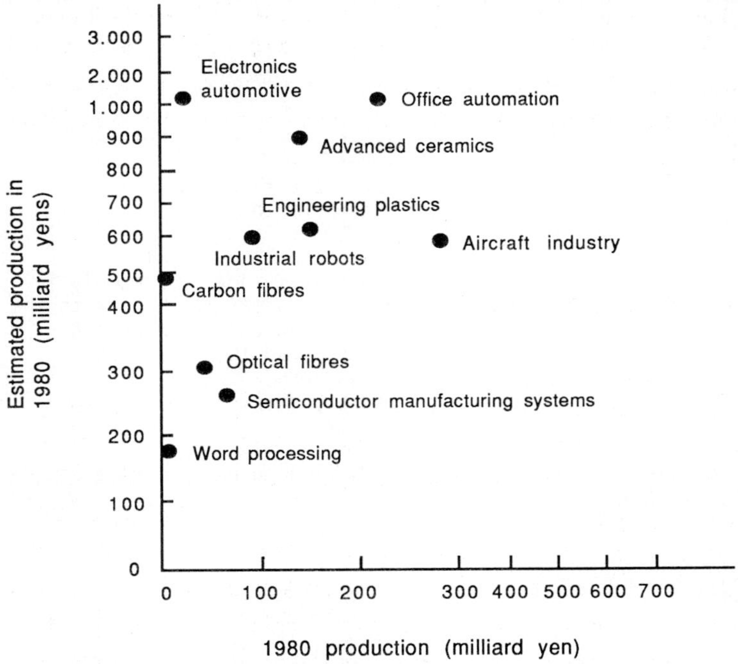

Source : Long Term Credit Bank of Japan, 1983 report (information taken from CPE Bulletin).

Table V.3.G (continued) : Predicted growth of different products between 1980 and 1990

pattern of competition in the long term between advanced plastics and advanced ceramics. One of the hypotheses that can be advanced concerning that competition is that, in an initial stage, plastics will dominate structural materials (except at high temperatures) and advanced ceramics will occupy an important position in the functional field. The following comment, however, suggests that a note of caution should be added in connection with the division between the two.

- Advanced ceramics are at present divided into structural ceramics (for structures - in engines, for example) and functional ceramics (insulating function in electronics, for example). But it is clear that with this kind of material the structural and functional aspects become increasingly hard to distinguish. Advanced ceramics are moving towards a modern materials concept which is both structural and functional. This property stems from the very high level of microscopic mastery needed in order to produce these materials. In the long run, then, they will be formidable <u>rivals for materials</u>, including plastics materials.

The reaction of the other principal countries to the challenge from Japan has been somewhat slower, especially in the case of the European countries, as can be seen from Table V.3.G.

Advanced ceramics in the United States

The authorities in the United States spend approximately 100 million dollars per annum on sustaining R & D efforts in advanced ceramics (it can be estimated that private research represents a more or less equivalent amount). The United States budget is thus more or less comparable with that of Japan, but one of the main differences lies in the fact that American research efforts are less integrated in an overall policy (including commercial policy) than is the case of Japan. One of the assets of the United States, however, is the importance of liaison between private industrial research and university research. The following instances of co-operation may be noted :

- MIT Ceramic Research Consortia, covering arrangements with Alcoa, General Electric, General Motors, IBM, ICI, Montedison, etc
- NTS Rutgers, in the field of ceramics research with Alcoa, Celanese, Corning, Dresser Inc, Ferro, IBM, etc
- Department of Energy/University of Berkeley, California, concerning advanced materials with Aerojet Tech Systems, Airesearch Casting, Alcoa, Coors, Corning, Dow, Ford Motor, Garrett Turbine Engine, General Motors, etc.

Private industrial research is carried out in three kinds of firms:

- Companies whose activity is essentially centred on ceramics (Norton, Ceradyne, Coors Porcelain, National Beryllia, Carborundum, etc). Among these companies, a firm like Ceradyne is developing rapidly, with its sales over the last few years up by some 100% per annum, with the aid in particular of substantial contracts in the field of military and automobile applications.
- Companies previously having a traditional ceramics sector which have made a "technological leap" into advanced ceramics (Corning, Owens-Illinois, General Electric, etc).

- Companies formerly lacking ceramics know-how but seeking to diversify in a growth sector : these include producers of miscellaneous materials (Sohio Engineering, Alcoa, etc), chemical firms (du Pont de Nemours, 3M, etc), end-users in the automotive sector (General Motors, Ford, etc) in electronics and electricity (IBM, Honeywell, GTE, Westinghouse, ATT, etc), in aircraft (Boeing, Mac Donnell Douglas, Hughes Aircraft, etc) or in the biomedical sector (Johnson and Johnson, etc).

Advanced ceramics in Europe

The ceramics research situation in Europe can be summed up as follows : Europe was distinctly late in reacting to the Japanese challenge. Only Germany (and most of all in structural ceramics for engines) seems able for the moment to occupy a competitive position (except in the case of certain slots where other European countries are competitive, notably the United Kingdom, which holds the Sialon patent and controls processes with a bright future such as the HIP processes, France, in ceramics for the field of optics, Italy in certain developments (4) , and it is worth mentioning outside the EEC the case of Sweden, which is the second country in Europe, after Germany, active in advanced ceramics). The EURAM programme has of course recognized the urgent need for a European reaction here, but it seems that an even greater awareness is needed before energies can be hardeness.

Many of the industrialists interviewed were sceptical about the attractions of advanced ceramics, pointing in particular to persistent technical bottlenecks (see Table V.3.H).

There is no denying their existence, but they do not appear sufficient to cast doubt on the strategic interest of ceramics. The want of a European strategy for advanced ceramics is particularly regrettable because there are centres in Europe in the field of conventional ceramics which could be ideal technological developments areas, even in the regions which are the most backward in Europe from the high-tech angle. For the metal-working firms most threatened,

---

(4) Advanced ceramics developments in Italy are studied more particularly in the report of the Instituto per la Ricerca Sociale, Nuove Materiali in Italia, R. MALAMAN, January 1987.

advanced ceramics can represent an ideal diversification as a means of recovering activities which produce substantial margins. Those which are tempted by a diversification of this kind will have to act quickly, as it is of course probable that the great European chemical firms will be tempted to venture into this field (they have actually begun to do so on a considerable scale) where chemical knowledge is a basic component (for the development of powders, notably).

The factors on which a European ceramics policy could be based are principally training policy (referred to in Chapter V.4), policy relating to norms (see section V.3.2.) and aid for the formation of a network of dynamic companies and laboratories.

One of the problems is that the web of small and medium enterprises in advanced ceramics is too under-developed in Europe, so that one of the essential links in the formation of a European network of companies is particularly weak.

The phenomenon is well illustrated in the case of France, where a mere five specialists dominate the branch : Ceraver, Desmarquest (taken over by Péchiney), Péchiney Matériaux, Société Européenne de Propulsion and Aérospatiale (highly active in composite materials and anxious to diversify into ceramics). Originally Ceraver turned towards ceramics for the nuclear field, which enabled it to acquire considerable expertise in heat treatment. Subsequently, in the early 80s, the company experienced cyclical difficulties with the decline in projects for the construction of nuclear power stations, and was taken over at that stage by Rhône-Poulenc. Here we have an example of a producer of ceramic parts (specializing in heat treatment) being absorbed by a major chemical group producing powders (special alumina, barium titanate, silicon compounds, etc).

This trend can be noted in other European countries. For instance, Ceramtech in Germany has been taken over by Hoechst, while the development of ceramics also regarded by Bayer as a strategic line. It is important to stress in this connection the especial dynamism of Germany, with companies such as Feldmühle (capitalizing 75% of the French market in ceramic tools). The group provides a huge range of engineering products in ceramics, occupying a position of virtual leader, mainly thanks to its considerable commercial know-how, the excellence of its products and the fact that it has been working in this field since 1957. It is also worth noting the dynamic strategy of a group like Rosenthal. These companies work in co-operation with Volkswagen and Mercedes, and have close links with sub-contractors such as Mahle, which manufactures pistons.

Hertel and Krupps-Widia, groups manufacturing cutting tools, have entered the field of ceramics, alongside carbide producing. It is interesting that Hertel buys black carbide ceramics from the Japanese manufacturer NTK, but produces its oxide ceramics in its factory at Ebermannstadt.

To sum up, it is safe to say that ceramics activities will develop in Europe around three major groups of industries :

- chemical industry : manufacturers of basic materials, powders;
- ceramic industry, which is mastering new forming and sintering technologies;
- user industries : electronics firms (between 40% and 50% of outlets) and motor manufacturers.

In any event, the European authorities must find means of injecting still more dynamism into the network of small and medium enterprises in advanced ceramics. The establishment of links with university laboratories, or with the research centre of the European Communities at Petten in the Netherlands, could be an important strategic element.

The production of a ceramic part generally takes place in four stages :powders, forming, sintering, finishing.

These methods of manufacture depend on the properties of ceramics. It is at the forming stage that an attempt is made to come closest to the definitive shape, because machining after the sintering stage is difficult and very costly.

* Powders

Powder quality is of vital importance. The finer the powder, the more uniform the blend of the different components. It is also a problem of porosity, joins between the grains, microscopic cracks. Fineness will be decisive in the choice of forming techniques, as will be seen later. The powder must also have an adequate degree of purity. It is of the utmost importance for the powder to be suited to the manufacturing process and desired properties of the part. It is also necessary to take into consideration the high cost and weight of the production tool to be employed for obtaining powders, and this is a considerable difficulty when it is remembered that the performances of materials depend on these original powders.

* Forming

It must first be noted that forming occurs before the material is sintered. Several techniques are possible, but depending on the process selected the characteristics of the part and production costs will be different. If there are possible techniques, other criteria will be used, such as possibility of automation, machining costs, amortization of equipment.

Forming techniques include the following : uniaxial pressing, isostatic pressing, thermoplastic injection and extrusion, slip casting.

- uniaxial pressing can easily be automated and permits the manufacture of simple parts in large quantities.
- isostatic pressing can be used for more complex forms, but machining has to be carried out before sintering, because the form of the parts is still too rough.
- thermoplastic injection and extrusion are ideal processes when highly complex forms are needed.
- slip casting is particularly suited for large parts of constant thickness.

* <u>Sintering</u>

There are considerable disadvantages connected with sintering. The cost of pressure sintering is high and cannot be contemplated for parts with complicated shapes. Hot isostatic pressing (HIP) is a technique being developed more and more at the moment; it is attractive economically and makes it possible to produce parts with a more complicated shape. It would have numerous applications. The most advanced studies here concern silicon nitrides and Sialons. It is important to note here that the Japanese and Americans have been equipping themselves extensively with HIP presses in the last few years. Strenuous efforts are being made in those countries to carry out application studies on ceramics. In the period between early 1983 and late 1985, 55 presses were installed in Japan; and by the end of 1986 90 presses will be ready for use; it is expected that ceramics will account for over one-half of the forecast occupation rate of presses. The Americans and Japanese seem to have reached a more advanced stage than the other countries.

* <u>Machining</u>

Machining is a very costly process, because the hardness of ceramics is such as to necessitate diamond tools. The need to reduce machining operations to the minimum, however, produces considerable savings.

The main difficulties in the ceramics field stem from the fact that these materials are extremely prone to defects which result from a variety of sources :

- forming is still poorly controlled and injection moulding is not completely perfected. The use of bonding agents for nitrides, carbides and oxides calls for caution.
- sintering may not be homogeneous : variation in volume when reaction sintering is effected, for example, or expansion differences.
- apparently finishing is the main source of imperfections, accounting for 50% of all flaws.

As will be seen, more knowledge is needed when it comes to quality control. At the moment there is no non-destructive test which accurately detects the size of flaws. Numerous studies need to be undertaken into the quality of powders and manufacturing processes. The problems have been identified and located, and what is needed now is a range of solutions, some of which are on the horizon.

Once the different solutions have been put in hand technically speaking,consideration will have to be given to the computerization of ceramics activities, so that ceramics production and processing can be enabled to cope with a very considerable range of applications in conditions which are economically acceptable. It is worth pointing out here that reliance on computer-assisted design is not particularly relevant, in the actual sphere of articles and their different production stages, as a result of the poor distribution of stresses within the article.

Mention must be made here of the existence of analysis by finished elements. This analysis makes it possible to calculate, with the aid of techniques derived from the decay of elements, the optimum form for the article bearing in mind production processes, so as to avert the problem of defects as far as possible. This analysis is of vital importance in the case of ceramics.

The ceramics factories of the future will probably be extensively automated. This automation, thanks to computerization, will be capable of producing a considerable array of products efficiently.

This poses very clearly, incidentally, the problem of the substitution and complementarity relationship between different materials for different technical articles.

Table V.3.H : Technological bottlenecks in ceramics.

V.3.1.2. Role of major research programmes and role of defence programmes

The major civilian or military programmes are among the factors making for the spread of sophisticated technologies, and their role in the development of materials is worth emphasizing.

A reminder was given in Chapter V.2, for instance, of the impact of space programmes on the development of composite materials (in particular carbon fibre). It is true that composite materials were conceived before the start of the first major projects, but this "available" technology had not been really tested or developed on an industrial scale at that stage. The space projects, consequently, acted as real innovation "catalysts" for composites. The process of disseminating space materials was, however, relatively important in this field because of the dynamic policy deployed by firms which has devised these materials for the space sector. Aérospatiale, operating in south west France, for instance, has not hesitated in the interests of dissemination to form small and medium enterprises in the region (including firms which for some outlets could be regarded as rivals of Aérospatiale). The result is a major contribution to the formation in that region of France of a network of SMEs specializing in composites. Thus a big firm in possession of technical know-how has promoted - at least at regional level - the dissemination of new materials towards other economic sectors.

Defence programmes frequently place considerable emphasis on materials as well. Some research is regarded as strategic for the army: in the case of structures, powder metallurgy, titanium alloys, improved performance of high-strength aluminium alloys, warm heading of steels with carbon, studies of superplasticity, epoxy resins, high-temperature ceramics and composites, notably carbon fibre. But a part of the dissemination of knowledge emanating from military projects may arise, especially in the case of structural materials, out of the improvement of control and maintenance processes (studies of degradation and protection of surfaces, studies of ageing of polymers, studies of breakdown phenomena, deformation at very high speed, toxicity and resistance to fire, non-destructive testing, etc). Among these processes, non-destructive testing is particularly important. It may concern either directly the materials themselves (ultrasonic testing, acoustic emission testing, checking depth of surface treatments by Foucault currents, stress testing by infra-red thermography, etc), or structures (measuring forms by photogrammetry, optical analysis of deformation, testing by laser holography, etc). These techniques will certainly become more and more widespread in industry, and - their mastery being particularly necessary in military projects (and in general in high-risk projects) - the problem arises how these processes will spread into the civilian sector.

For <u>functional materials</u>, military research programmes are particularly active in the field of materials for electronics, where research carried out into III-V compounds (gallium arsenide, especially) is strategic. But numerous other functional materials are concerned, for instance materials for infra-red in general or bio-sensors for detecting toxic emissions.

This brief outline of the interest of the military in the materials field is of course far from explicit, and probably represents an infinitesimal proportion of the actual work carried out in this area by all countries manufacturing arms throughout the world. The military equipment industry is an area where information is virtually unobtainable.

It is also important to emphasize a number of differences between "state" and "group of states" in facing this problem. As the United States has only one defence policy, by definition, it consequently has a critical mass of significant military research, and spin-off into the civilian sector is facilitated by this political unity. In Europe, a group of states which are independent when it comes to their defence policies, the situation is highly complex. But it must be pointed out that while France and the United Kingdom conduct their own military research programmes, which accounts for their progress or competence in certain materials (infra-red detection, high-temperature composites, etc), the FRG, which has no official programme, is nevertheless not behind in materials - far from it. This situation is incidentally similar to that of Japan. It is intimated in some quarters that the two countries in fact carry out military research via company-to-company agreements with the United States, while in others it is retorted that the savings effected on maintaining defence systems (nuclear defence in particular) can be invested in civilian programmes. Whatever the truth, or truths, of the matter, there is no disputing that the effect of military spin-off must be brought down to a scale that is more reasonable than that which some people would like to accord it, without of course dissimining it entirely. A good way of measuring this spin-off is to compare it with the spin-off from the space sector, which by and large is very close both qualitatively and quantitatively to the problems that the military are anxious to resolve, both in the concept of <u>risk</u> and that of <u>performance</u>.

In general terms, if the impact of major civilian or military programmes is to be truly positive for a country or group of countries, two pitfalls have to be avoided :

- The first is for major programmes to create a split in the economic fabric between the firms that are "protected" by the state and are self-sufficient, confident that public contracts will be renewed, on the one hand, and on the other a category of firms that are excluded from the narrow circle receiving government contracts. The risk here is that the first category will gradually cease to innovate while the second will have inadequate access to advanced technologies.
- The second is for major programmes to accept in an important technological branch the domination of one country or group of countries. An instance of this phenomenon was given in Chapter V.2, in connection with semiconductors, where early in the 70s European projects bowed to American domination so that the situation throughout the development of projects was perpetuated.

These thoughts on the role of major programmes lead, in connection with materials, to speculation about the case for developing major programmes in Europe for the production of materials in space (see V.3.1.3).

V.3.1.3 Major facilities

Industrial instrumentation

Investment in scientific apparatus is the area where states can directly intervene in the development of fundamental materials research. There are two scales, in the equipment necessary for the science of materials, generally termed major facilities and medium facilities.

Major facilities include synchotrons, high-flux neutron sources, for example, but "space" laboratories can also be classified in this category. The equipments costs for these apparatus are colossal, and beyond the investment possibilities of any one country small or medium in size. In fact only the United States and the USSR are able to finance this kind of operation on their own. Europe quickly realized this state of affairs, and the different countries joined forces to construct and utilize these giant facilities. Institut Laue-Langevin (ILL) in Grenoble (neutron flux) or European Space Agency (ESA), or the future European synchotron. (Strictly speaking neither CERN (Geneva) nor Jet (close to Oxford) can be regarded as facilities directly utilized by materials science). Their association put Europe well ahead in activities concerning neutron studies. Low- and medium-power synchotrons could be afforded by countries on their own, as in the case of LURE at Orsay (France), Daresbury (United Kingdom), Daisy in Hamburg or Bessy in Berlin (Federal Republic of Germany). The scale of investments (some 2 milliard French Francs) for the new high-power synchotron (6 GeV) necessitates association between the European countries concerned. But the project still comes up against stumbling blocks and no agreement has yet been reached on how funding is to be divided out.

In the United States, the Materials Research Society in 1985 presented a programme having absolute priority to the Federal Government for the development of four major items : a 6GeV synchotron, a stable flux neutron source, a small synchotron (1 - 2 GeV) and a high-intensity pulsed neutron source (5) . Total funding amounts to some $500 million, spread over 10 years (see Figure V.3.a). Here too it seems that the final decisions have not yet been taken, and that indecision in Europe is such that it is possible to delay matters slightly without running the risk of being overtaken, as was the case with ILL in Grenoble.

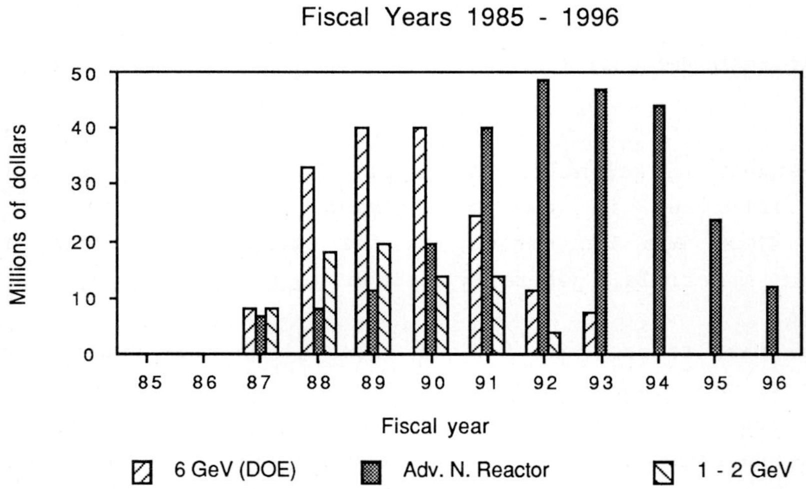

Figure V.3.a

The subject of materials and space is highly topical and deserves special attention; a more detailed study on the question appears at the end of this section.

The medium category includes all projects whose cost is not in excess of some 5 million French Francs. It concerns most spectroscopic analysis techniques (NMR, ESCA, electronic microscope, etc), materials preparation techniques (ion implantation, evaporation, furnace, etc) or calculators (mini-computers and medium computers). For many years the research policies of the different industrialized countries operated in favour of the acquisition of this range of instrumentation by research groups, generally in the form of joint services or partnerships between public and private research. The main advantage of this policy was to promote the dissemination of these highly sophisticated techniques from

(5) D.E. EASTMAN, MRS Bulletin, November - December 1984

physics to chemistry or to industry, both of which came to have greater access to specialists whose training was in physics, who contributed their knowledge and know-how. In an initial stage, the policy was amply justified. But it slowed down considerably the advance of research in a second stage, because the occupancy of the equipment (by definition highly fragmented), divided between different teams, precludes lengthy and systematic studies, and it is these that are the source of scientific progress. The Americans were fairly quick to grasp the significance of this negative aspect, principally through the example provided by the major private fundamental research laboratories. These could afford a "dedicated" apparatus which they entrusted to one specialist or one team, in this way enabled to exploit in full the area of discovery up by the new instruments. This aspect is emphasized particularly in the Pimentel Report on chemistry : "Much fundamental research today depends upon (.....) sophisticated instrumentation (.....) but whith such specialized application that shared use is quite impractical" (6) . The report suggests that approximately one-third of medium instrumentation should be dedicated. An analysis of this kind can of course be tranferred to all the advanced countries, and if Europe wants to maintain its excellent fundamental research standard, it must of course take a similar attitude. If the individual states tend to favour shared instrumentation mainly because of cost, an incentive policy on the part of the European Community should operate in favour of instrumentation in the form of dedicated apparatus. This attitude seems to be somewhat in conflict with the unifying and co-operative role that generally devolves upon the Community, which prefers to promote co-operation : the apparent conflict does not stand up to scrutiny, for the scientific reasons noted above, but also because good co-operation can be practised quite as well between teams or individuals each using a dedicated piece of apparatus, as between teams sharing the same piece of equipment - and often better.

(6) "Opportunities in chemistry", op. cit., page 305

Projects for making materials in space

Since the first experiments carried out during the joint American-Soviet Apollo-Soyuz flight in 1975, a new field for experimentation has opened up for the scientific, university and industrial community : the possibility of testing processes and manufacturing products in space.

The main contribution of the space environment is that it makes it possible to create conditions of reduced gravity, or microgravity, in a craft orbiting the earth, and such conditions modify radically the action of a number of physical phenomena. The manufacture of materials is one of the areas which could derive the greatest advantage from these new conditions, on several counts : more detailed knowledge of the fundamental properties of materials, higher quality of manufacture in some cases and the use of new fabrication processes.

Experiments carried out to date cover a very wide range of subjects, from the crystallization of semiconductors to obtaining exceedingly homogeneous glass, taking in on the way different metal alloys and the manufacture of latex spheroids. It is already possible to observe, however, that the diverse themes tackled reveal some fields of materials experimentation which could prove strategic in the new technical and economic system being set up. In each of these fields, the utilization of space could mean major technical advances, with undeniable benefits for their promoters, in terms of both commercial scope and technological independence. Evidence of this potential is provided by the greater participation by industrialists to be noted since the early 1980s, at the instigation of the United States, the Federal Republic of Germany and, more recently, Japan, whereas initially the experiments were mainly conducted by public or university research laboratories.

Nonetheless, if the attractions of microgravity are unquestionable for fundamental research, it does not seem that very many economic agents can reach a level where its commercial exploitation shows a profit, until space stations or platforms - manned or unmanned - become operational, which will taken at least 10 years. This is due principally to the technological risks and experimentation costs, both very high, which are features of the space sector.

It is for this reason, after giving certain details of a technical nature on the type of conditions prevailing in space and the products or processes that those conditions permit, that we shall be reporting on the questions which are central to the nexus of problems associated with the development of space as a research tool and means of production, so as to arrive at a better definition of what is at stake. In most industrialized countries, however, certain agreements are being concluded between firms or between firms and space agencies, statements are being made on future markets, or patents and sole rights are being acquired which lead on to programmes of experimentation in space, some of which are well under way (notwithstanding the temporary non-availability of the American space shuttle).

We shall then indicate, with the aid of certain examples, how these structurations reflect the strategies that those countries plan to develop, and what place Europe will be able to occupy on the international scene.

a).Technologies and products in space

a.1).Physical characteristics of manufacture in space

The space environment presents a number of features which radically modify the physical conditions of production : "microgravity", the term often used for defining all production activities in space, is not the only phenomenon present, even if it is perhaps the one with the most important practical consequences. Two categories of such phenomena may be distinguished :

- The first covers the natural conditions of the environment which are neither influenced nor disturbed by the presence of spacecraft. These are, principally, weightlessness (gravity is reduced between 1,000 and 10,000 times by comparison with earth's gravity), high vacuum (from 10^{-5} to 10^{-11} millibars residual atmospheric pressure, depending on orbital altitude and level of solar activity), with an infinite pumping capacity and exposure to electromagnetic and particle radiation from the sun and the cosmos in general (gamma, ultra-violet rays, nuclei and other high-energy particles). It does not appear, however, that in the short term a manufacturing programme stands to gain from this latter factor, mainly because it is unpredictable and we know too little about its interactions with manufacturing processes in use today.

- The effects classified in the second category are directly connected with the placing of spacecraft in orbit and the reduction of gravity that it affords (not weightlessness, which is an absolute ideal not likely to exist in the universe). The force of gravity is of course not entirely eliminated at a distance of several hundred kilometres. Microgravity is only the result of the equilibrium between the spacecraft's inertial energy when its engine is extinguished (which tends to send it on a free trajectory in space) and the force of gravity. In a manned craft, it is worth noting, microgravity does not remain stable. Apart from the disturbances due to the different on-board motors and instrumentation any movement within the craft may in fact provoke temporary micro-accelerations which can bring gravity up from 1.10^{-4}G to 3.10^{-2}G (7) and affect certain processes : it is one of the main elements in the argument between the advocates of manned stations and those who are in favour of fully automated production modules.

The virtual absence of gravitational effects has two consequences:

- First, it tends to eliminate the hydrostatic pressure which goes with convection and natural sedimentation, ie macroscopic movements in fluids induced by temperature gradients (heat convection) or concentration (convection in a solution). The effect of convection movements, on earth, in a liquid mixture of several substances, is to draw the densest or coldest materials downwards, and the least dense or hottest upwards. This is the main reason for the difficulty of obtaining alloys of heavy and light metals.

These fluids movements also have harmful effects in monocrystal growing processes. The irregular speed of crystallization and the heterogeneous nature of chemical compositions in the solid produce an increase in the density of structural micro-flaws; in the case of semiconductors, these flaws are often considered catastrophic.

The reduction of gravity can also make it possible to avoid flaws in the macroscopic structures which appear when, on cooling down and solidifying on earth, materials collapse and deform under their own weight.

Microgravity also makes it possible to do away with the constraint of using a container - the chemist's reactor vessel or the metallurgist's crucible. Numerous disadvantages are attached to

(7) The force of gravity at ground level is equal to 1G.

containers : chemical interaction with liquids, which is a very slight source of impurity but nevertheless latent in certain processes, the difficulty of keeping them at the same temperature as the liquids they contain, creation of crystallization cores outside the central part, which is the reverse of what is wanted. Under microgravity, it is no longer necessary to contain liquids - they can b suspensed or maintained in a state of levitation by various techniques of acoustic or electromagnetic positioning, etc, which are already mastered.

-BThe second consequences of gravity reduction is the new preponderance of physical effects which were hitherto secondary or hidden. They are essentially phenomena connected with diffusion (transfers of mass and heat tend to become purely diffusive), capillary forces (surface tension of liquids making it possible to realize larger "floating zones" or "liquid bridges") and electromagnetic interaction.

a.2).<u>Fields utilizing these new properties</u>

All these aspects, in isolation or combination, are relevant to the fabrication or production of materials. Exploration of the possibilities they offer is under way, but is still for the greater part in the sphere of public or private research. Some areas are already sufficiently mastered to permit or suggest market possibilities or future applications of a type other than the mere verification of space possibilities :

- solidification and crystal growth, in particular of alloyed semiconductors such as GaAs or metal alloys;
- fabrication of glass : absence of sedimentation and presence of a vacuum, permitting absence of contamination;
- phenomenon of diffusion : heat diffusion in alloys;
- fusion : soldering, welding metals;
- thin layer technology : polymer or metallic films;
- fluid science (better knowledge of effects of concentration, such as Marangoni effect; use of suppression of convection effects) relevant to manufacture on earth;
- manufacture of latex spheroids (perfect concentricity because of absence of gravity).

It is clear that the progress made by research to date is not the same in all the areas of experimentation on which a start has already been made, which we have just instanced. This particularly applies to work carried out on strategic materials, such as crystallization of semiconductors, knowledge of certain types of alloys, for composite materials, or manufacturing processes such as thin-layer technology or separation by thermo-migration. We shall be referring again later on to the strategic options for which states and firms in the main industrialized countries have settled.

In any event, however, the changeover to a real exploitation of microgravity, ie incorporating one or more operations carried out in space in the production process in the broad sense (from applied research to manufacture of finished products) still depends on a number of factors.

These development conditions are largely bound up with the specific nature of the space sector, but also the novel character of the undertaking and relations with research and production methods used on earth.

b).Technical and economic problems connected with the space laboratory

b.1).Specificities of space

The specificity of space is reflected essentially in terms of accessibility, ie flight expediency and cost, of course, but also in terms of risk. Even if it is possible to effect manipulations in microgravity for periods between several seconds and several minutes with the aid of shot towers, ballistic flight aircraft or probe rockets, only the American shuttle and Soviet modules at the moment are able to carry large numbers of experiments.

Europe's dependence on the American shuttle or Soviet spacecraft is seen in the procedure for selecting experiments, national priorities and waiting periods, which correspond to delays in the American shuttle programmes, but also the probable delay in construction of the space station. The costs of experiments are not confined to flight costs, with a flate rate per kilogram, but includes as well - and most of all - expenditure on infrastructure, the costs of preparation on earth or costs of miniaturization of instrumentation and weight reduction.

A two-fold concept of risks has to be added to this notion of cost as well : risks connected with the launch, on the one hand, and risks concerning the experiment itself, on the other (fragility and sensitivity of apparatus, but also in some cases actual validity of the experiment). So far this specificity has often been unfornately reflected in quality "compromise" effected on the space shuttle : most experiments have been carried out on a cohabitation basis. Sub-qualitative factors are worth mentioning : microgravity disturbed by orbit corrections or mere movements of cosmonauts, occasionally mediocre quality of vacuum, interference between certain experiments.

One of the prime consequences of all the factors just mentioned is the virtual impossibility of repeating experiments, which is one of the fundamental bases of scientific method. It is important, therefore, to stress the significance of future space stations or automatic platforms which will doubtless be able to guarantee optimum experimental conditions over long periods of time.

b.2).<u>Novelty of microgravity</u>

For most industrial sectors microgravity is a new possibility. This novelty may have various consequences.

- <u>Straightforward ignorance</u> : this lack of knowledge poses the problem of passing information on the scope and results obtained in space to potential users. In some cases this ignorance takes the form of a flat refusal to envisage the potential for development of a field regarded as too "futuristic".
- <u>Hesitation on the part of firms</u> to launch into space, since it means very high budgetary resources and also considerable risks, with the prospect of the venture paying off probably a very long-term one.
- <u>New rules with new partners</u>. Industrialists are confronted with new forms of relationships and work (relationships with space agencies and fabricators, sharing out of funding and responsibilities, procedure for access to space resources, problem of keeping experimental findings confidential, precise definition and selection of experiments, etc).

b.3).<u>Two-fold nature of relations with ground research</u>

This two-fold element is expressed both in terms of competition and in terms of complementarity.

<u>Competition</u> : Will materials produced in space still be in a position to compete, when they come to be marketed, with similar materials that have been improved during the period of space development ? This competition, which exists at the research level too,

may generate new techniques on earth, to replace those which are still programmed for space flights and consequently no longer have any justification.

<u>Complementarity</u> : this can be expressed in two ways :

- formalization of a space experiment or production often necessitates partial or total reconsideration of techniques, in the light of specific constraints (materials used, problems of volume, etc). The preparation of such experiments may thus lead to advances in knowledge of the physical phenomena involved.
- Some experiments are carried out in space merely for the sake of improving knowledge of certain products or processes, which will then be utilized on the ground.

The main features of this nascent activity of manufacturing in space are such that the conditions of its development can be apprehended in two ways :

- At the level of market potential, it is quite clear that industrialists will only commit themselves if the prospects of an adequate return are there, in other words if for a given product or process there is a commercial outlet on a sufficient scale for the "cost of space" to be recouped. The main candidates should therefore be the products or processes on which space manufacture confers a qualitative added value in excess of the inevitable extra cost (the example of latex spheroids), products with a very high price/mass ratio (like pharmaceuticals) and lastly those which have to be manufactured in space in order to attain a certain quality, and which accordingly present strategic advantages for firms or countries (the example of semiconductors, in particular GaAs).
- At experimental level, space is presented as an instrument of fundamental and targetted research which could well prove an unavoidable step towards knowledge of the intrinsic physical properties of materials close to the state of perfection or for mastering the preparation on earth of very high-performance materials.

It is important to stress that in most cases experiments - and particularly those where the ultimate objective is commercial - are in fact transpositions of techniques utilized on earth to space. It is certainly going to be necessary to build up a body of knowledge about the conditions of microgravity, and technological quantum leaps will have to be imagined. They could lead to a completely new view of scientific and technological use culminating in new products and processes of which we are perhaps unaware at present. In the main industrialized countries structures are currently being set up which will be able to exploit this research and commercial potential presented by space manufacturing.

c).National strategies and organization

On the basis of the analysis of experiments carried out in space to date, but also national claims or the purely economic of certain companies (hypothesis confirmed fairly abundantly by literature and scientists in the fields concerned), the eminently strategic space productions are those of III-V semiconductors, whose indisputable interest for future developments in photonics and high-speed electronics was demonstrated in Chapter IV.1. Little has been obtained so far in the way of results, but extremely promising forecasts for these markets, presenting a very high added value, should be confirmed as soon as the space market opens up. In 15 years' time it can be reasonably expected that there will be a $400 million market for space GaAs alone. In the field of private research, it is worth noting the interest shown in alloys and their manufacturing processes, modification of structures or knowledge of those structures.

Observation of the different agents present in each of the countries interested in microgravity makes it possible to indicate the different national organizations, although they are bringing into play the same kind of agents, national or international space agencies, space industries, present or potential user industries, the scientific community as a whole (public and university research centres, private research centres, etc), "centres of excellence" and service companies. The national organization of space research remains in the hands of governments : they are recognized as having a mission to promote research activities in certain areas and to direct a scientific community into strategic sectors or sectors defined as strategic. These state strategies are clearly linked with the economic need to control a certain technology or to strengthen predominance in one sector or another.

Strategic choices, not always clearly defined at present, can *a priori* be considered at two levels :

- the level of firms, which can opt for a diversification strategy or for maintaining and strengthening their position in one market. The mere desire to keep a "technological watch" may also be a motivating force.

- the level of nations, with the direction of experimentation in space possibly corresponding to a random (blind research) strategy, or else a strategy of centring on focal points.

c.1).<u>Outside Europe : the cases of Japan and the United States</u>

Outside Europe, two countries have exerted themselves considerably in the fields of materials research and production in space : the United States in the first place, with the setting up of the present infrastructure (space shuttle) and the medium-term outlook (space station, unmanned platform, etc), but recently also Japan, apparently anxious to catch up.

It is important to stress, in the case of <u>Japan</u>, the purposeful approach of the state and high-tech industry as a whole : Japan is so far without an operational launcher, but six Japanese technological groups including the most important industrial companies are currently participating in funding and scientific work in microgravity. Their efforts are co-ordinated by NASDA (Japanese space agency), under the authority of AST (agency for science and technology) and MITI, which is also taking an active part in accordance with the wishes of industrialists themselves. AST and MITI approved the establishment of the Japan Utilization Promotion Centre in 1986, to which space, steel, chemical, pharmaceutical and electronic industries are contributing. Its aim to encourage and promote the commercial utilization of space. The interest of the Japanese in the microgravity sector appears self-evident. The development of probe rockets (six flights in the TT 500 programme, between 1980 and 1983), future participation in space shuttle flights in co-operation with Europe and more particularly Germany (D2 Spacelab flight), on the one hand, and on the other with the United States (FMPT flight), and lastly participation in the space station, confirm a very marked advance in this area. For the moment, Japan is concentrating its preliminary efforts on semiconductor crystallization phenomena and composite materials. After a series of 18 experiments on their probe rocket programme (seven experiments in semiconductor crystallization, five in high-performance composites and alloys, one in ultra-pure glass, five in physics, including four concerning crystal formation), Japan is now proposing 22 experiments of the same nature on the FMPT flight.

The importance of this area of activity probably indicates a determination to acquire a further asset in the field of semiconductor production. Space appears to interest the whole of Japanese industry, which to date, it is true, had no relevant expertise. Efforts in this field are probably so many pledges of the strategic interest offered by this type of production.

The very pronounced determination of the United States to embark on the commercial use of space is reflected by intense efforts in the field of microgravity. In 1985 the total space budget amounted to $30 million, divided out between NASA, the universities, industrial research and centres of excellence.

The desire to associate industry as a whole more closely in the development of microgravity has led to the creation of the Space Business Center and the conclusion of different agreements (JEA, TEA, IGIA, see below) between NASA and industrialists, who are enabled in some conditions to benefit from experimentation ressources on the ground or carried on the shuttle, free of charge.

The infrastructure for the commercial use of space by NASA offers an array of possibilities that is sufficiently broad to permit the satisfaction of private demand. The technological advance of the American shuttle is such that American space research leads the way. The main advantages of this position, in the eyes of industrialists, are follow-up of experiments by regular flights and, in the short term, the setting up of unmanned platforms. The importance of the military sector in terms of investment gives industrialists access to a relatively inexpensive facility.

Because of the potential of its research and industry, not to mention its "seniority" in space, the United States is present in all the experimental spheres of microgravity. The research themes are of course too numerous to be listed : 127 objectives have for instance been listed by NASA in all fields, including 19 in electronic materials, 32 in solidification of metals, alloys and composites, 31 in fluid sciences, 12 in glass and ceramics and 13 in experimental technologies. The following among those objectives are particularly noteworthy :

- TEAs (Technical Exchange Agreements, ground experiments) between NASA and INCO, to study the influence of convection on electrodeposition, Honeywell's Electro-optical Division for the study of mercury cadmium tellurium, du Pont for the study of the catalytic properties of certain alloys, etc.

- JEAs (Joint Endeavour Agreements, flights free of charge) between NASA and MRA, for crystal growth of GaAs by electro-epitaxy, GTI for the construction of a metallurgy research furnace, Martin Marietta for studies on fluid dynamics and 3M Corporation for research into organic polymers, crystal growth and fine coatings.
- IGIAs (Industrial Guest Investigators Agreements), enabling members of industrial laboratories to be associated with NASA teams for carrying out experimental work.

It seems plain, then, that the United States and Japan have occupied positions in the microgravity field which are all the firmer since the "cost of entry" into these future markets is high. On the one side, the economic might of the United States and its superiority in space technology, and on the other the importance of the Japanese industrial groups which are involved, and their co-ordination by MITI, mean that the danger of Europe being outstripped is still greater.

Nonetheless, Europe does have distinct assets in the competition which in the long term can be expected to develop, first and foremost the pacesetting position held by the Federal Republic of Germany (and to a lesser extent by France) and the tradition of success in European co-operation, mainly through the European Space Agency programmes.

c.2).Europe

The Federal Republic of Germany, which of all the European countries is the most involved in microgravity, stands out for two principal reasons :
- first, its extensive participation in experiments of the DFVLR, the body which carries out German space policy;
- second, the creation of INTOSPACE, a private German company which however has a European purpose. This service company, the first of its kind, aims to be the interface between the space and the non-space sectors, managing for the latter the different procedures that have to be followed (technological, administrative, legal, etc) for the purpose of carrying out microgravity experiments or manufacture. Its capital is divided geographically between the Federal Republic of Germany (37.5%) and Italy (37.5%), the remainder among the other countries of Europe. Structurally, it is devoted to user industries (two-thirds) and space industries (one-third).

A further point is participation by universities, by tradition in contact with the world of industry, which is very considerable. Here it is important to mention the creation of university centres of excellence, which in the medium term should become financially independent.

The case of the Federal Republic of Germany also appears interesting because of the scale of private investments or holdings, notably on the part of industry. Concerns like MAN, KRUPP, BAYER, MTU, BATELLE have already carried out some experiments and have holdings in the capital of INTOSPACE. Its present leading role in pure physics enables the FRG to position itself in space as well. German industry, thanks to its research, is also concentrating its efforts in this area with the predominant role of the state as decision-making vector : the development and funding of the Spacelab structure, participation in European but also American launches, the TEXUS programme are so many guarantors of intense space activity, the learning phase of which is approaching its end.

German research is apparently concentrating, apart from classic experiments, on crystal growth for semiconductors and relatively specific physical experiments for which FRG is now a world leader. The Federal Republic of Germany is also very seriously interested - through essentially industrial research, it is true - in metal alloys.

In France, experiments in microgravity were for a long time carried out by the principal public research centres (CEA, CNET, etc) under the aegis of the CNES, which master-minded flight opportunities, notably with the USSR. Experiments have been conducted in this framework, particularly on III-V semiconductors.

Over the last two or three years, industrialists seem to have been committing themselves rather more, and at the moment an overall structuration of French activity is being carried out, co-ordinated by CNES, which is encouraging the creation of groupings of the GIE or GIP type covering space and user companies, the above-mentioned research centres and itself, in the field of inorganic materials. GRAME (Groupement de Recherche sur les Applications de la Microgravité aux Matériaux et à leur Elaboration - group carrying out research into the applications of microgravity to materials and their preparation) is the result of an agreement of this kind concluded between CNES and the CEA, which should however be opened to companies. The main areas of interest concern semiconductors, homogeneous and high-purity glass and the manufacture of space equipment (furnaces). A number of French companies are also members of the German INTOSPACE. Academics are also "invited" to participate and take advantage of space, and research workers are preparing certain highly specific experiments, albeit still in isolation.

As in most areas relating to materials, and in particular those of a strategic nature, it appears that the only way in which the countries of Europe can play a significant part in the future is through European co-operation. The efforts that have already been made in space, whose success is undeniable, would in this way have a chance to expand through the establishment of a coherent strategy - in two ways. Firstly, from the point of view of strictly space activity, with the complementarity between the new space transport vehicles of the Hermes type and a space station that would have long-term autonomy, and part of whose justification would lie in microgravity manufacturing and experimentation. Secondly, because the possibility of experimenting in microgravity could have implications for numerous strategic sectors which Europe cannot afford to neglect.

It appears, in any event, that any strategic decision in the field of space materials production is closely bound up with the nature of the materials development scenarios which have been described in these pages.

V.3.2. AUTHORITIES, NORMS AND STANDARDS IN MATERIALS

Any reflection concerning norms is indissociable from speculation about the changeover from old production system based on "standardization" to a new system based on "variety". Norms in fact define the "rules of play" for a system. They lay down for a component, product or article the conditions in which industrial operations are carried out in the context of the general economic environment. At the same time, however, they offer the possibility of transmitting information about industrial products to the rest of the economic system. In this way they provide a guarantee, through the overall compatibility of the economic system, that outlets will exist for the products concerned.

Because of the far-reaching changes noted in the previous chapters, it is clear that as the technological system moves away from the regime of product standardization to a new system where the scope offered by new materials widens the range of new production possibilities, the "rules of play" have to change. There must be "deregulation", so that a new system of norms can be defined. In the case of materials, the lines of thought suggested below show that it is necessary to pass from a system of static norms based on the macroscopic properties of materials to a new system of "dynamic" norms based on microscopic properties.

It is the actual viability of the variety-based system which is at stake, since the multiplication of interdependences in the industrial fabric, as variety expands, creates an increasing need for compatibility among information items circulating in the system, and particularly those concerning the nature of industrial products and articles. This point is so crucial that it produces a shift from traditional inter-firm competition towards a quest for the imposition of a standard. The standard is not just the condition for the operation of a real market for a given product - it is also the genuine norm for the spread of innovation. Only when it has been recognized and accepted can the machinery for the spread of innovation spring into motion.

Naturally, when we think of norms we have to look at the problem of the attitude of the authorities. There are public bodies at European level (CEN - Comité Européen de Normalisation - European Committee for standardization - or CENELEC - Comité Européen de normalisation dans l'électrotechnique - Europe Committee for Electrotechnical Standardization) which act as arbiters and watchdogs of the public interest. These institutions are backed up, in each European country, by national public agencies such as AFNOR - Association Française de Normalisation (French normalization association) which, after studying preliminary projects prepared by industrial branches and after requesting a public enquiry, established a recognized French norm - the NF norm - and which also acts to establish and verify the norms it lays down, by testing products in the field. These organizations all face a two-fold problem : as they depart from the former standardization system (where a gradual move towards relatively stable norms had been achieved), the problem of questioning former norms presents itself, together with the problem of defining new norms, but also the problem of the place of the authorities when new rules come to be defined.

At this stage particularly it is advisable to make a clear distinction between the unavoidable need for "de-regulation" of norms and questioning the principle of intervention, even on the part of states. Unless special care is taken, there is a risk of even greater exposure to international competition, because it is obvious that in this field the United States and Japan, the leaders, are applying a vigorous industrial policy on which authorities and industrialists have concerted their efforts. The quest for major international standards, the establishment of deterrent norms for imports or the definition of major international norms on the environment are all examples which point to the strategic role, at international level, of the problem of norms, and which have provoked the intervention of the American or Japanese authorities. The problem of de-regulation cannot be boiled down to a principle of the "less state" kind, but its raises the question of the new manner in which the state must intervene so as to guarantee that a group of countries maintains its international competitiveness.

It also raises the question of the necessary "outside the state" determination of certain norms. There are professional associations of producers or users, in fact, but very few consumer associations (in the sense of the public at large) which lay down the quality norms, without state intervention, which are necessary to the correct functioning of a sector - and which in general ensure reproductibility in the manufacture of a product, making it possible to check that use complies with laws that can be economically controlled.

For the European Community, more perhaps than for any industrialized country, it is essential to reflect on the issue of norms if European industrialists are to have a guarantee of conditions conducive to the dissemination of innovation and industrial competitiveness. To take just one example : until now office furniture was an activity involving essentially structural materials. Despite certain differing norms, the articles produced (tables, chairs, etc) could be relatively easily exported to all European countries. Now, with structures embodying more and more functions (electric plugs are being incorporated in tables, lighting strips in cupboards, electronic equipment is being made an integral part of desks, etc), the sector is liable to be faced with a multiplicity of norms, differing from one European country to another, and so to be faced with a shrinking market and reduced potential for innovation. With this simple example, which is repeated in numerous sectors, it is not only the problem of the existence of a genuine European market which is posed, but also that of the conditions for spreading innovations throughout European industry.

V.3.2.1 Specificity of norms applied to materials

Norms defined in relation to traditional materials

Traditional materials can be classified, in connection with the problem of normalization, in three main families :

- ferrous and non-ferrous metals, conventional alloys;
- materials for construction in the broad sense, including concrete, bricks, wood, glass, traditional ceramics, etc;
- miscellaneous intermediate products, including paper, glue, paints and varnishes, natural fibres, etc.

The features common to all these materials, from which the norms regulating their manufacture and use will naturally derive, are first relative technological simplicity of manufacture, involving generally a homogeneous microscopic composition, and second virtually infinite stability in time, on the human time scale (of course, the existence of certain highly aggressive environments such as sea water, for example, means that this statement has to be qualified).

In most cases these materials are utilized in a traditional environment (air - water - known chemicals) which has macroscopic consequences for the appearance and properties of materials (corrosion, wear, etc) that can easily be detected by the manufacturer or user. In most cases the physical norms will be macroscopic norms.

The norms for chemical composition and heat and mechanical treatment during manufacture are designed to "help" the manufacturer. Norms for physical properties (ductility, resistivity, impact-resistance, malleability, transparency, fluidity, viscosity, density, etc) benefit the user in the manufacture of products (aircraft, motor cars, buildings, machine tools, clothes, etc) which are reliable and can be reproduced. These two kinds of norms, together with certain norms relating to utilization, are also intended to protect the final consumer, both for his personal safety and for environmental protection. Lastly, observance of all these norms permits normal and regular commercial operations, thanks to the establishment of straightforward controls.

It is already possible to glimpse, through this brief account of the utility of norms applied to materials, certain deviations to which a system of this kind is liable : chemical impurities not having any influence on properties but disastrous for the environment at the point of utilization, manufacturing norm imposed during production, also causing considerable environmental damage, concealed protectionism, refusal to sell, etc.

Are these traditional norms applicable to new materials ?

The principal characteristics of new materials, contrast with traditional materials (be they materials in the electronic field, silicon, etc, composites, super-alloys or ceramics) is the fact that they necessitate an extremely complex manufacturing technology, particularly when it comes to the reproducible nature of the edifice on an atomic scale; incidentally, some so-called traditional materials which incorporate in the course of their production these new reproducibility technologies becomes new materials (example : certain special steels or cast irons, ceramics, etc). Mastery of this microscopic structure confers on these new materials most of the properties for which they have been prepared (example : single crystals of Al-Ni alloys or silicon; doping of semiconductors; cross-linking of polymers, etc). If the traditional physical properties which serve to measure traditional materials still apply when new materials take the place of old (density, hardness, tensile and impact strength, etc), the new properties due to microscopic mastery will result in the elaboration of new norms which will measure their limits and reproducibility (shape memory, heat resistance, purity grade, etc). What is more, the environments in which these materials are going to evolve may be highly aggressive towards this microscopic structure, the secret of their success. Apart from the constraints on materials exposed to radio-activity and neutron flux, which apply to both traditional and new materials, and which have necessitated a whole range of new norms, sunlight (principally UV) which considerably degrades certain plastics or other polymers by depolymerizing them, gases in general, oxygen in the air in particular, which diffuse through the mass of materials causing chemical migrations, phase changes, premature doping, will all necessitate the establishment of new and often unknown norms. On closer inspection, it is highly probable that, when these norms come to be defined, they will apply to traditional materials too, the norm in that case producing an effect that goes beyond its original attributions by modifying the manufacture of the actual products.

The time variable

There is one point, however, which seems to be confined to the field of new materials, ie the taking into account of the time variable, because many of these new materials "grow old without working". (We are not referring in this paragraph to ageing due to work, generally referred to as wear). Needless to say, traditional materials become corroded, but both in construction (buildings, public works, bridges, roadways, etc) and in the manufacture of articles (motor cars, trains, household appliances, machine tools, etc) most of these materials age less fast than the product (not to say the user himself). On the other hand, new materials under environmental stress (see above) or structural stress (depolymerization due to adjuvants, to monomers caught up in the mass, to catalyst residues; demixing or unsticking of composites because of differing dilatation coefficients, etc) age far faster than the object they make up. This phenomenon has shown its full effect in the aircraft industry, which has played and continues to play (with the space industry) a key role in the establishment of norms which take this time dimension into account. A new dimension is thus added to the concept of a norm : the boom in new materials, coupled with the boom in new products, whose construction is possible thanks to these new materials (example : the fighter aircraft transparent to radar microwaves), will lead to the establishment of evolutive norms as opposed to static norms. A material will no longer have to have rigidity x, yield stress y and dilatation coefficient z when it leaves the workshop with consistency x, y and z throughout the period of its utilization, but rigidity x1 at time t1 (ditto for y and z), x2 after 1,000 hours, x3 after 5,000 hours and so on, with the proviso that x1, x2 and x3 - and so on to xn - will have to remain at least equal to a limit x min guaranteeing the safety and reliability of the object constructed. The consequences of the establishment of such norms are considerable and numerous : elaboration and measuring of norms, integration of the time variable in a product made up of different materials (necessity of introducing detachable parts), application of a servicing schedule in the light of the age of the product (since their inception airlines have practised this new form of work), and so on.

V.3.2.2 Establishment of norms

How are norms to be established, technically, for new materials ?

The adequate homogeneousness that is easily obtained in synthesis of traditional materials means that in most cases conformity with the norms laid down can be achieved on the chemical and physical analysis of a sample, in general an excellent statistical representative of a whole production batch (of cast iron, concrete, glass, etc). These analyses are simplified technically because they can be destructive.

The general rule concerning new materials which demand perfect microscopic homogeneousness is that sampling is no longer statistically reliable. Checking electronic components for specification is sometimes the most costly stage in the manufacturing chain. Each part coming out of the workshop has to be tested and verified; of course, it goes without saying that any destructive method of analysis has to be ruled out. This has produced, and still produces, analyses of very great complexity (mainly spectroscopic analyses) and the development of a branch of science which is developing in parallel with developments in new materials, namely metrology. A second difficulty in analysis stems from the microscopic origin of the properties for which a norm is wanted. The torsional strength of a conventional steel is obtained, in general, from macroscopic elements such as chemical composition, annealing temperature or quenching speed, and the corresponding norm is generally measured only on the basis of the thickness of the part; the torsional strength of a single-crystal Ni-Al/jet engine blade will depend on the number of defects in the atomic structure and the direction of the shear planes, and it can only be measured on the basis of highly sophisticated spectroscopic analyses.

Apart from these difficulties stemming from the microscopic origin of the desired properties of new materials, the establishment of norms for new and unkwon properties poses essentially new problems demanding a solution. Furthermore, the rapid development of technologies leaves technical bodies little time in which to draw up preliminary proposals which in turn are liable to become obsolete before they have been sent on to the decision-making bodies. Lastly, the problem of ageing and the establishment of evolutive norms as defined in the previous paragraph, apart from the intrinsic difficulties of integrating the time variable, comes up against the limited possibility of speeding up the test and so reproducing in a short space of time (which is both economically reasonable and technically significant) the "damage" produced by real time.

Who establishes these norms - and who should establish them?

It seems logical that the power to lay down different norms for new materials which are relevant to the fields of health, safety, the environment or even the utilization of products resulting from these materials or their commercial use, should be placed in the hands of states or groupings of states (EEC, Council of Europe, GATT, the UN, etc) through public procedures. If this governmental role in theory seems natural in our democracies, it is by no means self-evident that our states have the technical means of laying down these norms (example: use of catalytic converters). It may be too that they exceed their powers under pressure from lobbies, by laying down norms with a protectionist aim (the same example of catalytic converters, plastics in the food industry, etc). It frequently happens that professional groupings, bringing together manufacturers, users and sometimes consumers (with the state generally relegated to the role of host), with a more corporative spirit (experts!) will find norms that are technically reliable and acceptable in the sense of free trade (example : catalytic converters, acid rain) which governments - taking a more general and hence more political view - would have difficulty in establishing.

When it comes to the technical norms necessary for the correct functioning of the materials industry, it seems logical in a liberal economy to allow a dialogue to grow up between manufacturers and users which will make it possible to work out norms. In the course of our investigations, however, we met different groups of manufacturers or users who are calling for intervention on the part of the state or multinational organizations when the imbalance between partners is too great, with the risk of a monopoly (a giant chemical group which imposes its sole quality norms on a multitude of small users which need more technically feasible specifications, consortium of users or large-scale systems users imposing extremely rigorous norms on small manufacturers of the SME type) or when competition is very keen within a profession, and has produced, for the sake of personalizing each firm's product, a multitude of incompatible utilization norms (mostly in electronic applications) or manufacturing norms making it possible to rationalize processes further down the chain (polymer processing industry, for instance).

A study of the system of establishing norms in the United States might be of value here, as the concern for technical efficiency which is highly characteristic of the United States, coupled with a "democratic" economic philosophy (anti-trust legislation, very powerful federal organisms : Food and Drugs, National Bureau of Standards), have led to a procedure for establishing norms which is probably exemplary. **It is important always to remember, however, that the actual concept of a norm includes a concept of limit (upper or lower), and that this concept should not be expected to define perfection, but rather to set limits on the acceptable.**

V.3.2.3 Norms and networks : role of the authorities

Over and beyond the actual problem of determining technical norms for materials, there is the critical question of the role of states in the formation of the full-scale information circulation networks which forms the backbone of the future technological system (of which technical norms are only one element). This network of information presupposes the establishment of infrastructures (ISDN networks, local networks, satellites, etc), data processing modes and data storage nodes. Here too competition is extremely keen, and the precise place of governments still has to be defined. In the matter of materials development, there is also a pressing need to define the role of the authorities.

Establishment of information networks and the role of states

The establishment of an integrated information network on a world-wide scale is one of the major problems facing society as the century draws to its close. These different "value added" (8) networks perform different functions : carrying simple data (primary telecommunications), communication between different systems and materials, information through access to on-line data bases, and processing by access to in-network software. It is possible to analyze the different networks in accordance with a technico-economic typology that is particularly useful because it makes it possible to differentiate the attitudes of the principal agents (9) .

- "The packaging switching networks" which "telecommunicants" are best able to supply in order to satisfy regional demand (local networks of ex-subsidiaries of Bell in the United States, for instance, national demand (Transpac in France) or international demand (GTE-Telenet).
- Horizontal VANs (value-added netwroks), which include communication, information and processing for users, regardless of their location, size and activity. The best placed to supply them are computer manufacturers, such as IBM, whose position is already dominant.
- Vertical VANs, whose users belong to the same activity sector (bank networks for the electronic transport of funds, for example). Their design and implementation are thus generally demanded by their users. For instance, the bank association in Great Britain has announced that a national system is to be set up for the electronic transfer of funds, which puts a question mark over an agreement previously signed between IBM and British Telecom for provision of the same service, whereby British Telecom is relegated to the role of sub-contractor.
- Services providing general information (stock market, scientific, holiday industry, etc) from very large data banks - the case of

---

(8) The following definitions were supplied by P. AUDOIN, on the occasion of the Transpac Forum on value added networks held in Paris, 25 - 26 November 1986.

(9) The following passage is an extract from "Télécommunications, innovations et produits nouveaux" (telecommunications, innovations and new products), A Jour Editeur, n°5, 15 December 1986; the issues features a considerable proportion of the conclusions of the statement by P. AUDOIN (see previous footnote).

teletext, in which the holder of the information will generally occupy a dominant position (eg Reuters).

For all these networks, the over-riding trend will be integration: integration in single structures providing the whole range of VAN services, integration of regional, national and international markets through the clientèle of major multinational firms, and integration of suppliers either through takeovers (IBM) or the formation of consortia. Entry thresholds are going to rise very rapidly and the early arrivals will be in the most favourable position.

The establishment of these networks necessitates inevitably a profound deregulation movement. It is clear, in fact, that in some countries the VAN market has been hampered to date by public telecommunications monopolies. A world-wide tendency can be discerned, consequently, towards liberalization of the market. But this does not mean that the egal environment will be the same everywhere. Numerous adjustements will have to be made in the light of local situations.

In countries where there is a vast public data transmission network that is reliable, charging attractive rates, a substantial diversification of supply will be possible (see teletext in France). On the other hand, in countries without an infrastructure of this kind, those supplying the service will have to construct their own network, or to rely on another company acting as carrier, which may however also be its rival. Depending on cases, pricing and strategy on the part of suppliers cannot be identical.

But VANs also suffer from an inextricable diversity of norms and standards both from the technical and from the organizational viewpoint. They are currently putting their house in order. It will be important to avoid "private norms" imposed by suppliers in a dominant position, as exemplified by the unhappy situation of informatics. But nothing is less certain, since there are excellent reasons for thinking that IBM is going to carry through its own strategy. For suppliers, the only possible means of parrying this move would be to turn to informatics companies which are independent of manufacturers, either by sub-contracting the construction of their network or by absorbing them, as in the case of General Motors.

The lack of standards is of course a major cause of uncertainty, prompting the different users to defer their choice. This is the technological risk inherent in the variety-based system, and it can clearly measured in the field of computer software. The main world producers (10) have recently (January 1987) co-ordinated their position in agreeing on a common standard, POSIX (portable operating system for computer environments), which will make it easier to utilize software written for one manufacturer's computer on the machine of another manufacturer. Is is noteworthy that the choice of POSIX has been reached as much as a result of consumer pressure as because of the fear that IBM would impose standards on its own (it is also, at the same time, a relative failure for ATT's UNIX standards).

In these conditions, it is believed that one of the roles that the authorities can play is that of "guarantors" of the communicability of the overall system. This role can consist, among other things, of helping to build up the great "data motorways" (ISDN) and ensuring, in conjunction with industrialists in the country concerned, that failures to keep pace due to the quest for standards do not irremadiably favour international competition. Apart from these roles, other problems are posed, notably who is to manage the information networks.

Importance of networks for materials

The development of networks holds a twofold interest for materials. On the one hand, the role of materials is considerable in the construction of networks (optical fibres, connectors, semiconductors, etc), as was seen in Part IV, and on the other the development of materials data banks is bound to expand and to play an increasingly strategic role. This phenomenon naturally pre-supposes a very broad strategic coherence among the countries of Europe. Increasing importance must in fact be attached to the utilization of scientific and technological information for the sake of improving the manufacture and processing of materials, as perceptible in computer-assisted materials design (CAMD) (11) .

---

(10) IBM, ATT, Digital Equipment (USA), ICL (UK), BULL (France), Olivetti (Italy), Philips (Netherlands), Nixdorf and Siemens (FRG).

(11) The comment on CAMD comes from P.F. GOBIN.

Most technological and scientific information is at present spread over a very large number of bibliographical data bases.

These bibliographical systems, be they specialist or multi-disciplinary, are not however in a position to supply statistical or factual information on a precise phenomenon (flow behaviour or machining conditions, for instance).

It is extremely important for all industrialists to have data banks on the technical properties of materials, and this need is still more acute for small-scale concerns which do not have their own banks or specialist documentation services.

The Americans are currently building up a National Materials Properties Data Network, with the aid of the NBS. Their ambition is to carry out an operation comparable to the Chemical Abstracts operation. Still more than in the case of bibliographical data, an operation of this kind in the field of actual data banks could make Europe "technologically dependent on the United States". Products whose properties were not included in such a bank would in fact be excluded in the long term from the market.

It seems that the Japanese have a considerable number of banks and are beginning to study their structuration. A European endeavour is under way in this area, and a considerable development of such digital data banks is expected in the next few years. In a second stage, their combination with suitable software should permit access to libraries of "behaviour laws" (in the broad sense, response to a physical, chemical and engineering need). The formation of European data banks is under way in particular at the EEC research centre at Petten in the Netherlands.

Lastly, their association with expert systems will incorporate the experience for processing, assembling, protection, etc operations and will permit logical choices to be made against a complex background.

These highly developed data banks will then enable a decisive step forward to be made, as the 21st century is glimpsed on the horizon, with the appearance of a new activity : CAMD (computer-assisted materials design).

It will be possible, on the basis of the definition of a function (transmission of electrical, photonic, optical energy, support of a structure, etc) performed in a given environment (applied stress temperature, radiation, etc) in conditions of imposed limits (dimensional stability, optical, acoustic absorption, conditions of incorporation in an assembly, reliability, cost, etc) to identify in a preliminary stage the materials already existing which constitute an approximate solution to the problem posed. In a subsequent stage, the system will be able to suggest the development of a known material or the devising of a new material, by altering its composition (theoretical forecasting of phase diagrams), its processing conditions (programmed by behavioural laws of a semi-product at a given temperature, for a given deformation, etc), potential assembly, price and so on.

A system of this kind, which will pay off in the long run since it will avoid costly testing by making it possible to focus efforts on highly probable solutions, necessitates total command of the complex chain of a material's production.

The undertaking is thus unrealistic as a whole today, but over the next few years blueprints centred on very special materials, with homogeneous behaviour patterns, should make their appearance.

V.3.2.4. Materials, regulations and environment

The need for the authorities to define regulations relating to the environment is unquestioned. It is necessary, however, to make a distinction between two kinds of action : on the one hand, it is important to ensure that at no stage of their lives (production, application, consumption) shall materials create environmental problems that are intolerable for society; on the other hand, the development of materials designed for use in resolving or attenuating environmental problems (development of catalytic converters or biosensors, for example) needs to be encouraged.

Environmental norms applied to materials

Environmental problems are liable to be encountered at any stage in the life of a material. Without referring to the actual winning of commodities, the production of certain materials may engender health problems or environmental damage. The risks of pollution by solvents during a curing cycle, the health risks that can be caused by the preparation of certain powders (the manufacture of certain ceramics or composites can cause particle dispersion, and although the carcinogenic effects such particles on humans are not yet clearly established, (it has been shown that certain ultra-short fibres utilized in the manufacture of advanced ceramics have caused lung tumours in animals), the risks presented through inhaling fumes in units producing reinforced plastics are all so many examples underscoring the risks inherent in the production of certain materials. These risks must prompt all the more vigilance when it is remembered that the materials in question are frequently "new" and that knowledge of their real effects is imperfect.

The phase of utilization of materials is also sometimes the source of health or environmental problems. The use of asbestos in buildings is perhaps one of the most preparative examples of the utilizations limits that must be set for materials when it is demonstrated that they have harmful effects on health. There is another source of massive pollution, whose consequences have been inadequately assessed but which are probably very serious : tarred surfaces (roadways, car parks, airports, roofing, etc), which are expanding very rapidly. Tar contains a broad range of products (asphaltenes, aromatic nitrogen derivatives, heavy metals, etc), all potentially dangerous, tar being in fact the garbage-can of the petrochemicals industry. Evaporation, air pollution and the action of rainwater, water pollution, all effect considerable quantities and are even beginning to affect the untouched soils of developing countries. These tars, coupled with gravel, have structural properties which are suitable for motor traffic and are thus difficult to replace (concrete consumes too much energy). An attractive solution would be to enclose the tar and gravel in a kind of high-strength polymer pouch, whose surface properties could be controlled - anti-skid, anti-aqua-planing, anti-freeze by introducing heat resistances at critical points - and which at the same time could incorporate coloured road markings-central and roadside lines - conspicuous colours at the approach to danger areas. The introduction of a process of this kind in Europe might lead to a significant up-turn for the major polymer producers and public works enterprises, and at the same time find favour with oil producers, who have the problem of eliminating residual tar to best advantage. The innocuous nature of the solution could be put to good use with the addition of solid residues to tars and gravels (dust after reprocessing) from chemical industry residue incinerator plants.

A large number of examples were also encountered, in studying the applications of structural materials (Part III), of intervention on the part of the authorities to regulate the production of materials deemed harmful for health (example of PVC in bottles in the United States). Here it seems advisable that a harmonized regulatory policy should gradually appear in Europe on materials and their effect on health. Otherwise barriers will once more be placed in the way of the spread of innovations and expansion of markets, with at the same time the risk of a sense of incomprehension growing up among Europeans, who will suspect that they are not subject to the same system as their neighbours.

Lastly, there are still the problems of recovery and recycling, which are liable, among other things, to create problems of pollution or deterioration of the environment. We endorse fully here the findings of the French materials task force, which reported that new materials are increasingly complex (composites, multi-layer) containing less and less noble, rare and costly raw materials, and are more and more difficult to recover in good conditions, that is to say without disastrous deterioration of materials (the final stage in the process being incineration, which technologically speaking is a desperate solution). Furthermore, these materials are far from being neutral vis-à-vis the environment, because often they do not decompose or are harmful in the long run, which makes it essential that they shall be eliminated and neutralized.

Fibre-based composites cannot therefore be discarded after use, as they are, with impurity. But the technologies enabling the matrix and fibre to be properly re-utilized do not exist today, and for a long time to come the sole economically viable solution seems to be for them to be burned.

Materials from scrapped cars are less and less attractive for recovery purposes, and more and more difficult to separate. In 1950 cars contained substantial quantities of copper, lead and chromium; this is no longer so, and in 10 years' time plastics, whether reinforced or not, will have taken the place of a considerable proportion of ferrous metals. Washing machines and dishwashers still contain stainless steel - but for how much longer ?

The pattern suggested above compels us to reflect on the future of the recovery branch, which is several years late (between 10 and 15) in anticipating the problems; it would do well to open its eyes to the difficulties now. In the interests of survival, ways will have to be found of processing industrially ever increasing tonnages, through costly processes which do not exist today. These harsh constraints will in this way necessitate powerful, high-performance industries which will have to have substantial budgetary and research resources.

It would be advisable here for private or state research laboratories to take part, but we come up against an obstacle at the moment which is not only psychological : the study of problems of this kind is regarded by research workers as beneath their dignity, and to devote themselves to them would represent a standstill in their careers. Thincking here will have to evolve.

It would also be helpful if designers could realize the problems they create : planning for recycling, from the drawingboard stage wherever possible, is therefore desirable. On the other hand, for goods of an ephemeral nature (disposable packaging, for instance), this concern could be more present and indeed spelt out and codified by the authorities in some cases.

Lastly, research and development work for new recovery processes (in particular for household refuse and ordinary industrial waste) is clearly of international interest. Europe-wide co-ordination on this subject could be strengthened, so as to rationalize efforts and cut costs.

Development of materials to resolve environmental problems

The development of materials in order to resolve environmental problems is perhaps one of the most urgently needed policies that the European authorities could establish. The built-up of veritable environmental perils (acid rain, river pollution, air pollution in Europe's great cities, etc) could well provoke the general public more and more, and if allowed to become worse could be the number one problem in the latter part of this century. Materials can help provide solutions. As a rule solutions presuppose the integration of the development of a material as part of a systems approach (an approach in terms of function). For this reason, the materials concerned are generally complex materials, or materials incorporated in complex systems (catalytic converters for motor cars are an excellent example for this latter category, see Chapter IV.2).

Emphasis was laid in Chapter IV.3, among other things, on a group of functional materials which for many reasons could be regarded as strategic : <u>biosensors</u>. In the case of the environment, biosensors can if developed play a significant role by warning local authorities <u>in real time</u> of developing environmental problems. The difficulty with biosensors is that there is no really guaranteed market (apart from the army) unless the authorities get together. Because of their advantages (and also apart from the technological potential they present), these materials certainly merit an effort on a European scale.

The growing safety requirement

Thinking on the growing safety requirement has to be brought into line with the trend emphasized above in connection with environment control. Recent disasters have opened the eyes of the public at large to the general problem of the reliability of major structures (dams, buildings, jumbo jets, trains, etc). The community is less and less inclined to regard accidents of this kind as inevitable, and as the idea of the "right to health" develops (refusal to accept mistakes, reflected in the growing number of lawsuits against doctors and hospitals), so we discern the ascent of a "right to reliability" in the technological universe.

The evolution of technologies and knowledge suggests a response to this expectation : continuous non-destructive testing (12)

Active research is currently being carried out into forecasting the lifespan on the basis of models of behaviour in which account is taken of the concept of "damage sustained". It is essential in fact, for vital structural parts, to have at least a rough idea of their lifespan so as to arrange for maintenance. This is the objective behind current laboratory tests which stimulate actual utilization conditions as realistically as possible.

The development of micro-electronic technologies is going to make it possible to envisage the cheap production, and accordingly large-scale production, of various sensors. We shall have sensors for displacement, effort, vibration, acoustic emission, etc. On the basis of the information so obtained it is reasonable to extra-polate what we see slowly taking place before our eyes : low-cost sensors will be placed at the construction stage at sensitive points inside structures,

---

(12) The comments on continuous non-destructive testing originated with the Prospective 2005 research group, CNRS - CGP.

and they will communicate, either on demand or on a continuous basis, according to the degree of risk (the risk is not the same in the case of a motor car, an aeroplane or a nuclear power station), local information on the behaviour of the material. The information will be incorporated by means of software into a behavioural model making it possible to indicate continuously what useful life still remains, and thus to operate - if the model is correct - in perfect safety. This will really mark the arrival of continuous non-destructive testing.

CHAPTER V.4

Materials, Employment and Skills

A statistical study of the way the number of jobs is evolving in Europe in the field of materials is not really conceivable, for a multitude of reasons suggested below. On the other hand, and because materials determine a considerable proportion of the organization of industrial work, a thorough analysis of the evolution of skills in the field of materials, in the light of work carried out to date may be extremely fruitful. Thanks to a new "visual angle", it affords an opportunity to comprehend better the pattern of skills and the mechanisms whereby jobs are destroyed and created, and so to pinpoint the blocks and difficulties in changing over to a new mode of industrial organization. In the background of the activities reported on in earlier parts of this work, the importance of knowledge and know-how have become manifestly apparent. The following analysis confirms that skills are indeed the key to the possible transition to a new production system.

V.4.1. MATERIALS AND EMPLOYMENT

Materials constitute a "transverse" technology, since they are more or less closely involved in all economic activities. True, there are sectors which produce materials (steel, chemical, metal-working, etc), but very numerous jobs are allotted to the manufacture, processing and utilization of materials in the European Community. A rough and ready examination (see Table V.4.A) of just the materials production and processing sectors does, however, enable us to put the number of jobs in this sector in the EEC in 1981 at 7 million (compared with almost 9 million in all the other sectors). Within these latter sectors, the definition of jobs connected with materials working is virtually impossible. It is for this reason that the precise statistical observation of the number of jobs concerned by materials in the European Community cannot be contemplated. A rough and ready examination (see Table V.4.A) of just the materials production and processing sectors does, however, enable us to put the number of jobs in this sector in the EEC in 1981 at 7 million (compared with almost 9 million in 1974). It is probable that the overall number of jobs will continue to fall away up until the end of the 80s, but in the absence of precise statistics these figures must be treated with the utmost caution. In fact the figures merely reproduce, in proportion, the drop in the number of jobs observed for the whole of the active population throughout the secondary sector.

Country Year	FRG	France	Italy	Nether-lands	Bel-gium	Luxem-bourg	UK	Ire-land	Den-mark	TOTAL
1974	2770	1711	1269	382	374	34	2132	30	101	8803
1981	2378	1445	1057	324	275	30	1487	52	85	7133

Source : BETA.

Table V.4.A : Estimated number of persons (thousands) employed in the production and processing of materials in Europe.

Any attempt (even if it were statistically possible) to make a distinction between "new materials" and "old materials" (in accordance with the definition rejected in Part I) would doubtless lead to an impasse, or in any case to a very restrictive and distorted view of industrial reality. It would be found, in fact, that "new materials" (composites, ceramics, etc) accounted for only a very small proportion of industrial employment (certainly no more than 5%) and that the jobs it was hoped to create in these sectors did not really offset the substantial number of jobs lost in "old materials" (particularly steel).

This view, which suggests a simple replacement mechanism between old and new materials, says nothing about the importance of the mechanisms whereby production tasks are redesigned, the dynamics of a new pattern of skills or the central role of know-how. Even assuming that the statistics were really available, in fact the "balance" between the jobs created with "new" materials and those eliminated with "old" materials can only present an advantage if the reference measurement is <u>stable</u>, that is to say if the <u>whole production structure remains unchanged</u>. If the production structure evolves at the same time as materials evolve (which has been demonstrated throughout this work), then an <u>argument in terms of jobs makes no sense</u>. <u>What becomes fundamental, however, is the way in which the manner of working changes</u>. And this means that we must direct our gaze first and foremost to the evolution of skills.

V.4.2 EVOLUTION OF MATERIALS AND EVOLUTION OF SKILLS

<u>V.4.2.1 Mechanisms whereby jobs are eliminated and skills transformed</u>

The changes affecting materials lie behind far-reaching movements whereby jobs are eliminated and skills evolve. The mechanisms at work can be broken down into two separate categories : the first concerns the conventional principle of replacing labour by machinery, generally automated machinery. The second is based on a mechanism operating actually "inside" production processes : through the phenomena of process redesign and materials manufacture - particular mention may be made of integration processes - which concern structural materials especially.

The first suggests the phenomena - which have been studied elsewhere in detail - of the introduction of information technologies into production. The second, in contrast, is more specific to the evolution of materials. But if we are really to understand the movements under way and their repercussions on skills, we must take into account the two types of mechanisms described below at the same time.

"Traditional" processes whereby jobs are eliminated

The most "traditional" process whereby jobs are eliminated is the replacement of manpower by machinery. Naturally, this process is at work in the materials sector - from their production to their processing - as it is in any other sector of the industrial system. Examples abound wherever work on materials can easily be replaced by machinery (digitally controlled machine tools, robots, etc) which can reproduce, by automated means, the movements involved in human work. This replacement has gone particularly far in repetitive, arduous and dangerous situations. Among the operations relating to materials, those of welding, machining, painting, handling, etc, best lend themselves to automation. Although extensive and real, this replacement mechanism is not peculiar - as has just been pointed out - to materials : it can be observed (and is very widely analyzed) in all industrial activities, and for this reason we shall not dwell on it. (It is to be noted that the changeover to automation, however, can have an impact on the choice of materials : for instance, the transition to automated laser machining is a definite handicap for the choice of aluminium, which "reflects" light). In the case of materials, however, an effect is found which is worth recalling : as manufacturing tasks become automated, so human work is relegated to the fringes of manufacturing, in the direction of supervision, preventive maintenance and verification, etc. If, at the level of actual manufacture, the work needed is whittled down and becomes <u>more abstract</u>, ancillary tasks are reinforced : new kinds of know-how are harnessed; they can generate needs for new skills (automation and robotization experts, etc). In any event, the organization of work must be looked at anew : the evolution of tasks requires at least a new sharing of tasks, and at best a redefinition of the roles of each, including those in command.

- Another mechanism which eliminates jobs and transforms skills is the entire replacement of one technological solution by another. The most representative example here is doubtless that of the replacement in electronics of vaccum tubes by semiconductors. In this specific case, a whole technological area was suddenly uprooted by another, and in the replacement process a whole body of skills rapidly became obsolete (it can be remarked that in this case it was firms other than the producers of vacuum tubes which made the first semiconductors). This case of sudden substitution is extremely rare in materials. The very great majority of changes affecting materials take place slowly, through gradual processes which take place inside existing firms, notably in downstream industries. This in particular is why integration mechanisms, which correspond precisely to this kind of evolution, need to be studied in greater detail.

Effects of integration processes

The other processes which eliminate jobs and transform skills are integration phenomena considered in Chapter II.5, a feature of which is a partial or complete reconception of production. The effects on qualifications and skills depend on the precise nature of the integration process.

- In the case of processes whereby production phases are integrated, the partial reconception of production is accompanied by a reduction in production steps and hence elimination of jobs. Generally, however, these presuppose the conservation of skills. The integration of production phases so corresponds, in some ways, to the conventional logic of the gradual acquisition of a technological solution. The improving of the production process as a general rule concerns a single type of material, in fact, whose design are has been bettered. Recourse to high-precision casting, for example, may make it possible - among other things - to avoid numerous steps in the welding or refining of a given metal. But the utilization of that metal is not as a rule affected. It is to be noted, however, that in the simplification of the operational management resulting from such a process, the elimination of the welders' or refiners' jobs is not the only effect. Besides savings in freight and handling, there are savings on intermediate parts or particular treatments which have become pointless (additives, materials protection, etc), which in some cases means eliminating sub-contracting and associated jobs.

- In the processes whereby parts are integrated, the reconception of production is generally far less partial than in the case of integration of phases. In most cases, parts integration processes provide an opportunity, through a new solution, for introducing a new material into the fabrication of an industrial product. In these cases, then, we observe not only a trend towards the elimination of jobs but also a gradual trend towards a "shift" in skills. Old qualifications tend to disappear (both in the company and among sub-contractors), at the same time as an old form of work organization comes under fire. The creation of new trades becomes necessary, leading to the need for new forms of relations between skills. What is more, these new solutions frequently entail recourse to new machinery designers and manufacturers. In the automotive sector, for instance, the redesigning of bodywork parts based on plastics or composites (with a reduction, by comparison with the previous solution based on metal, in the number of parts) inevitably spells the end of intermediate tasks (assembling, welding, etc). This latter phenomenon presents firms with substantial problems concerning the management of skills. It is the occasion for a profound renewal of learning curves in the company, but also for a renewal of relations with sub-contractors and machinery manufacturers. The greater technical complexity of the solutions employed necessitates recourse to the forms of "partnership" referred to earlier. The forms of knowledge which appear to be more and more necessary within these forms of partnership include chemistry, since the new technical solutions made necessary are still based on an increasingly fine mastery of the microscopic structure of materials, which looks to chemistry at all stages of production : design, production and control of materials.

V.4.2.2 Reconception of know-how

In all, the mechanisms which eliminate jobs and skills relate to work on materials (milling, drilling, trimming, refining, painting, etc) and transfer operations which consist of connecting, physically, in the old organization, the various stages of production together (for instance, handling).

This tendency for the direct relation in production between man and material to become gradually more remote (corresponding to a kind of "dematerialization" phenomenon in production) is not only encountered in the utilization of structural materials. In the production of certain functional materials, it is progress in the mastery of the microscopic structure of the actual materials which necessitates recourse to automation and abandoning manual work on the material. The example of the etching of semiconductor circuits shows clearly that the miniaturization of chips, made possible by progress in the mastery of the microscopic structure of matter has necessitated recourse to computer-assisted technical solutions in place of tasks effected by skilled etchers.

In this way, recourse to information technologies coupled with the mastery of the microscopic structure of materials for the sake of reshaping production, ultimately produces the conditions for the changeover from a traditional kind of skill (based on metalworking and engineering) to a new kind of skill, based on mastery of the integrated organization of production and detailed knowledge of the properties of the material.

It is important to stress the very slow and very gradual nature of changes since it is a wholly step-wise process of change which is at work, arising out of the multitude of minor individual modifications in materials processing and production processes.

It is also important to stress the existence of phenomena of inertia, which naturally impede the changes under way, and which must be integrated by any forward-looking approach to skills. These phenomena may have two main effects. Either they help "block" an enterprise or series of enterprises in over-rigid industrial traditions (for instance, certain machine-tool manufacturers in whom the steel tradition is deeply ingrained cannot think in terms of composites, although composites may provide an efficient, up-to-date solution for making robot arms that are both light and rapid), and then such enterprises are liable to be left behind by their rivals. Or they may encourage materials designers to counter the technical evolutions under way by suggesting competitive solutions which "respect" the traditional skills. To some extent, the introduction of aluminium-lithium in aircraft structures is a way of countering the complete redesign (based on composites) of structures and components, and offers the advantage of fitting in with the traditional skills of the existing workforce.

The transition taking place naturally generates the need for new jobs and skills in design, control, maintenance and also the multi-disciplinary capacity to integrate other forms of knwoledge. It is becoming increasingly necessary to have men available in the company who are capable of managing complex solutions based on partnership with other companies.

As a general rule, the management of <u>interfaces</u> is being confirmed as one of the prime company requirements. The development of materials is consonant in fact with the determination to integrate different properties, not to say different functions, in a single material or, at least, in a single component. This concern for integration poses <u>interface problems</u>, that is to say problems concerning the compatibility of properties at all levels : material, component, assembly of parts and so on. For example, the development of composite materials depends on mastery of the interface between the matrix and the reinforcement, the quality of semiconductors is based on the proper stacking of thinlayers, the assembly of plastic and metal parts demands high precision and regularity of parts. All this entails more and more dove-tailing between skills, compelling the specialists to work together to resolve their common difficulties.

In more general terms, the aim of present advances is to find the shortest path between intermediate materials and the final product, at the same time defining the properties and functions to be performed and the materials and components to be produced. This objective poses the problem of the interface between industries producing materials and industries using materials. The dialogue that is needed here must in fact be set up between personnels engaged in industries of very different orders. Producers of materials have to design and turn out materials that match user industries' needs better and better, and the user industries must be familiar with the array of available materials and their potential, if they are to express their requirements to the best possible effect. Interdisciplinary team work is needed , then, as the dominant form of work, in order to be able to manipulate certain materials or components but also in order to reconcile the constraints on one side, and needs on the other. This means that each professional must possess a technological culture which is adequate to enable him to communicate using a minimum common language.

V.4.3 EFFECTS ON THE TECHNOLOGICAL SYSTEM AS A WHOLE

Given this reconception of production, which we have placed at the centre of the changes under way, the effects on employment and qualifications extend to the whole of the industrial system. As has already been emphasized, the proper functioning of production units demands the continual development of research activities on the one hand, and marketing activities on the other. In terms of personnel numbers, actually <u>within</u> existing enterprises (producing materials or using materials to produce industrial articles), it is probable that at least over the next ten years the number of jobs eliminated will be greater than the number created. Over and above the net effects noted within existing enterprises, however, production reconception mechanisms are likely to produce significant indirect effects.

For instance, the dissemination of new technical solutions in materials to the other sectors of economic activity may give rise to new activities which create jobs in industries producing materials for leisure activities, sport and so on. Lower production costs must of course be the foundation for the spread of these new technological solutions and their utilization throughout economic activities. The indirect effects in terms of job creation in some sectors are likely to be considerable. Emphasis has accordingly been laid on the fundamental importance of the development of basic functional materials (optical fibres, semiconductors, etc) for the development of the telecommunications sector, and more precisely for the creation of jobs <u>in services</u> tied to the establishment of a telecommunications network that is rapid and accessible in cost terms. More generally, any new industry derived from materials generates a procession of services either directly around the product (see chemistry of the product - chemistry of the function (13) , for example) or indirectly for managing new activities (sport, leisure activities, travel and holidays, education, health, etc). These services are as a rule employment-intensive since they need to offset, by the physical presence of men and women, the dehumanizing aspect of technology and products of new industry.

---

(13) "La Chimie en Europe" (Chemistry in Europe), Economica 1984. Collective work edited by P. COHENDET.

The revelation of the mechanisms currently now makes it possible to identify better the implications for employment and skills arising out of the two development scenarios utilized in this work.

V.4.4 EMPLOYMENT AND SKILLS IN A SCENARIO DOMINATED BY CONSTANT VARIETY

The hypothesis of a system based purely on constant variety (in which it has been shown that materials play a decisive part), marked by a continual reconception of production and a growing ability to master complexity, poses considerable problems when it comes to skills. Such a system breaks once and for all with one of the main aspects of the previous system, based on the standardization of products and processes, and on specialization of tasks : it shatters the traditional mechanism of skill acquisition. In a "perfected" Taylorian system, in fact, there is a clear division of labour between design and execution which brings with it a negation of the notion of a learning curve for operatives : the latter simply have to carry out repetitive and specialist tasks for the sake of achieving standardized production. But the permanence of the production system and accumulation of experience inherent in specialization of tasks permits successive improvements and, paradoxically, makes for the appearance of a body of skills specific to the labourforce.

This development is no longer possible in the case of a variety-based system : with this hypothesis, in fact, in which depending on relations between supply and demand conditions it is possible to move on rapidly from one kind of technological solution to another, these traditional learning mechanisms no longer apply. We come back, in this way, to the sources of Taylorism where the whole of knwoledge and skill is situated at the level of production, conception and organization.

This situation appears to be in complete contradiction with the fact that, in a system based on continual variety, competition takes place mainly in terms of mastering complexity, so that the workforce needs to be able to master the management of this complexity itself, by means of appropriate instruments (expert system, flexible workshop, etc).

More generally, difficulties arise out of the fact that in a variety-based logic, no level of complexity that is attained can be regarded once and for all as stable. The workforce at all levels is thus encouraged to update its skills all the time, through a suitable training system. This kind of adjustement will be the easier or more difficult to effect depending on the age of the persons concerned : young people having undergone basic training of a good standard seem to be in a more favourable position than the other age groups. Their training is recent and relatively appropriate; they have not yet acquired working habits, so that they can be trained to be all-rounders, to have a comprehensive knowledge of materials. The oldest, however, will be less adaptable, whatever their level of responsibility, and they too will have to widen their range of skills if they want to keep their jobs in the long run. A conventional mechanical engineer, will have to become a manager, become familiar with computer science and use both computer-assisted design and computer-assisted manufacture. In this way we have a measure of the crucial - and to some extent cruel - weight of age in companies' future organizational tables.

The example of the manufacture of single-crystal silicon wafers illustrates one aspect of this heightened personnel involvement : this material needs an extremely high degree of zero defect, a norm imposed by semiconductor producers. But it is only possible to verify this quality effectively at the end of the "customer chain", when the component is completed. The necessity of the mastery of the microscopic structure of manufactured materials in fact requires a continuous, no longer a statistical system of control by sampling. Of course, it is not economic to analyze each wafer, atom by atom, by some spectroscopic process. It is better to optimize the manufacturing process and leave the task of detecting microscopic defects to the user of the materials. However, optimization of the silicon manufacturing process presupposes feedback from the customer to the supplier (so that the latter can move towards the zero defect stage). One of the few ways of achieving this feedback is for an identification number to appear on each wafer, whose technical characteristics are computerized. This method permits total traceability : identification of the history of the wafer and the worker(s) having participated in its production. The success of such a system also presupposes product control at several successive work stations. In this way, workers (highly skilled personnel) in fact become responsible for the quality of their work and can imagine or devise realistic solutions for defects. They are no longer mere "operatives" - they also take part in the design stage, or at least in the accumulation of experience.

If the functional materials industry is alone in gradually adopting this process because of its absolute requirement in microscopic mastery, the tendency of all materials, in particular structural materials, to encourage this microscopic advance will sooner or later bring all these industries towards this phenomenon whereby a sense of responsibility is created.

The difficulties inherent in the functioning of a system based on permanent variety are thus two-fold at least :

- a system of this kind presupposes considerable ambition in the matter of training; on this hypothesis, most skills would in fact have to be substantially re-appaised, and so adequate training resources would have to be found;
- a variety-based system naturally poses the problem of social relations in the firm, and more generally that of the support of the social corpus for such a system.

If variety is imposed without consultation and without training, there is an immense risk that a general pattern of society will grow up with two groups of "agents". On the one hand, there will be a group of "designers and organizers" creating the added value, highly paid and holding real power in numerous areas. On the other, there will be a group made up of "operatives", poorly paid and in reality without any genuine responsibilities in their work and without any possibility of taking part in working out future solutions. Of course, this risk may become general throughout a whole country (aggravation of the divide between so-called high-tech zones and the rest) or internationally (aggravation of the gap between the highly developed countries and the rest). As it is not really possible to contemplate this solution, it seems in short that a variety-based system demands, if it is to be viable, that people shall be highly trained and highly responsible.

V.4.5 EMPLOYMENT AND SKILLS WITH A SCENARIO FEATURING THE IRREVERSIBLE ASCENT OF PLASTICS

The "plastics" scenario has implications in terms of the "cost to society" which differ from the scenario featuring variety. This hypothesis fits in with a traditional logic of production, in fact, and in particular maintains traditional learning mechanisms, suggesting a much more progressive reconception of the industrial system. It is true that there are extremely complex forms of technical solutions in the "plastics solution", bringing with them a need for qualifications at the highest level, but on the whole the "plastics solution" goes along with the conventional pace of technological change. As has been seen, it consists of imitating, in a preliminary stage, the traditional solutions before progressively replacing them. It is also based on a body of knowledge that is relatively accessible for the skills present.

Even if the "tailor-made" kind of production suggested by the "plastics solution" (or organic matrix composites) may today seem inaccessible for certain enterprises (lacking the technical level needed to carry out the processes required, or lacking the financial resources), even the most elaborate plastics or composites can provide very advanced technical solutions which preserve the traditional forms of production. The evolution of thermoplastic composites is significant here, because they combine the characteristics of composites with the possibility of presenting the material in traditional intermediate form (sheet, bars, etc). The recent introduction on the market of the APC (aromatic polymer composite) by ICI is a good example of the development of these "new" materials in "old" shapes. Similarly, the recent development of thermoplastic pre-impregnated fibres, produced by extrusion or filament winding (or in the form of braided or woven fabrics) enlarges the scope for processing thermoplastic composites. It is possible to shape these fibres at will, before heating them to arrive at the final composite. These counter-examples, which among other things rehabilitate the concept of the semi-finished product, are indicative of the current evolution in processes : while it is certain that present progress gives pride of place to a kind of production which is more integrated and more flexible, this trend must not be systematically interpreted and does not even appear expedient in some cases.

In other words, the "plastics solution" seems to respect a more "graduated" evolution of work, even if some aspects of the solution demand mastery of the most complex techniques.

Basically, the "plastics solution" could become established because it appears to be the best overall response to the need for a more and more advanced mastery of the microscopic structure of materials. It presupposes, naturally, the acquisition of chemical know-how, but this is something that can be attained relatively gradually. For instance, if we look at the gradual transition in assembly industries from certain assembly processes involving welding or riveting to bonding processes (which are integration processes too), it is possible to observe the way in which very gradually chemical know-how makes its way in, becoming increasingly necessary in a decisive stage of production.

The acquisition of chemical know-how can of course take a number of different forms. For instance, a motor manufacturing firm was compelled, in order to produce plastic bodies, to integrate provisionally, in the economic sense of the term, one of its sub-contractors so as to "train" it in the new techniques. The training completed, the firm is thinking of giving the sub-contractor back its independence.

V.4.6 OUTLINE CONCLUSION

As has already been emphasized, scenarios are only an ideal vision of reality, and one that is inevitably provocative. They over-state trends, and their only merit is to serve as a frame of reference for reflection. For problems of employment and skills, they make it possible in any case to see just how far the role of skills is central to the evolution of the technological system. It also appears that the problem of skills and working conditions does not depend on technological developments alone. It is at the heart of the objectives or strategies of a country or group of countries. The fundamental problem, which extends beyond the limits of this study, is perhaps whether the state of international competition permits the co-existence, in different countries, of different systems of skills.

It can be imagined, for instance, that Japan presented a challenge to some extent in seeking to show that it was capable of managing a virtually permanent variety-based system. On the other hand, it is clear that Europe would appear more inclined, for various reasons, to adopt a system assuming the "irresistible ascent of plastics" (weight of the chemical industry, rejection of excessive mobility, slow pace of adjustement, etc). Supposing that such a dichotomy could exist, would the European "project" be valid in an environment marked by cut-throat competition and markets on a world-wide scale ?

Far more pragmatically, examination of the two scenarios shows that, regardless of the trends that are emerging, a number of "underlying trends" seem to be affirming themselves in Europe in connection with the problems of employment and skills in materials :

the importance of the forward management of jobs, the importance of the concept of evolutive skills and most of all the importance of the concept of training based on alternating periods of work and study.

V.4.6.1 Forward management of jobs

The forward management of jobs at all hierarchical levels, which so far has been neglected and is still in its infancy, should now move into the foreground : establishment of skills records for everyone, planning of staff needs, organization of in-house training courses, setting up quality circles, progress groups, more rigorous recruitement. The question is no longer one of adapting human resources to the new technologies, but rather of imagining the future conditions of production so as to have a labourforce that is ready to tackle those conditions when the right moment comes. For example, when Montedison decides to develop a new area of activity, it begins by addressing the problem of skills. For this purpose it sends young graduates to train in laboratories where important work is being carried out. Having in this way acquired the appropriate expertise, it then proceeds to buy the machinery and at the same time, perhaps, to retrain the production personnel concerned.

V.4.6.2 Evolutive skills

The concept of evolutive skills is a response to the need for the accelerated development of products, to the diversity of the materials utilized and the variety of materials processing techniques which will be manifested in any event - even in the case of the "plastics" scenario. Skills acquired become obsolete more and more rapidly. It is therefore essential to renew them at regular intervals. The most immediate solution is of course to take on staff with the required skills, and if necessary to get rid of the staff who no longer fit the bill by early retirement schemes and other incentives. But this staff turnover can only be partial, and the operation cannot be renewed indefinitely : it also creates a heterogeneous element in the work community because of the co-existence of young graduates and older members of staff, who are left on one side and become embittered. Flexibility on the labour market, something ardently sought at the moment in all European countries, is no better solution. While it appears satisfactory for employers at micro-economic level, in the long run it fails to resolve the macro-economic problem of renewing the stock of available skills. It is thus essential to take into account the constant evolution of skills needs in training courses, as in careers, and so to move in the direction of the establishment of evolutive skills. This concept needs to be seen in conjunction with that of evolutive norms, considered in the previous chapter : new materials will age much faster than the article they are built into so that it will prove indispensable to define evolutive norms as opposed to static norms. Evolutive skills in parallel with static skills, will have to be defined in collective agreements. So the question arises, in particularly acute forms, of how skills are to be recognized.

Already a great many large and medium firms have agreed to make unparallelled efforts in the provision of continuous training. Although this option is costly, it has the two-fold advantage of maintaining a social climate - which outright dismissals would have affected irremediably - and committing the enterprise to harnessing its human resources. The objective is no longer to train exclusively for a new technology, at a given work station. Rather, training relates to the means of best fulfilling a function (instead of a task), if need be regardless of the technology and materials concerned : training to grasp abstract concepts, to analyze any problems and handle them, creating quality-consciousness. What was to date a feature of the working pattern of executives is currently spreading to the other levels in the hierarchy.

It is clear, then, that the problems of evolutive skills relates most of all to a problem of training, and so gives pride of place, among the different methods, to permanent training which includes periods of work and periods of study.

V.4.6.3 Training in which work alternates with study

The policy of training combining periods of work and study presupposes a certain continuum between the initial training of the young and that of adults, both based on the fundamental idea of work alternating with study. As has been seen at a number of points in this section, the extent and speed of changes in the materials field make exchanges between the educational system, where initial training takes place, and the enterprise absolutely essential. There is no curriculum that can provide students with a collection of skills that is adequate, that will enable them to meet the requirements posed by the continuing development of materials, especially since national educational systems are often very slow to react and to consider new educational forms : from seven to 10 years, between identification of needs and passing out of the first graduates. It appears pointless, then, for educational systems to train on a large scale and at considerable cost large numbers of specialists in one or other material, since there is every chance that their knowledge will fairly quickly become obsolete because the relevant skills change so fast. Similarly, it is clear that experts in different specialities who are incapable of team work or a dialogue among themselves are of limited usefulness. Lastly, training that is too centred on historic achievements and too remote from industrial realities can only give rise to difficulties in adjustement following recruitment.

What is needed rather, in the matter of initial training offering outlets in the field of materials, is first to promote a high standard of general education in which sufficient place is set aside, whatever the predominant option, for basic scientific subjects : physics, chemistry, computer science, materials engineering. These "specialists in general subjects" will in this way have acquired sufficiently general training to enable them to make on board subsequent adjustements in the best possible manner, and to be easily transferrable from one company to another. This theoretical training should be supplemented by practical training in firms.

Allowance must be made in the training currently being dispensed for the need to think more in terms of "multi-materials" both at the basic training stage and in training for technicians and engineers. Basic training must be founded on the science of materials, situated at the crossroads of chemistry, physics and engineering. The objective of this science is an understanding of the properties of materials : mechanical, chemical, electrical, magnetic properties, and so on, which are not intrinsic properties but which stem from the whole thermal and mechanical past of the material. It has for that a guiding line, or if preferred a unifying concept (since it applies to all categories of materials) : properties/internal structure relations. It is this structure, in fact, from the atomic scale to the macroscopic level, moving through all the degrees of micro-structures studied by the different spectroscopic techniques, which governs properties. This kind of research is very largely the work of university laboratories. Training for engineers and technicians will have to leave more room for materials engineering, which is designed to master relations between, on the one hand, the basic properties of materials and, on the other, their suitability for application and the quality of their performance (reliability, resistance to deterioration, etc) which incorporates follow-up and quality control (non-destructive testing, for instance).

This training needs to be extended and enriched throughout working life by means of permanent training, retraining that is both theoretical and practical. In some cases the intervention of university-type teaching could be a response to the obsolescence of theoretical training.

The system poses the problem of recognition of skills, since periodic retraining of this kind does not necessarily mean that acquisition of a degree or recognized qualification - in contrast with the continuous occupational training system practised in the Federal Republic of Germany, which constitutes the basis of the "model" presented - what it confers is an extra skill. This genuinely continuous form of training would make it possible to return to the initial objectives of the continuous training systems that have been current since the early 70s throughout Europe : occupational advancement, and most of all requalification to meet the needs of the production system.

The piecemeal nature of so-called continuous training rarely makes it possible to anticipate projects, to make allowance for low levels of the initial skill and technological culture. Such training generally constitutes a measure to accompany changes, instead of preceding them.

If the objective of permanent training seems to be such that it can be envisaged in certain conditions for large and medium firms, it appears still difficult to realise in the case of small concerns. It may be noted here, however, that Belgium is carrying out an interesting experiment at the level of small and medium firms : Royal Decree N° 123 promulgated in January 1983 is designed to further the entry into the world of work of young unemployed persons (industrial and civil engineers, graduates and those with postgraduate qualifications in whatever subject) so as to enhance the general scientific level of the SMEs concerned. The state pays the trainees' salaries in full or in part for two years, and employers' federations may sponsor them (following the progress of the trainee and possibly intervening in the event of problems). To date, most trainees have been recruited by the firm, although the experiment is on an extremely limited scale for the moment.

Implementation of a permanent training policy of linked work and study raises in pressing terms the problem of training for instructors; it will have to be both theoretical and practical, and most of all it will have to be constantly updated. This bolsters the idea of a permanent dialogue between the education and the production systems.

Conclusion

The first, and the principal, conclusion of this survey is as follows : the development of materials is going to have an impact on society that is at least as great as that of the information technologies. This entails major consequences for the strategy of firms and of the authorities, too inclined in the past to think solely in terms of the computer revolution. All the lines of argument developed in this report in fact confirm that the reconception of products and processes, that the satisfaction of quality and reliability requirements, that energy savings, the creation of new industrial articles or new services, and the decentralization of human activities in space, are just as dependent on the mastery of materials as on that of information technologies.

This need to master the evolution of materials and at the same time the evolution of the information technologies appears to be the more urgent as the outlines of the new industrial environment gradually emerge. However audacious or provocative the scenarios worked out in a prospective study, one hypothesis becomes more and more credible : that of a profound upheaval in the manner of producing, the way in which the requirements of demand are met, and types of relationships between the different economic agents. This hypothesis leads to the establishment of a production regime that is rich in information, where a greater range of technological solutions becomes possible - provided that one fundamental constraint is observed : the mastery of complexity at two levels. First, there is control of the complexity inherent in the material itself, through mastery of the microscopic structure of matter, and second, there is control of the growing complexity of industrial production through the quest for material and organizational solutions which simplify the management of processes.

The present importance of materials for the production system arises first of all out of a key phenomenon : progress in the mastery of the microscopic structure of matter. It is in fact out of this mastery that derive the new possibilities available to the industrial system, and the capacity to find valid and appropriate responses to the requirements of demand. Progress in the mastery of the microscopic structure of matter offers a special opportunity for judging how far the development of materials and that of the information technologies have become indissociable. As we move forwards in this mastery, the amount of information to be processed in order to control the production of materials and their industrial application in fact expands with ever increasing speed. Here materials generate their own complexity, and the only route to viable industrial production lies through having recourse to powerful data processing resources.

Thanks to advances in the mastery of the microscopic structure of matter it is possible for materials to make a decisive contribution to the formation of a new production system. On the one hand, they help respond to the new types of demand, by constantly enlarging the range of industrial solutions and opening up unsuspected possibilities for creation, and on the other hand they unceasingly offer solutions to what companies can supply, subject as they are to the phenomenon of growing complexity, by making it possible to simplify application processes, permitting constantly greater integration of functions and pruning production costs. Materials are in this way bringers of both variety and integration, and in this sense they lie at the heart of the far-reaching changes currently taking place.

The report sets out not only to ponder the question of actual materials, but also to look closely into the probable development of the production system. A deliberate comparison has been drawn between the transition from a former production regime based on standardization of products to a new production regime based on variety. With the advantage of the more detached vision that we can afford to take in this conclusion, we may advance the hypothesis that since the start of the industrial era the production system has constantly evolved towards higher and higher levels of complexity, generating successive phases of integration (in order to make production profitable) and variety (in order to enlarge markets). The period dominated by the Henry Ford philosophy can thus be described as a preliminary phase, in which machines were lined up and integrated in a single production units so as to make it possible to satisfy demand marked by a relatively low level of variety, in profitable conditions. The endeavours of General Motors round about the mid-20s to diversify its range through technical, economic and financial integration so as to achieve a dimension great enough to permit financial resources to be set aside for the development of new models, open up a second period : the period that can be described as planned variety, which lasted until the mid-70s. We are now apparently entering a new phase, that of permanent variety of products, their viability guaranteed by efforts to achieve integration early on in the production stage, by simplifying processes, reducing the number of components, and overall cost control. In this way it can be observed that as complexity expands, and variety increases further on in the production system, so the need for integration gradually shifts towards the earlier stages in the system. These comments underscore still further, then, the role of materials and in particular that played in the early stages of the mastery of the microscopic structure of materials in order to find appropriate responses and guarantee the viability of the new production regime.

The second conclusion reached in this study concerns rivalry between materials. Two scenarios have been worked out : that of absolute variety of materials (with each retaining a chance to prevail) and that of variety restricted to the field of plastics only (more specifically organic polymers). It became plain that what was at stake (and the industrial and social consequences here are considerable) was at present still very much an open question and that the trend towards one or other of the scenarios will very much depend on the projects and decisions of the agents concerned, and particularly what the authorities want. It is in fact essential, if variety is not to become indecision, for the authorities to lay down the necessary rules of the game in areas where their role is indispensable. It is probable in the case of industrial firms that their movement towards mastery of a greater variety of materials will only be gradual : in an initial stage variety will probably test its paces in the elaboration of "intermediate forms" (composites, fibres, functional layers, etc) - concepts through which firms will learn to combine materials, to try the whole gamut of minor differences and to master new problems. It is probably only at the end of this learning process that a clearer choice between the development scenarios will be made.

Irrespective of the scenario that prevails, the over-riding tendencies have been indicated. Among them, the convergence of structural materials and functional materials appears decisive. Not only are structural materials destined to integrate functional materials more and more, but a still more radical hypothesis suggests itself - that structural materials are gradually going to take a back seat to functional materials. More precisely, the rule of structural materials could be summed up as constituting the protective membrane - the "skin" of industrial objects (governing relations between such objects and their immediate environment, mainly), while functional materials will be more and more incorporated in those industrial objects (linking together, in particular, industrial objects - even when they are separated by a considerable distance - by means of information networks). The idea of "well-informed" or "smart" objects frequently aired in connection with the vehicles or buildings of the future is a faithful reflection of this reversal being perceived in the logic of materials evolution. Up until now, structural materials were in fact predominant (they came under a form of production generally based on efficiency of scale, while the incorporation of functional materials took place at a larger stage, being more or less improvised). Conditions now appear right for giving priority to functional materials (covered more markedly by the logic of efficiency of scale) at the expense of structural materials (which will be a vital element in differenciation and variety). In concrete terms, the growth of the functional aspect in the supply of industrial objects should take on two main forms : on the one hand, a greater proportion of functional materials in industrial objects, and on the other the gradual incorporation of functional materials in actual structural materials (perhaps culminating in a concept of variable "informed composite" in which the functional component cannot be separated from the structural component). This "swing" between the two types of material corresponds to the aim to satisfy a fundamental principle : that of matching a type of demand that transmits its requirements increasingly in terms of functions to be performed (safety, comfort, respect for the environment, insulation, etc) to supply, which is going to incorporate more functional materials in industrial responses, so as to furnish the material basis essential for the satisfaction of demand and the development of multiple services connected with industrial solutions.

This convergence between the functional requirements of demand and the enlargement of the functional potential of supply leads to several important corollaries : first, it confirms that the industrial object is destined to <u>become more complex</u>. More precisely, the materials incorporated and the functions integrated will become more and more complex. Now, this phenomenon means at both ends of the production system that the <u>cost of errors</u> will become more and more prohibitive. Early on in the production stage high-quality components and elements have to be produced that are thoroughly reliable, because as the final object becomes more complex, so the cost of an error at the early stage in production becomes greater. This is a further reason for developing a form of production that is rich in information, and ensuring progress in the mastery of the microscopic structure of matter. In the later stages of the production system, we need only think of the higher cost today of a dented car wing (it was enough at one time to change a steel bumper, but nowadays an integrated and complex set of parts has to be replaced) in order to gauge the consequences of the greater complexity of industrial objects. At the later stages there arises the question of the recovery of such objects. How in fact are we to recover complicated parts which combine several parts which combine several materials, in an integrated manner, with the risk of causing further environmental damage if we fail to resolve the problem ? One solution which might be considered, already applied in electronics, for instance, is for producers themselves (or through sub-contractors) to be responsible for recovering industrial objects, even if this means possibly recycling some of the components that could be reutilized.

Lastly, a comparison between variety of demand and variety of supply raised the question of how the movement of information is to be effectively realized between the early and later stages of the production system, and how supply is to be tailored to demand in practice. It is here that the role of the designer, as a veritable "functions manager" may appear decisive and so means destined to expand considerably.

<u>Our third conclusion concerns the strategy for Europe</u>. In the field of materials, the study has shown that Europe clearly has sizeable assets with which to resist competition from the United States and Japan. Emphasis has repeatedly been placed on the strategic role than can be played in this competition by European chemicals. But if it is to remain really competitive, Europe will have to overcome two major obstacles. The first is its excessive dependence on the United States and Japan when it comes to functional materials. How is it possible in fact to counter competition unless we can control the

materials around which the industrial responses of the future will take shape . How are we to move in the direction of a society that generates jobs in services if we fail to master the material basis of the technological system ? The second obstacle is a purely internal European obstacle. The Europe of materials, in contrast with the situation that can be observed in other sectors, is not operating on a two-speed basis (the most highly industrial regions, and the others) but on a three-speed basis (Germany, the most highly industrialized regions, and the rest). For a great many reasons (linked work and study forms of training, existence of a robust chemical industry, tradition of mastery of production processes, etc), the authors conclude that Germany, more than any other European country, is satisfying the conditions for international competitiveness in this area. Further consideration will still have to be given to this conclusion, of course, but in any event it poses the problem of how European efforts are divided out and points to the case for stepping up co-operation among the different countries of Europe.

Lastly, as Europe enters further into an inevitably desperate form of competition with the United States and Japan for the sake of confirming its industrial competitiveness, affirming its independence and defending its jobs, another problem - just as fundamental - clearly appears : Europe is liable to cut itself off still further from the least advanced countries, to the point of generating profound malfunctions in world equilibrium. This is because, with the production of new materials incorporating a steadily increasing proportion of grey matter and a steadily declining proportion of noble materials, the industrialized countries gradually break with the need to import commodities and materials from the least developed countries. A durable decline of a few per cent in certain strategic materials can in fact help trigger a sudden and far-reaching downturn in certain developing countries' trade balance, and worsen the debt crisis. This phenomenon, coupled with others, is thus liable to cause a further deterioration in the situation of countries that are dependent on only a few sources of specific raw materials (Bolivia, Malaysia, etc). Some people will argue that in the long term the disappearance of this situation (which implied excessive dependence) may prove beneficial. The fact remains, however, that it is urgently necessary to envisage solutions so that an acceptable transition can be affected. This problem, which must be looked at squarely, naturally goes hand-in-hand with a vast problem of global financing which stretches beyond the limits of the present study (in the same way as a problem arises in the rich countries, in financing terms, of bringing financial innovations into line with technological innovations).

Taking into account the arguments that have just been developed, the evolution of materials is a pointer to the far-reaching changes foreshadowing the world of the future. Perhaps more so than with other technologies, the message that can be discerned from the evolution in materials is that of a subtle combination between alliance and competition. It is perhaps this phenomenon which characterizes the quest for partnership solutions or international agreements at company level, and it is perhaps this phenomenon too which, at national level, spells the necessity for Europe to strive to cope with its rivals and at the same time seek forms of technological and commercial alliances with them. Europe needs to balance its strategy here with care. While competition with the United States and Japan is of course necessary and something to be advocated, there may be a risk of an excessive number of individual agreements concluded between European firms, on the one hand, and American or Japanese firms on the other (at the expense of the establishment of an adequate number of intra-European agreements) ultimately leading to a dislocation of Europe. Were this to be the case, it is clear that Europe would not even be able to go on acting as the privileged intermediary for developing countries, a role which it has traditionally and consistently sought to play.